ENVIRONMENTAL INDICES
Theory and Practice

by

Wayne R. Ott, Ph.D.
Senior Environmental Engineer
Monitoring Technology Division
Office of Research and Development
U.S. Environmental Protection Agency
Washington, DC

This book was written by the author in his private capacity. No official support or endorsement by the Environmental Protection Agency or any other agency of the Federal Government is intended or should be inferred.

ANN ARBOR SCIENCE
PUBLISHERS INC
P.O. BOX 1425 • ANN ARBOR, MICH. 48106

Copyright © 1978 by Ann Arbor Science Publishers, Inc.
230 Collingwood, P.O. Box 1425, Ann Arbor, Michigan 48106

Library of Congress Catalog Card No. 77-85082
ISBN 0-250-40191-6

Manufactured in the United States of America
All Rights Reserved

To Patricia

PREFACE

In the last two decades, many indices and statistical approaches for interpreting and presenting information on the state-of-the-environment have been developed. Prior to this book, a person interested in these techniques had to spend much time and effort searching through professional journals, research reports, and various governmental publications. He may have found the sheer diversity of these approaches confusing, and he may have had difficulty understanding how each technique related to the others. Thus, I have attempted to bring together, in a systematic fashion, all existing environmental index systems, along with principles for their design, application and structure.

Chapter I introduces the reader to environmental data, presenting simple communicative approaches such as environmental quality profiles. It also describes the national monitoring activities that generate these data and discusses the difficulty of constructing meaningful environmental damage functions. Chapter II presents a new conceptual framework that is designed to embrace nearly all existing environmental indices, allowing the behavior of different index structures to be compared and probed in detail. As part of this system, I have developed the (I_1, I_2)-plane, which permits such problems as "ambiguity" and "eclipsing" to be presented graphically. Chapter III concentrates on air pollution indices, using the conceptual framework introduced in Chapter II to analyze and compare published air pollution indices. Chapter III also gives a detailed summary of the historical evolution and scientific basis for the Pollutant Standards Index (PSI), which has been developed for uniform application throughout the United States. Computatonal aids (equations, tables and nomograms) for applying PSI to actual air quality data are included. Chapter IV covers water pollution indices, using the theoretical framework and concepts from Chapter II to examine today's water pollution indices; it also presents design principles for an ideal water quality index and discusses a candidate index structure. In both Chapters III and IV, the current air and water index usage patterns in the United States are described in detail. Finally, Chapter V presents conceptual approaches, such as Quality of Life and environmental damage functions, that extend beyond the traditional fields of air and water pollution. In this chapter, I introduce the "damage space," a graphical system designed to facilitate analysis of environmental damage functions. Mathematical criteria are proposed for the properties of damage functions (for example, antagonism, superposition and synergism), and it is shown that these criteria also can be applied to today's environmental indices to gain insight into their behavior.

This book should serve as a basic reference for users wishing to apply indices to analyze environmental data. Thus, each index is presented in sufficient detail to enable the reader to apply it directly, and numerous examples are included. This book also is intended to familiarize members of the general public, regulatory officials, and environmental specialists (environmental planners, engineers, statisticians, research scientists, etc.) with the basis for, and limitations of, today's environmental indices. Because of its breadth of coverage, it is hoped that this book also can be used as a college text on the measurement of environmental quality and the quantitative interpretation of environmental data. To assist the student in understanding the techniques presented in this book, problems for study are included at the end of each chapter.

Because measurements of environmental variables give man the most objective and precise means available for determining the state-of-the-environment, and the impact of his activities upon it, I hope that this book will stimulate greater activity in the development and use of quantitative techniques for interpreting environmental quality.

I wish to thank the following individuals, who are employees of the Environmental Protection Agency and close personal friends, for taking their own time to review various chapters of this book and make many thoughtful suggestions: Dr. Lance Wallace, William Sayers, Dr. David Mage, William F. Hunt, Jr., and Victor Randecker. I also wish to thank Lorraine Deavours, technical editor, for her thoroughness and meticulous care with this manuscript. Without the excellent assistance of these persons, I could not have finished this book in a timely fashion.

Wayne R. Ott

**Wayne R. Ott** has been with the Environmental Protection Agency and its predecessor agencies for more than 12 years, working primarily in the areas of evaluation of national environmental policies and research and development. He is a Commissioned Officer in the U.S. Public Health Service, on detail to EPA, with the rank of Senior Sanitary Engineer. His areas of technical specialization lie in the development of new mathematical techniques for the interpretation and display of environmental data, development of quantitative methods for environmental decision-making, environmental statistics, and quality assurance. He has written more than 40 papers and technical reports covering both practical and theoretical aspects of environmental modeling, environmental data analysis, systems analysis, projections of environmental trends, computer techniques, air and water pollution indices, siting of monitoring stations, statistical methods, monitoring survey design, and the analysis of national environmental regulations and policies. In 1976, he served as program co-chairman of the national EPA Conference on Environmental Modeling and Simulation.

Dr. Ott conducted the initial research which showed the feasibility of developing a nationally uniform air pollution index. His study findings identified the basic structure of a candidate uniform index, and he participated in EPA's efforts to select a final index design and to implement it nationally. His research on indices also helped stimulate passage by Congress of amendments to Section 319 of the Clean Air Act, which mandate use of a standardized index in cities throughout the United States. In 1977, Dr. Ott received the Commendation Medal of the U.S. Public Health Service for his work on air pollution indices. He has served as an international consultant on environmental data analysis problems, and he assisted the government of Mexico in the development of an air pollution index now in routine use.

Dr. Ott received his PhD degree from Stanford University in civil and environmental engineering in 1971 and also holds a Stanford MS degree in engineering science (1965), a MA degree from Stanford in communications (1966), a Stanford BS degree in electrical engineering (1963), and a BA degree from Claremont Men's College in business economics (1963). He is a member of the American Statistical Association, the Air Pollution Control Association, the American Society for Quality Control, Kappa Mu Epsilon (mathematics honor society), Tau Beta Pi (engineering honor society), and Sigma Xi (scientific research society).

Drawing on his extensive familiarity with the literature and on his work in developing indices for the federal government, he has created this basic reference book on environmental indices. For the first time, a theoretical basis is presented for a variety of index structures, and the mathematical implications of alternative forms are probed in detail. This rigorous new approach is intended both to give insight into the performance of existing indices and to provide general principles for the design of future environmental indices.

CONTENTS

I INTRODUCTION 1
   Role of Indices 2
   Language of Indices 7
   Environmental Quality Profile 9
   The Search for Damage Functions 27
   Environmental Monitoring 36
   Problems for Study 45
   References 46

II STRUCTURE OF ENVIRONMENTAL INDICES 49
   Mathematical Structure 50
   Subindices 51
      Linear Function 51
      Segmented Linear Function 54
      Nonlinear Function 59
      Segmented Nonlinear Function 64
   Aggregation of Subindices 66
      Additive Forms 66
      Maximum Operator 76
      Multiplicative Forms 79
      Minimum Operator 89
   Summary of Index Structures 91
   Problems for Study 93
   References 95

III AIR POLLUTION INDICES.	97
Uniform Administrative Limits	97
NAAQS.	98
Episode Criteria	101
Significant Harm Level.	103
Historical Background	104
Indices in the Literature	105
Indices in Use.	122
Criteria for a Uniform Index	135
Compendium Findings.	138
A Nationally Uniform Air Pollution Index	140
Development of PSI.	143
Structure of PSI	147
Data Used in PSI.	154
Calculation of PSI	156
Reporting PSI.	168
Performance of PSI	171
Adoption of PSI	184
Conclusion	190
Problems for Study	190
References.	191
IV WATER POLLUTION INDICES.	197
Indices in the Literature	197
General Water Quality Indices	198
Specific-Use Water Quality Indices	222
Planning Indices	247
Statistical Approaches	254
Biological Indices	263
Comparison of Indices	264
Indices in Use.	273
Indices Developed by Agencies.	284
Prospects for a Nationally Uniform Water Quality Index.	294
Criteria for an Ideal Water Quality Index	300
Conclusion.	300
Problems for Study	301
References.	304
V CONCEPTUAL APPROACHES	309
Quality of Life	309
QOL in Metropolitan Areas.	312
Other QOL Approaches	322

Other Environmental Indices	323
Solid Waste	323
Noise	323
Total Environment	324
Species Diversity	325
Environmental Damage Functions	326
Univariate Damage Functions	326
Multivariate Damage Functions	329
Discussion	345
Conclusions	348
Problems for Study	349
References	351
APPENDIX A	353
INDEX	357

CHAPTER I

INTRODUCTION

In this era of technological development, man has collected vast quantities of data and information about himself, his society, and the physical world around him. This large body of data has grown so rapidly that it challenges man's ability to understand and assimilate it. The same technology which made it possible to create this large data base also has produced the automatic computers which make the task of storing, analyzing and processing the data more reliable and efficient. The computer, however, is just a tool, a slave to the programmer's will, and there still remains the task of extracting from the data the pertinent information required to answer questions of importance. Not only must the data be manipulated and reformatted in a way that is understandable to the user, but exactly the right information must be extracted that is relevant to the questions that are being asked. Furthermore, in democracies such as the United States, the user of the information—a Congressman, a member of the public, a businessman, a student—often is relatively remote from the information and unaware of its technical subtleties. Yet, it is this user, often with his limited technical background, who must rely on the data to draw conclusions about governmental policies and actions. Indeed, it is often this same user who has responsibility for formulating governmental programs and regulations. Furthermore, members of the general public, who support these programs through their taxes, have a right and a responsibility to determine the impact of these programs on the world around them. Consequently, in democratic societies, the need for translating vast quantities of technical data into a form that is readily understood by the layman takes on special significance.

In the environmental field, an interested member of the public, a representative of a citizens' group, or a governmental official typically may seek to determine whether a particular environmental problem is becoming "better" or "worse." These questioners usually will seek answers in the simplest possible form. The environmental scientist or professional working in the field may feel, on the other hand, that the answer to the question is complex, requiring the interpretation of hundreds of thousands of measurements of different pollutant concentrations and other variables, sometimes compounded by missing data, inconsistencies, and quality control problems, and often giving vague or uncertain results. Unfortunately, however, the questioner usually will not be satisfied by a 500-page "telephone book" full of raw data, time series plots, statistical analyses of pollutant concentrations at different locations, and other complex findings. He wants a simpler answer.

The questioner could, of course, hire a consultant already familiar with the data to go through the book of numbers to determine a simple answer to the question. This sometimes happens. Another common but unfortunate event is for the questioner to be told that the problem is "too complex," that his question cannot be answered unless he is willing to learn more about the technical details of the problem. Usually, the fault does not lie with the person asking the question but with those in the technical and scientific communities who may be unwilling or unable to take the trouble to express the answer in terms that the layman will understand. One reason, of course, is that technical specialists often do not feel comfortable with simple answers to complex questions; they see many nuances of the questions and possible areas for misunderstanding. They prefer to give no answer rather than an imperfect answer that could lead to misunderstanding. Yet the layman usually prefers an imperfect answer to no answer at all.

Here is where "indices" can play a potentially important communications role. Ideally, an index or an indicator is a means devised to reduce a large quantity of data down to its simplest form, retaining essential meaning for the questions that are being asked of the data. In short, an index is designed *to simplify*. In the process of simplification, of course, some information is lost. Hopefully, if the index is designed properly, the lost information will not seriously distort the answer to the question. Unfortunately, however, one may not know in advance which question will be asked. This situation creates the hazard that the index will be used for purposes other than those for which it was designed. This book takes the position that index development must begin with a carefully defined concept of the purpose of the index, and the original purpose must be respected when the index is being used.

ROLE OF INDICES

Various authors, governmental officials, and committees have emphasized the desirability of developing and utilizing environmental indices. The role that these indices are to play usually is linked to the basic reasons for which environmental monitoring data are collected. Environmental monitoring data consist of routine measurements of physical, chemical and biological variables that are intended to give insight into environmental conditions. These data often provide an important yardstick by which to judge the effectiveness of regulatory programs in improving environmental quality, as noted by Russell Train[1] in his 1973 address before the National Conference on Managing the Environment:

> Accurate and timely information on the status of the environment is necessary to shape sound public policy and to implement environmental quality programs efficiently. It is virtually impossible to develop effective programs and to monitor their implementation without good monitoring data. Very detailed data are necessary for certain types of planning and enforcement.

From a purely conceptual point of view, environmental monitoring data serve as a "feedback loop" by which to evaluate the effectiveness of regulatory activities. Once the environmental monitoring data are collected, there is a further need to translate them into a form that is easily understood. Train[1] sees environmental indices playing an important role in the translation process:

> For top management and general public policy development, monitoring data must be shaped into easy-to-understand indices that aggregate data into understandable forms. I am convinced that much more effort must be placed on the development of better monitoring systems and indices than we have in the past. Failure to do so will result in sub-optimum achievement of goals at much greater expense.

Once the indices are developed and applied, they should serve as "tools" to examine trends, to highlight specific environmental conditions, and to help governmental decision-makers evaluate the effectiveness of regulatory programs as noted by Train[2] in *Science*:

> A limited number of environmental indices, obtained by aggregating and summarizing available data, could be used to illustrate major trends and highlight the existence of significant environmental conditions. These indices could provide measures of the success of federal, state, local, and private programs in coping with environmental problems that must be solved.

Environmental indices, of course, are not the only source of information that is brought to bear on environmental decisions. Decision-making will be based on many other considerations besides indices and the monitoring data on which they are based. However, Train[2] sees environmental indices as performing an important role in making the technical data available to managers:

> Policy-making neither can nor should become totally "scientific." Vital decisions will always depend ultimately on the values we hold and on the way we express these values through the political system. But we must

also strive to make maximum use of the scientific evidence available to us, and development of environmental indices is one important way of doing this.

A report by the Planning Committee on Environmental Indices of the National Academy of Sciences (NAS)[3] concluded in 1975 that there has not been sufficient progress toward developing methods to evaluate environmental conditions:

> Environmental indices provide an important method for evaluating the state of the environment. Despite strong statements of need from all three branches of government, progress toward the development and use of methods for evaluating environmental quality has not been satisfactory.

The NAS report saw environmental indices performing an important role in four areas:

- assist in formulating policy
- provide a means for judging the effectiveness of environmental protection programs
- assist in designing these programs
- facilitate communications with the public concerning conditions of the environment and progress toward its enhancement

The Committee recommended that the federal government take an active role in developing environmental indices:

> A program to develop and use environmental indices (EI) should begin immediately. The Congress and the Executive should assure that this effort is implemented by directing and encouraging adequate programs and by providing appropriate stimuli and resources.
>
> The Committee has concluded that the development of EI is an important and urgent matter. Indices are needed now for a variety of reasons, and the need for indices will become increasingly apparent and pressing in future years. A useful system of EI can and should be initiated now. Programs to use existing indices and to develop additional EI should begin immediately. Thus, both short- and long-term goals can be identified for the development of environmental indices.[3]

Within the federal government, the President's Council on Environmental Quality (CEQ), an executive staff office, is charged with coordinating the development of methods to evaluate environmental conditions and trends. Under Section 102(2)(B) of the National Environmental Policy Act,[4] all federal agencies are required to ". . . identify and develop methods and procedures, in consultation with the Council on Environmental Quality . . ., which will ensure that presently unquantified environmental amenities and values may be given appropriate consideration in decision-making along with economic and technical considerations . . . ." In Section 204, the Act further directs the CEQ to ". . . gather timely and authoritative information concerning the conditions and trends in the quality of the environment, both current and prospective . . ." and ". . . to document and define changes in the natural environment. . . ."

The Environmental Protection Agency (EPA), which conducts environmental research and carries out a variety of regulatory activities, has a more "operational" role than the CEQ in environmental indices. EPA's functions in this area include developing the technical basis for indices, evaluating them, applying them to large data banks, and using them in environmental decision-making. The NAS Committee[3] recommended, for example, that EPA develop a standardized system of air pollution indices:

> The Committee recommends that a uniform national system of air quality indices be developed and adopted. The EPA appears to be the appropriate agency to have operational responsibility for EI concerning air and water quality.

Since the time of the Committee's report, the EPA has come forward with a recommended uniform air pollution index for use by state and local air pollution control agencies (Chapter III).

The author's review of the literature has identified six basic uses of environmental indices. The uses listed here are not necessarily unique to a given index, because indices sometimes are applied for more than one purpose. Nevertheless, one can find examples in the literature where an index has been developed or proposed for each of the following purposes:

- Resource Allocation — Indices may be applied to environmental decisions to assist managers in allocating funds and determining priorities.

- Ranking of Locations — Indices may be applied to assist in comparing environmental conditions at different locations or geographical areas.

- Enforcement of Standards — Indices may be applied to specific locations to determine the extent to which legislative standards and existing criteria are being met or exceeded.

- Trend Analysis — Indices may be applied to environmental data at different points in time to determine the changes in environmental quality (degradation or improvement) which have occurred over the period.

- Public Information — Indices may be used to inform the public about environmental conditions.

- Scientific Research — Indices may be applied as a means for reducing a large quantity of data to a form that gives insights to the researcher conducting a study of some environmental phenomenon.

In each of these applications, the index helps convey information about the state-of-the-environment. Because the questions being asked are different in each application, however, the index may differ in terms of the variables included, the basic structure, and the manner in which it is applied. Because different users have different data-reporting needs, identification of the users should be a critical part of the development and application of any environmental index, as suggested by Coate and Mason[5]:

> It is absolutely critical that the user be identified. The scientist, administrator, elected official, and general public cannot usually be satisfied by the same environmental measure. The administrator needs to see the resource allocation implications and the scientist needs to see the cause and effect implications. Who the user is will also affect geographical or political aggregation of data and the decision to highlight or obscure interjurisdictional comparisons.

Although a great many environmental indices have been developed and appear in the published literature, their use in evaluating governmental programs or in making management decisions appears to be very limited at the present time. In fact, the most successful use of environmental indices has been not in management but in the issuance of daily reports of air pollution levels to the public (public information). Even in this area, the findings reported in a published compendium of air pollution indices[6] show that very few of the indices originally proposed in the scientific literature ever found their way into routine use by air pollution control agencies to report daily air quality levels. Rather, state and local agencies tended to develop their own, individualized indices. Consequently, a great variation in air pollution index structures emerged from city to city, ultimately creating a need for the federal government to recommend a standardized air pollution index for public information reporting.

The process of developing environmental indices is itself a controversial area. The debate seems to center primarily around the amount of information that is lost in the simplification process made possible by the index. Persons who are very familiar with the complexities of making environmental measurements generally view the potential distortion occurring in an index as unacceptable. In contrast, persons who are more removed from the measurement process have greater willingness to accept the distortion for the sake of obtaining an easy-to-understand, although perhaps crude, picture of environmental quality. Coate and Mason[5] sum up the argument as follows:

> Much of the discussion as to the appropriateness of constructing and disseminating environmental information in the form of indices may be reduced to the following: The advocates say "something is better than nothing" and the adversaries say "something defective is worse than nothing." Both sides agree that better information is needed for decision-making

or public information purposes and both sides agree that perfect measures of environmental quality are impossible to construct except perhaps for very limited topical areas. The advocates say the best way to proceed is to develop indices, test their usefulness through experience, and refine the process through use.

This argument illustrates a "classic dichotomy" of views toward environmental indices. One view holds that the raw, undoctored data give the best means of evaluating environmental conditions. The other view holds that the raw data are too complex, that a simplification process is necessary, and that some distortion is acceptable in the process.

A book by Inhaber[7] compares environmental indices with economic indices, such as the gross national product (GNP), and argues that environmental indices would not be more difficult to produce than economic indices. He attributes the failure of environmental managers to use indices to the narrowness of their professional field, noting that economists have "over the past half century...built up a strong theoretical base for the indices they now use." He fails to point out one important difference between economics and the environmental sciences, however. GNP—the dollar sum of all the goods and services produced in the United States in a year—is a relatively simple concept, and members of the public can readily develop an intuitive feeling for its meaning. It is only slightly more complex than the concept of the birth rate or the total U.S. population.

A combined water quality index that includes dissolved oxygen, pH, temperature, suspended solids, chloride content, coliform organisms, color, hardness and turbidity would be much more complex than most economic indices and would be very difficult for the layman to understand intuitively. It would even be difficult for the environmental scientist to understand intuitively. What, for example, does an index change from "50" to "100" mean about water quality? Would it mean that the water is not drinkable, not swimmable, not capable of supporting aquatic life, or not suitable for irrigating crops? Would it mean that the water is "twice as dirty" as before, and in what way would it be twice as dirty? Index values might have changed simply because of seasonal temperature changes, or because of increased turbidity due to heavy rainfall. Regardless of the reasons, the layman has no frame of reference by which to intuitively gain a feel for the meaning of the index. Thus, the designer of environmental indices faces problems very different from those in economics.

LANGUAGE OF INDICES

Some common language has developed to describe environmental pollution data. In the area of environmental indices, however, not all of the terms are uniform. To avoid confusion, we shall introduce some definitions of terms which will be consistently used throughout the book.

In the mathematical sciences, the term "variable" usually refers to some attribute of interest which takes on different values. In the environmental professions, however, the word "parameter" has come into common usage as a substitute for "environmental variable." It means some measured environmental quantity. To avoid confusion between the mathematician's term and the environmental language, we shall use the term "pollutant variable" to denote *any physical, chemical or biological quantity intended as a measure of environmental pollution.* For example, reduced visibility due to atmospheric particles, sulfur dioxide concentration in the atmosphere, acidity of a stream, or the mass of pollutant emissions discharged from a smokestack per hour are all pollutant variables.

Environmental indices occasionally include pollutant variables reflecting the quantity of a pollutant released into the environment—mass of pollutants emitted from a smokestack or effluents released from a wastewater treatment plant—and not the quantities actually present in the ambient environment after diffusion and mixing have occurred. Because these pollutant source variables pertain only to pollutant quantities initially discharged into the environment, they do not directly reflect the state-of-the-environment. Variables which depict the state-of-the-environment, called *environmental quality* pollutant variables, are measures of actual ambient conditions—the content of pesticides in the soil, the concentrations of gases in the atmosphere, the quantities of toxic substances present in a stream. In this book, the term pollutant variable encompasses both environmental quality variables and pollutant source variables.

The term "environmental indicator" refers to a single quantity derived from one pollutant variable and used to reflect some environmental attribute. For example, the *number of days* that observed atmospheric concentrations of sulfur dioxide exceed some fixed ambient air quality standard represents an indicator of sulfur dioxide pollution levels. Similarly, a dimensionless number that varies between 0 and 1 and reflects the quantity of dissolved oxygen in a stream is an environmental indicator of dissolved oxygen content.

Environmental indicators can be presented individually or they can be mathematically aggregated in some fashion to form an "environmental index." An "index" is a single number derived from two or more indicators. In computing an index, the first step is often to compute the individual indicators, one for each pollutant variable; thus, the indicators also are referred to as "subindices." In summary, the main difference between an indicator and an index is that an indicator is derived from only a single pollutant variable, while an index is derived from more than one pollutant variable. A number of indicators presented at the same time to give a picture of

INTRODUCTION 9

environmental conditions (but not aggregated together) is referred to as an "environmental quality profile."

ENVIRONMENTAL QUALITY PROFILE

The Environmental Protection Agency's Seattle Regional Office (Region X) brought together data on a variety of environmental topics—air pollution, water pollution, radiation, pesticides, solid waste, and noise—when it published its first Environmental Quality Profile in 1976.[8] In preparing this report, it sought to communicate information on the state-of-the-environment through a series of easy-to-understand bar graphs and charts. Although the publication does not go too deeply into the technical details of each environmental topic, it is impressive for its breadth of coverage. Thus, it is useful to discuss this material in some detail to give the reader who is unfamiliar with environmental variables an introduction to environmental quality data. The original charts were in color. The black and white version presented here, which uses intermediate shading to reflect the colors, is somewhat less dramatic than the original version.

The data reported in this Environmental Quality Profile relate to the 833,000-square-mile area represented by Alaska and the Pacific Northwest States of Idaho, Oregon and Washington. The intended audience of the report includes members of the public and government officials:

> This document is directed to both the public and to Congressional, State, and local officials. It is intended to help develop an overall perspective on environmental issues as well as to assist in policy analysis, program management, and program evaluation. This report is to be the first annual report to the people of Region X and to their elected and appointed officials on the quality of our environment. Is it improving? Where are the problems? And what can be done to solve them?[8]

Figure 1, which is from the Profile, presents an overall synopsis of environmental conditions in the four-state area. For reporting water quality violations, two indicators are used: (1) river miles in the regional area not meeting ambient standards, and (2) severity of the violation of the standards. Two similar indicators are employed to report air pollution: (1) number of days during which the ambient air quality standards are violated and (2) severity of the levels on those days in which the standards are violated. The indicators used for reporting radiation and noise are based on the number of persons exposed, while the pesticide indicator is based strictly on the pesticide concentrations measured in food, air and water. Although none of the eight indicators shows a "satisfactory" condition in 1976, six of them show that the trend in the environmental problem is toward improvement. The indicator of environmental noise implies that the noise problem, already serious, is "worsening."

10 ENVIRONMENTAL INDICES

COMPONENT		INDICATOR	TREND
WATER	RIVER MILES NOT MEETING STANDARDS		IMPROVING
	SEVERITY OF VIOLATION OF STANDARDS		IMPROVING
AIR	DAYS OF STANDARDS VIOLATIONS		IMPROVING
	SEVERITY OF POLLUTION IN DAYS IN WHICH STANDARDS VIOLATED		IMPROVING
RADIATION	NEAR-TERM EXPOSURE		NO CHANGE
PESTICIDES	CONCENTRATION IN FOOD, WATER, AIR		IMPROVING
SOLID WASTE	PERCENT POPULATION SERVED BY SANITARY LAND-FILLS		IMPROVING
NOISE	NUMBER OF PERSONS EXPOSED TO UNACCEPTABLE NOISE LEVELS	SERIOUS PROBLEM	WORSENING

LEGEND

☐ SATISFACTORY CONDITION   ▒ AREAS OF CONCERN; MORE ACTION NEEDED   ▓ SERIOUS PROBLEM

Figure 1. Synopsis of environmental indicators in 1976 for the Environmental Protection Agency's Region X, which includes Alaska, Oregon, Washington and Idaho.[8]

While Figure 1 gives a useful general overview, it does not offer insight into specific details of the problems. Thus, the Region X Environmental Quality Profile then discusses each subject area separately, offering more detailed indicators for each area.

Water Quality

The sections of the Environmental Quality Profile dealing with water pollution discuss general water quality first, followed by drinking water quality. Within the 833,000-square-mile area of the four states, the report notes that there are 35,855 miles of tidal shoreline, streams and inland bodies of water, with 38,340 miles of riverbank and shoreline.

Although these states are relatively free of industrial water pollution, the majority of the principal river waters in the Pacific Northwest do not meet all water quality standards. Figure 2 shows the percent of "stream miles" meeting the water quality goals of the Federal Water Pollution Control Act,[9] according to criteria recommended by the National Academy of Sciences.[10] The names of the Region's 16 principal river basins are listed at the bottom

INTRODUCTION 11

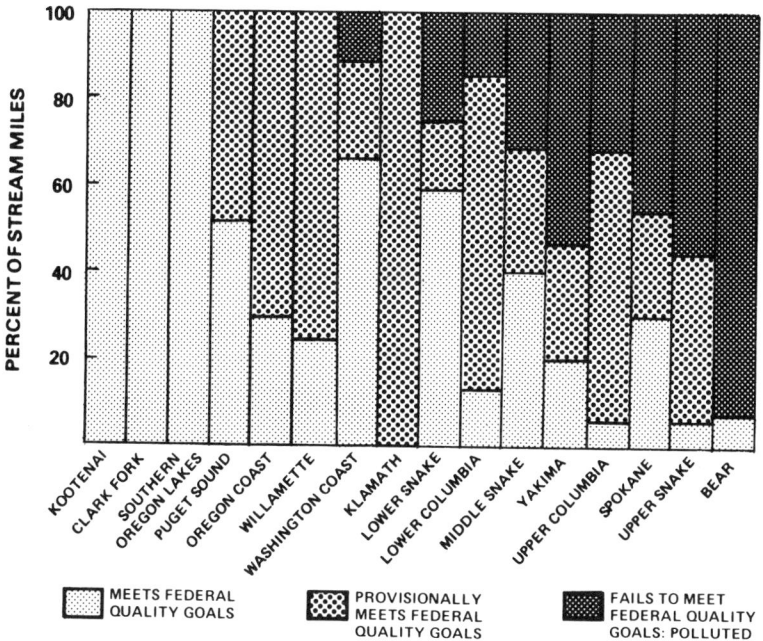

Figure 2. Relative water quality of Region X's 16 prinicipal river basins.[8]

of the chart. A common reason for violating water quality standards in this part of the United States is high bacterial concentrations or thermal pollution. The Environmental Quality Profile also notes a disturbing increase in the presence of organic and inorganic toxic substances. Although these increases may be traced to increasing industrialization of the area, there was too little sampling in the past to firmly establish a trend. Thus, toxic materials such as polychlorinated biphenyls (PCB's) may have been more common than was suspected before active water quality monitoring began.

To illustrate water quality trends, the Environmental Quality Profile presents summary charts for the time periods from 1966 to 1974 (Figures 3 and 4). Over this time period, state and federal pollution control programs have emphasized control of municipal and industrial dischargers, or "point sources." The effectiveness of these programs is evident in Figure 3, which shows criteria violations for point-source related pollutants. Between 1966 and 1974, the number of observations not meeting these criteria declined from 31% to 21%, a 33% reduction. If, however, the water quality evaluation is broadened to include variables that are not point-source related,* such as

---
*A "nonpoint source" is one that does not discharge from a particular point, such as the runoff of rainfall from crop land.

12 ENVIRONMENTAL INDICES

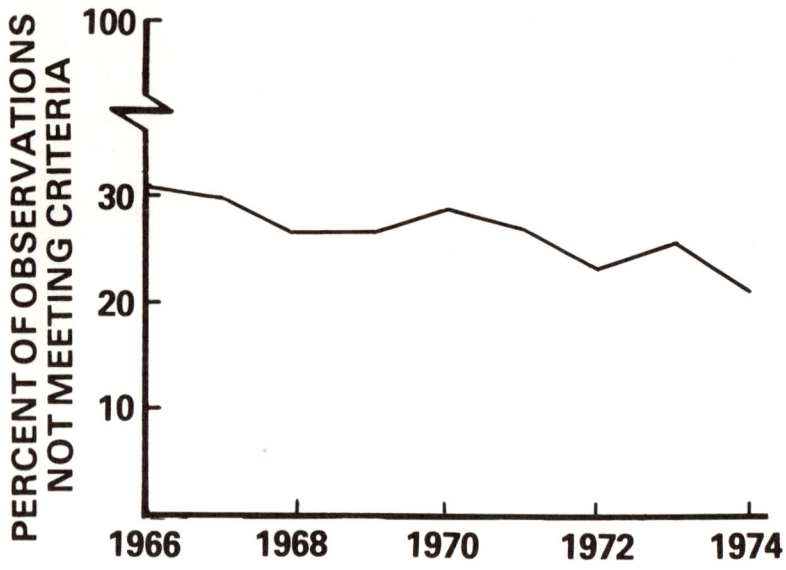

Figure 3. Status of point-source related water pollutants from 1966 to 1974 in Region X.[8]

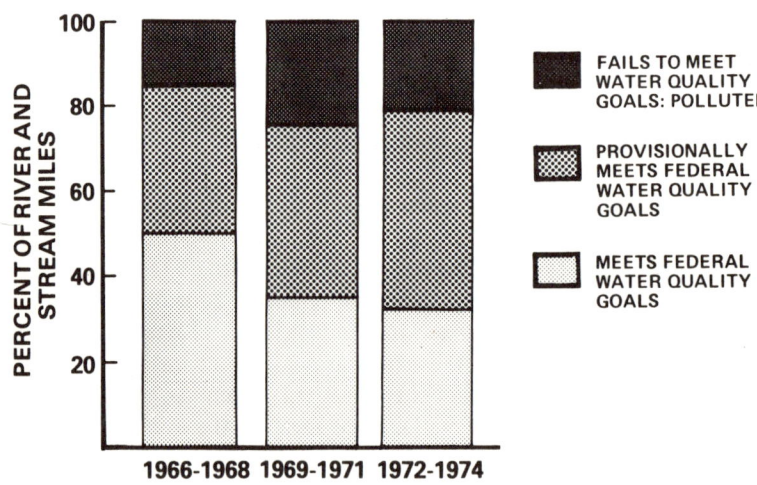

Figure 4. Stream miles meeting federal goals in Region X, 1966-1974.[8]

INTRODUCTION 13

pesticides and gas saturation, a different picture of water quality trends emerges. As shown in Figure 4, the proportion of total stream miles that consistently meets federal water quality goals (darkest shading) is decreasing, indicating a perceptible decline in water quality in the Northwest during the past decade. This change is offset to some extent by increases in the total stream miles that provisionally meet federal goals. The reasons for this decline apparently are related to the intensity of water and land use in the Northwest. Most of the apparent overall decline occurs in rivers east of the Cascades, which are heavily used for power generation and irrigation. The Environmental Quality Profile attributes this change, in part, to a lack of emphasis on nonpoint source problems by federal and state pollution control programs over the past 10 years. In the future, nonpoint source control programs will be receiving greater emphasis in the Northwest.

The Profile also presents a more detailed set of indicators for violations of water quality standards and criteria for each of the 16 major river basins (Figure 5). Simplified definitions are provided to explain each characteristic of the water associated with violations of the standards (Table I), and these definitions are reproduced here to assist the reader in understanding some of the more common water quality variables of importance. In the figure, an upward pointed arrow indicates that the pollution characteristic is worsening; a downward arrow indicates it is improving; and a horizontal arrow indicates no change in the recent past. A lack of shading implies that the characteristic is not a contributor to violations of standards; light shading indicates that the characteristic is a minor contributor; and dark shading indicates that it is a major contributor to violations of standards.

Air Quality

Under the Clean Air Act,[11] the federal government has established air quality standards, which apply on a national basis, for six air pollutants: carbon monoxide, hydrocarbons, photochemical oxidants, nitrogen oxides, sulfur dioxide, and particulates.[12] These National Ambient Air Quality Standards (NAAQS) are further divided into two categories: *primary standards,* which are set at levels designed to protect public health, and *secondary standards,* which are set to protect against the many other, nonhealth-related effects of air pollution (damage to materials, plants and animals, reduction in visibility, economic losses, and other factors). The NAAQS are discussed in greater detail in connection with the analysis of air pollution indices (Chapter III). The characteristics of these six pollutants are described briefly as follows:

14 ENVIRONMENTAL INDICES

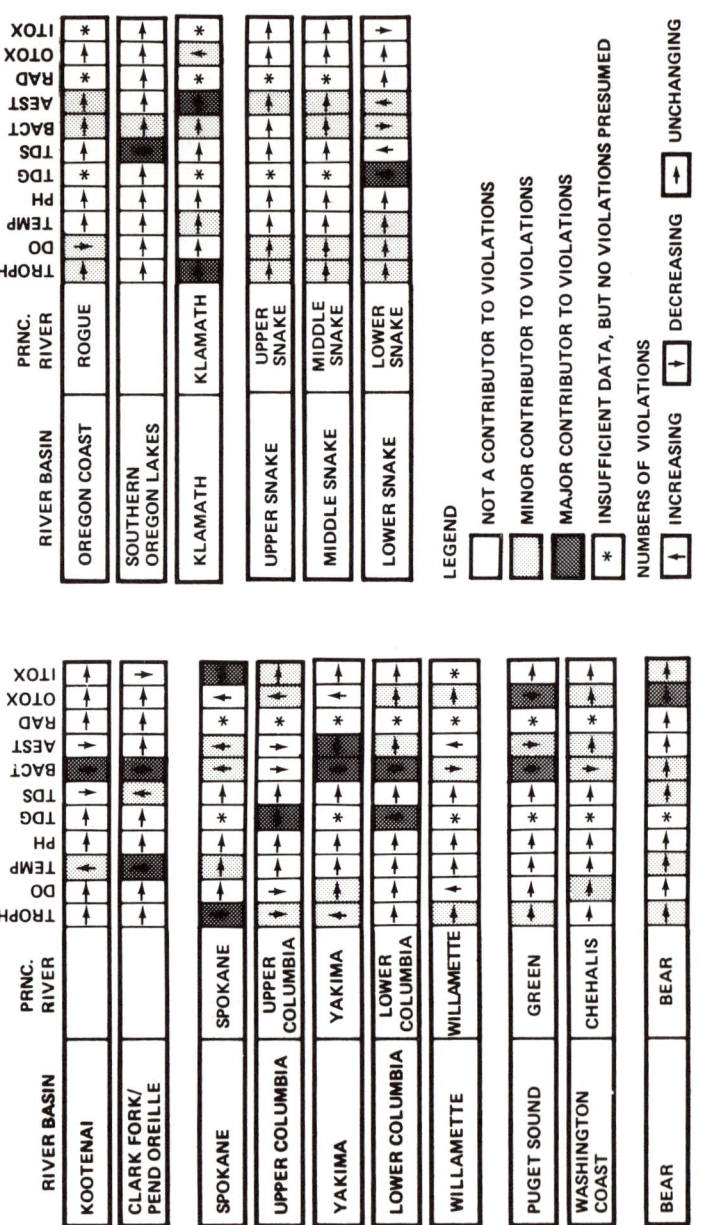

Figure 5. Trends in 10 principal water quality characteristics for Region X's 16 major river basins.[8]

Table I. Definitions of Water Quality Variables Included in Region X Environmental Quality Profile[8]

TROPH:	(Trophic Condition) This refers to the intensity of biological activity. Excessive biological activity is characterized by very murky, turbid water and nuisance-level growths of algae and aquatic weeds.	TDS:	(Total Dissolved Solids) This is the measure of nongaseous minerals in solution in water–its relative "saltiness." Excesses affect water taste, plant and animal life, and cause mineral build-up in pipes and appliances.
DO:	(Dissolved Oxygen) Oxygen dissolved in water or wastewater. Adequate dissolved oxygen is essential to the life of fish and other aquatic organisms. Discharge of excessive organic solids generally is the cause of low DO concentrations.	BACT:	Bacteria indicating the probable presence of disease-causing organisms and viruses not natural to water. They come from the intestines of warm-blooded animals, including man.
TEMP:	Temperature of water governs both the nature of life forms and the rate of chemical reactions. In general, higher temperatures are undesirable for the types of fish and shellfish found in the Northwest and Alaska.	AEST:	(Aesthetics) This refers to detectable oil and grease, sediment, and similar considerations.
		RAD:	Radioactivity may be present in water as a result of discharge of radioactive wastes or fallout. Its slow decay presents a direct threat to aquatic life and potential increase in the cumulative dose rate for other life forms.
pH:	This is a measure of acidity or alkalinity of water. Extreme levels of either can imperil fish life and speed corrosion.	OTOX:	Organic toxicants include pesticides and other poisons that have the same effects and persistence as pesticides.
TDG:	(Total Dissolved Gases) This is a measure of the concentrations of gases in solution in water, and it can affect the metabolism of aquatic life forms. High concentrations of gases in excess of 120% saturation can cause high mortalities in migrating fish.	INTOX:	Inorganic toxicants are the heavy metals and other elements. Although many are naturally found in water and are essential to life in low concentrations, excess concentrations are poisonous.

Carbon Monoxide

Carbon monoxide (CO) is a colorless, odorless gas produced by incomplete combustion. Its principal source in metropolitan areas is the motor vehicle. Carbon monoxide, when inhaled, is rapidly taken up by the bloodstream, where it combines with the hemoglobin molecule. By replacing the oxygen normally carried by this molecule, carbon monoxide impairs the ability of the bloodstream to supply oxygen to the brain. Carbon monoxide thereby can affect mental function at relatively low concentrations and can reduce visual perception and alertness. CO also impairs heart function by weakening the contractions of the heart, which decreases the supply of blood to various parts of the body. In healthy persons, the result is reduced ability to perform exercise. In persons with heart disease, who may be unable to compensate for reduced oxygen, the result is to decrease the likelihood of recovery from heart attacks. Carbon monoxide also may adversely affect persons with lung disease, anemia, or cerebral-vascular disease.[13]

Nitrogen Oxides

The oxides of nitrogen usually originate in high-temperature combustion processes and, to a lesser extent, in chemical plants. Although measurement of this pollutant in the atmosphere has proved difficult due to problems with the measurement technology, experience has shown that, in various forms, the oxides of nitrogen can affect humans as well as materials and vegetation. Instances of extremely high occupational exposures involving firemen, welders, silo fillers, miners, chemists and other industrial workers have shown that nitrogen dioxide (NO_2) can be fatal to humans. At lower levels, it can cause acute bronchitis and pneumonia. The group of pollutants known as nitrogen oxides also can affect lung tissue and lower the resistance of laboratory test animals to influenza. Nitrogen oxides participate, along with hydrocarbons, in atmospheric chemical processes which, in the presence of sunlight, produce photochemical oxidants, or "smog."[14] This class of air pollutants irritates the respiratory tract and causes a stinging, watery reaction in the eye.

Hydrocarbons

The term "hydrocarbons" refers to a class of compounds whose molecules contain just two elements, hydrogen and carbon. The hydrocarbons commonly found in urban atmospheres include such compounds as methane (CH_4), ethane (C_2H_6), propane (C_3H_8), ethylene (C_2H_4), propylene (C_3H_6) and acetylene (C_2H_2). Motor vehicles are the chief source of hydrocarbon emissions, with the remainder coming from evaporation of industrial solvents in painting, dry cleaning, and other processes and from the marketing of gasoline. Although no adverse effects on human health are directly attributable to hydrocarbons, this pollutant does participate in the chemical reactions that produce photochemical oxidants.[15]

Photochemical Oxidants

Photochemical oxidants are not emitted directly into the atmosphere but arise from a complex series of chemical reactions initiated when hydrocarbons and the oxides of nitrogen are exposed to sunlight. Ozone (O_3), peroxyacyl nitrate (PAN), formaldehyde, acrolein, nitrogen peroxide, and organic peroxides are all formed in this manner. Photochemical oxidants can affect the eyes, causing symptoms of tears and inflammation, and can cause changes in lung function. The principal constituent of photochemical oxidants, ozone, is a severe irritant to all mucous membranes. At certain atmospheric concentrations, photochemical oxidants have been shown to impair the performance of athletes and to affect persons with asthma.[16]

Particulates

The terms "particulates," "particulate matter," or "total suspended particulates (TSP)" all refer to the same thing: the total mass of liquid and solid particles in the air, such as soot, smoke, dust, mist and sprays. This pollutant class includes a wide variety of nontoxic materials such as fine sand and dirt, as well as other materials known or suspected to be toxic, such as beryllium, lead, asbestos, cancer-producing organic compounds, and suspended sulfates and nitrates. Atmospheric particulates often arise from the combustion of fossil fuels, such as coal and oil, which contain metals and other substances. When burned, these substances form small, solid particles. The bulk of the suspended particulate matter in the atmosphere lies in the size range between about 0.1 and 10 μ. Particulates often are small enough to penetrate deeply into the lungs, where they can cause either temporary or permanent injury. Injury may be confined to the surface lining of the lungs or throat, although particulates may affect the entire body either by weakening resistance to infection or by introducing chemicals which cause cancer or other diseases. Inhaled lead particulates, for example, may cause lead poisoning—manifested by nervous and blood symptoms—while causing little damage to the lung itself. Analysis of epidemiological studies indicates an association between air pollution, as measured by particulate matter accompanied by sulfur dioxide, and health effects of varying severity.[17]

Sulfur Oxides

The oxides of sulfur include sulfur dioxide (SO_2) and sulfur trioxide (SO_3), which are transformed in the presence of moisture and catalysts to sulfurous acid (H_2SO_3) and sulfuric acid (H_2SO_4). In the United States, over 95% of the sulfur oxides emissions arise as an unwanted by-product of the combustion of coal and oil for generation of heat and electricity. These fossil fuels usually contain sulfur, often in the range of 1 to 3%, in the form of inorganic sulfides and sulfur-containing organic compounds. The sulfur is difficult to remove from the fuels before they are burned and difficult to scrub from the stack gases after combustion, where it is in the form of gaseous sulfur dioxide and sulfur trioxide at high temperatures. In major U.S. cities, increases in the number of deaths have been observed when both atmospheric sulfur dioxide concentrations and particulate concentrations were high, suggesting that the particulates may carry sulfur dioxide and its related compounds into the lungs, where they can do considerable damage to the respiratory system. The tendency for two pollutants to work together in such a way that the health effects are magnified is referred to as "synergism." In addition to adverse health effects, high concentrations of sulfur dioxide are associated with the discoloration of buildings, corrosion of materials, and damage to statuary and other surfaces.[18]

In dealing with air quality, the EPA Region X Environmental Quality Profile uses four of the six air pollutants for which NAAQS exist—carbon monoxide, oxidants, particulates and sulfur dioxide. The Profile identifies a total of 20 counties which comprise the principal industrial and populated areas of high air pollution potential in the Northwest. The concentrations of various air pollutants are routinely measured at selected sites in each county. When more than one site existed in a county, the site with the highest concentration for each pollutant type was used. Thus, the data are based on "worst case" conditions rather than "representative" conditions.

The indicator selected to characterize the air quality problems of each county was the number of days that the NAAQS were violated and the "severity" of the violations. The top chart in Figure 6 shows the number of days during the year that the NAAQS were violated for each pollutant. For these counties, the greatest number of violations of the standards can be attributed to carbon monoxide, with particulate matter playing an important role in the Alaskan counties. Photochemical oxidants come after carbon monoxide and particulates, with sulfur dioxide making a negligible contribution. However, sulfur dioxide may still be a significant problem in these counties. Many localized points are very close to particular sources where measured sulfur dioxide concentrations show up a problem, but these concentrations are not reflected in the present data which are collected at locations in the general urban area.

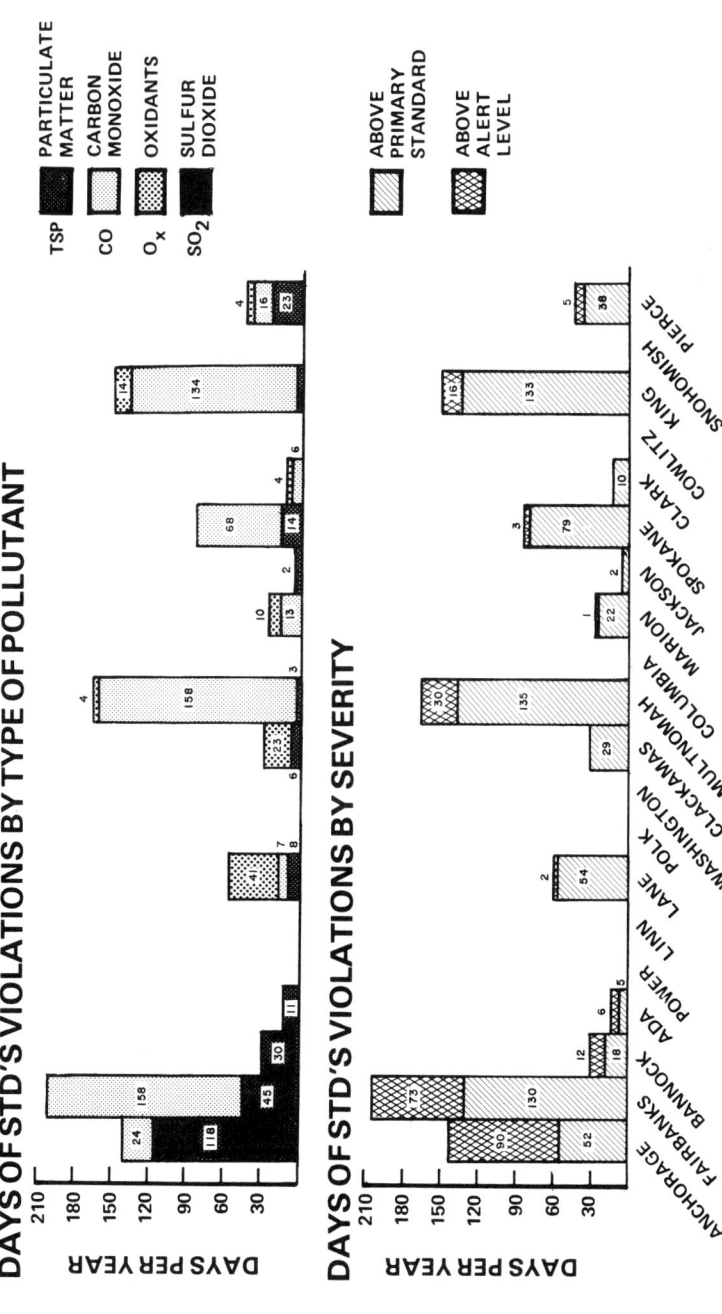

Figure 6. Air quality levels in 20 Region X counties in terms of the number of violations of ambient air quality standards and the severity of these violations.[8]

In addition to the NAAQS levels, the federal government has published a recommended system for avoiding air pollution episodes.[19] Three levels are defined—Alert, Warning, and Emergency. These levels, which are slightly higher than the NAAQS values, are designed to trigger short-term administrative control actions such as reducing traffic or curtailing industrial activities. Usually action is taken only when adverse meteorological conditions also show signs of persisting. The Alert level is the level at which the public is notified that a serious air pollution problem exists. The Profile uses two measures of the severity of air pollution (Figure 6): the number of days in which the concentration is above the Alert level (dark shading) and the number of days in which the primary NAAQS is violated (light shading).

The two severity indicators give a somewhat different picture of environmental conditions than an indicator based on the NAAQS alone. For example, there were more days in violation of the NAAQS (but not the Alert level) in Fairbanks (130 days) than in Anchorage (52 days), while there were more days in violation of the Alert level in Anchorage (90 days) than in Fairbanks (73 days). Thus, air quality reaches undesirable levels in Anchorage less frequently than in Fairbanks, but, when it does, it reaches more severe levels. These results illustrate two separate, important characteristics of environmental data which must be dealt with when making comparisons of environmental quality: (1) *frequency* of adverse conditions, and (2) *degree of severity* of adverse conditions. It is not unusual to find one location that ranks above another in terms of the frequency of adverse conditions, while the ranking is reversed if just the severity of these conditions is considered. Which location, then, has the most serious environmental pollution problem? This difficulty in interpreting the results, along with many differences in the ways measurements are made and variations in quality control practices, makes comparison of air quality levels in different cities a formidable technical challenge. As discussed in greater detail in Chapter III, these difficulties contribute to a "ranking paradox" characteristic of environmental quality measurements.

The air quality trends over time in Region X can be further described on a county-by-county basis by pollutant (Figure 7). On the figure, the shading indicates the severity of pollution levels while the arrows indicate the trend—whether the number of days of standards violations is increasing (upward arrow), decreasing (downward arrow), or remaining the same (horizontal arrow). The narrative with the Profile describes the overall air quality conditions as follows:

> The Seattle and Spokane metropolitan areas still have significant carbon monoxide problems too many days of the year. Carbon monoxide (CO) is a major threat to the residents of Fairbanks, Alaska, during its long winter and is becoming an increasing problem in other Alaska cities such as

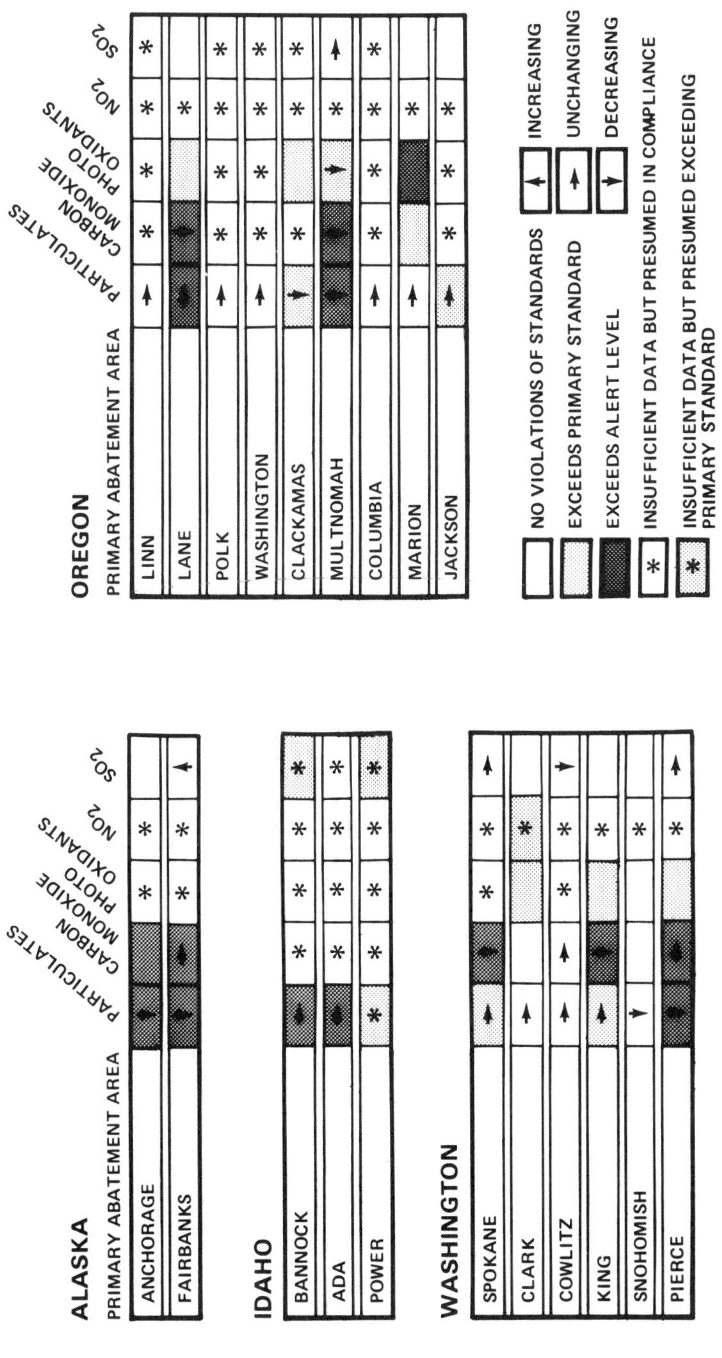

Figure 7. Status of air quality levels in 20 counties in Region X, by pollutant.[8]

Anchorage. Developments associated with the Alaska Pipeline are expected to compound this latter problem.

Portland, Oregon, has frequent CO problems and other air quality problems. In other communities, such as Tacoma, Washington, and Kellogg, Idaho, the problems stem from industry. Heavy metals and sulfur oxides emissions from smelters have long been difficult problems in both areas.

EPA is working closely with the States of Alaska, Washington, and Oregon to establish mechanisms to reduce the CO levels in problem areas. This includes measures to reduce emissions from vehicles as well as measures to reduce the number of vehicle-miles traveled in urban centers having high levels of CO.[8]

These narrative excerpts show that the detailed presentation of air quality data possible in the Profile permits administrators to set priorities regarding the geographical distribution and importance of various air pollution control actions.

Radiation

The Region X Profile treats radiation in less detail than air or water pollution. The chart summarizing the radiation problem (Figure 8) considers two radiation sources—fallout and industry. Radiation from atmospheric fallout, attributable to nuclear weapons testing, was relatively serious in the past, but it has assumed reduced importance and currently is below applicable radiation standards. By contrast, the number of nuclear power plants in Region X is expected to increase by the year 2000, and the increased handling and processing of nuclear fuels raise the potential for increases from this source over the next 5 to 50 years. To place the problem of radiation exposure in its proper perspective, the Profile lists the exposure contributed by various radiation sources (Table II). Although the largest potential source of radiation exposure is from natural background sources, there is a significant, continuing dose rate from medical uses of ionizing radiation, and it is believed that this rate will continue.

Pesticides

The pesticide problem can be divided into two general categories: "persistent" and "nonpersistent" substances. Persistent pesticides have chemical properties that are comparatively resistant to natural breakdown processes. This group includes most of the chlorinated hydrocarbon insecticides (DDT, aldrin, dieldrin, etc.) and some of the mercury-containing fungicides. The nonpersistent group consists of naturally derived as well as synthetic substances used to control pests. Pesticides of this type usually are either more degradable (such as organophosphates and carbamates) or relatively more selective as to the target pest affected (such as pyrethrins, biological controls, etc.) than the persistent substances.

SOURCE OF RADIATION EXPOSURE	PAST	CURRENT	NEAR-TERM (1-5 YR.)	LONG-TERM (5-50 YR.)
FALLOUT	ABOVE STANDARDS	BELOW STANDARDS	DECREASING	DECREASING
INDUSTRY	IN COMPLIANCE	IN COMPLIANCE	MAY INCREASE BUT STILL IN COMPLIANCE	POTENTIAL FOR INCREASE

Figure 8. Status and trends for exposure to radiation, from Region X Environmental Quality Profile.[8]

Table II. Radiation Dose Rate from Various Sources[8] (Source of Exposure in Millirems[a])

	1960	1970	2000
Natural	130.0000	130.0000	130.00
Occupational	0.7500	0.8000	0.90
Nuclear Power	0.0001	0.0020	0.20
Fuel Reprocessing	-	0.0008	0.20
AEC Activities (excluding open-air weapons tests)	0.0100	0.0100	0.01
Open-Air Weapons Tests	13.0000[b]	4.0000	4.90
TV, Consumer Products, Air Travel	1.6000	2.6000	1.10
Diagnostic Radiology	72.0000	72.0000	72.00
Total	217.0000	209.0000	209.00

[a] The millirem is a unit of biological dosage related to the amount of energy deposited in tissue by various kinds of ionizing radiations.
[b] 1963 data.

The Profile presents the general status and trends of pesticides in the environment somewhat more qualitatively (Figure 9) than air or water pollution. Because of their residual properties, persistent pesticides build up in the systems of animals as one moves up the food chain. Little is known about the possible consequences of long-term exposure to even small amounts of these substances in our everyday existence, and the use of many of these persistent substances recently has markedly declined. This decreasing trend is best illustrated in the case of DDT, which is now prohibited except under very special conditions. The organophosphates (nonpersistent), which have been substituted for the chlorinated hydrocarbons, are increasing in use. Because of their short life, however, their concentrations do not significantly increase.

24 ENVIRONMENTAL INDICES

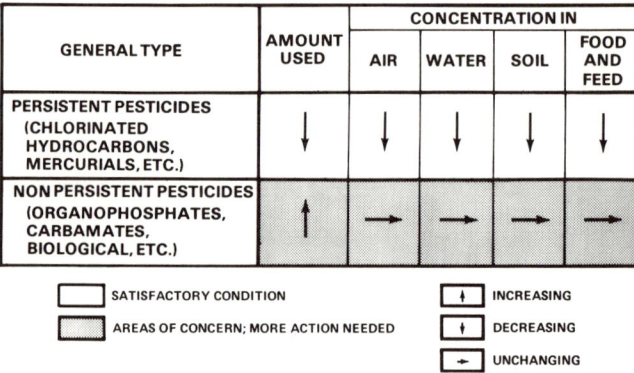

Figure 9. Trends in environmental concentrations for the two general categories of pesticides, as presented in the Environmental Quality Profile.[8]

Solid Waste

The Profile notes that waste management activities present a spectrum of problems, from extreme health and environmental hazard to efficiency of collection operations. The diverse nature of the wastes (dead animals, mercury-rich industrial sludges, dredge spoils, abandoned cars, septic tank pumpings, residential solid waste, infectious hospital wastes, demolition debris, feedlot wastes, etc.) and the diversity of their occurrence make the problem of waste management very complex. As an indicator of progress in this area, the Profile lists the percent of the population served by waste disposal facilities that meet state requirements in Washington, Oregon, Idaho and Alaska (Figure 10).

Noise

The Noise Control Act of 1972[20] gives the Environmental Protection Agency authority to set noise standards for new products that are major sources of noise (cars, trucks, etc.) and existing noise sources that require nationally uniform treatment (interstate railroads, trucks and aircraft). However, as with much other environmental legislation, the primary responsibility for control of noise rests with state and local governments. The federal government provides technical assistance to states and communities that need help in writing laws and establishing noise control enforcement programs. The Region X Profile presents the status of noise control efforts in terms of four basic sources of noise (Figure 11). Noise from aircraft and ground

INTRODUCTION 25

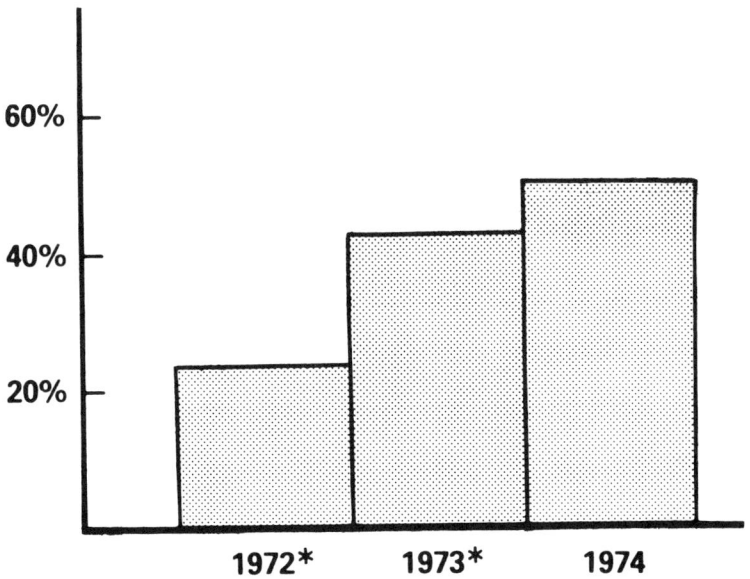

Figure 10. Percent of Region X population served by solid waste disposal facilities meeting state requirements.[8] *Excludes Alaska.

PRINCIPAL AREAS OF NOISE CONTROL

TYPE OF CONTROL	FEDERAL	STATE	LOCAL
INTERSTATE VEHICLE STANDARDS	EXIST		
CONSTRUCTION NOISE STANDARDS (ORDINANCES)			
BUILDING CODES (INSULATION & LOCATION)			
AIRPORTS/AIRCRAFT	EXIST	IMPLE-MENTATION NEEDED	IMPLE-MENTATION NEEDED

TRENDS IN NUMBERS OF PERSONS EXPOSED TO UNACCEPTABLE NOISE LEVELS

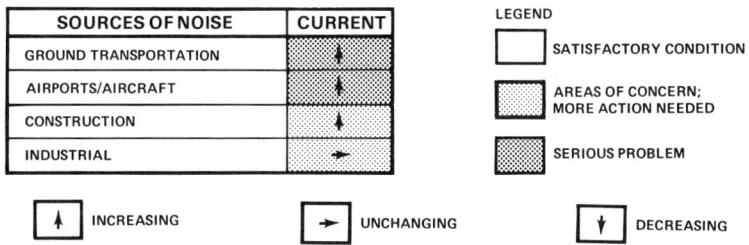

Figure 11. Status of noise control efforts, as presented in the Region X Environmental Quality Profile.[8]

transportation, already regarded as a "serious problem" (dark shading), is increasing. Noise from construction activities, while not yet regarded as a serious problem, is an area of concern (light shading) and is also increasing. Industrial noise, also an area of concern, remains relatively constant, however. Although adequate federal noise controls exist for interstate motor vehicles and aircraft operations, the Profile indicates that further implementation is required at the state and local level.

Discussion of the Profile

The Region X Environmental Quality Profile appears to provide a useful format for presenting environmental quality data to managers and to the public, enabling them to better understand environmental conditions and changes in these conditions over time. Although the raw data are not presented, each indicator comprising the Profile bears a close relationship to the raw data. Where possible, the pollutant variables are presented in such a fashion that they can be compared with the legal and administrative levels (existing standards and criteria).

In a profile, the observer's eye scans the separate, individual indicators, and his mind is asked to aggregate the indicators to form an overall impression of environmental conditions. Because the mathematical aggregation of different pollutant variables to form a single number does not occur, proponents of profiles see them as giving less chance for misinterpretation or misunderstanding than aggregated indices. For example, an indicator reporting the "number of days exceeding the CO national standard" is very specific and leaves little room for ambiguity, unlike, perhaps, a dimensionless (aggregated) environmental quality index whose value has just changed from $I = 110$ to $I = 140$. Each indicator is just what it claims to be, and the possibility for misunderstanding is reduced.

In an index, on the other hand, the aggregation process is carried out by a mathematical equation and not by the observer. Development of the initial equation almost always requires more assumptions and arbitrary decisions than design of an environmental quality profile. Thus, aggregated indices are frequently criticized by scientists, engineers and public health specialists familiar with the data, who feel that the assumptions can introduce serious distortions. They argue that the distortions can cause the observer to misinterpret the data.

Critics of environmental quality profiles usually argue, conversely, that the presentation is still too complex, that the reader must possess considerable knowledge about the chemical and physical properties of the individual pollutant variables to understand the meaning of the indicators, and that so much information is presented on the charts and tables that it is difficult to absorb. In short, these critics would prefer to see the profile simplified down to a single number, an "index." They look to the scientific community to construct an index that will give the necessary simplification without introducing the distortion usually inherent in the aggregation process.

This discussion once again illustrates the "classic dichotomy" of views toward indices. One viewpoint prefers the data in the most complete form possible, but is willing to accept the resulting complexity, while the other viewpoint prefers the data in as simple a form as possible, but is willing to accept the distortion introduced in the simplification process.

THE SEARCH FOR DAMAGE FUNCTIONS

The process of creating meaningful indicators and indices of environmental quality would be simplified if valid mathematical functions were available relating pollutant variables to their effects on man and his environment. If an accurate equation were available, for example, which could predict the number of additional deaths in New York City as a function of the measured concentration of each air pollutant, this function would provide a useful guide for designing air pollution indicators. If a second function were available for accurately translating the measured observations of all pollutants into an "estimated death rate," a general air pollution index could conceivably be based on these estimates. Such a relationship, called a "damage function" or a "dose-effect function," is an equation or a set of curves relating pollutant variables to observed effects on materials, plants, animals, biological organisms, human health, and the aesthetics of man's surroundings.

A literature review by Hershaft, Morton and Shea[21] defines a damage function as follows:

> A damage, or dose-effect, function is the quantitative expression of a relationship between exposure to specific pollutants and the type and extent of the associated effect on a target population. In plotting such a function, the ordinate may represent either the number of individuals affected or severity of effect, whereas the abscissa indicates the dosage in terms of time and ambient concentration. In reporting a damage function, one needs to specify the pollutant, the dose rate, the effect, and the target population.

These authors make a distinction between a biological damage function and an economic damage function:

> The damage can become manifest in a number of ways and can be expressed in either physical and biological, or economic terms. If the effect is physical or biological, the resultant relationship is known as a physical or biological damage function, or a dose-effect function. In an economic damage function, on the other hand, the effect is expressed in monetary terms. Economic damage functions can be developed by assigning dollar values to the effects of a physical or biological damage function, or by direct correlation of economic damages with ambient pollutant levels.[21]

Various investigators have made initial progress toward developing damage functions. Examples in air and water pollution include development of a relationship between SO_2 concentration and the corrosion rate of zinc panels exposed to the atmosphere,[22] curves relating oxidant exposure levels to injury in the tobacco leaf,[23] data showing a relationship between total suspended solids in streams and trout and shellfish density,[24] and functions expressing a relationship between total suspended solids in tap water and life expectancy of household plumbing fixtures such as toilets.[25]

Because of the importance of the impact of air pollution on human health, a great deal of health-related research has been undertaken in the air pollution field. A variety of studies documenting the effects on health and welfare of the various air pollutants are summarized in the air quality criteria documents which have been published for each of the major pollutants.[13-18] Because of the complexity of establishing meaningful dose-effect functions over the entire range of a pollutant variable, most published studies have focused on determining the nature and character of an effect, or the minimum concentration at which adverse effects occur rather than a full dose-effect function. The problem of establishing valid dose-effect functions in air pollution is extremely formidable from a scientific standpoint, because so many other factors affect human health in ways that are not yet well understood. Establishing a more general damage function that includes the entire range of effects—sickness and increased incidence of respiratory illness, impaired physical activity and reduced coordination, soiling of buildings, corrosion of materials, reduced plant growth, reduced visibility, and other factors—is even more formidable.

Liu and Yu[26] have attempted to combine data on the effects of air pollution on the deterioration of materials, household soiling, and human health (morbidity and mortality) to produce physical and economic damage functions which can be used for making regional damage estimates. Applying their approach to 40 Standard Metropolitan Statistical Areas, they found the largest aggregate air pollution damage, approximately $1 billion, in both New York and Chicago, with the smallest damage, approximately $15 million, in each of two Pennsylvania cities, Johnstown and York. For 40 metropolitan areas, they estimated the damage to human health from air pollution as ranging from $1.5 to $2.2 billion, with material deterioration damage at about $0.7 billion, and household soiling costs at about $3 billion. Their study should be treated as a crude "first step," however, due to many methodological problems and the limited understanding of basic cause-effect relationships. Like other investigators in the field, Liu and Yu note that the significance of their findings must be qualified by various theoretical and empirical limitations:

The major difficulties often encountered in estimating air pollution damage involve the lack of knowledge regarding the shapes of functions describing the relationship between air pollution and various receptors, and the lack of a satisfactory theoretical model specifying the way air pollution affects various receptors. The impossibility of accounting for all major factors which might affect various receptors, the lack of reliable formulations used for translating physical damages into monetary terms, and the presence of numerous econometric problems have also caused concern to investigators.[26]

A paper by Hunt *et al.*[27] concludes that there are not, at the present time, sufficient scientific data to create valid damage functions on which to base an air quality index. Arriving at the existing NAAQS for a given air pollutant—which can be viewed as just one point on a damage curve—has been an extremely difficult process, requiring extensive research, evaluation of vast quantities of environmental data, detailed epidemiological studies, medical advisory committees, and many other involved steps. The complexity of the problem can be seen by examining any of the published air quality criteria documents.[13-18]

In the water pollution field, the process of documenting effects is also very complex, as can be seen by examining any of the water quality criteria publications.[10,28]

Theoretical Damage Functions

If valid damage functions were to be developed, what would they look like? From a purely theoretical standpoint, we might expect them to exhibit a "threshold" phenomenon. A threshold phenomenon implies that some minimum threshold value exists below which no damage occurs. Above the threshold value, effects increase rapidly with increases in the pollutant variable. One hypothesis for the existence of a threshold level, also called a Threshold Limiting Value (TLV), is based on the concept of adaptation.[29] Adaptation is the tendency of man and other living organisms to develop tolerance to low concentrations of toxins. Many common air pollutants, for example, arise from natural as well as man-made sources, such as forest fires, marine organisms, and photochemical reactions with certain organic compounds produced by plants.[30] Presumably, continued exposure to these natural and man-made pollutants at low concentrations causes a certain tolerance to develop, which is reflected by existence of a threshold level.

A second argument for existence of a TLV concerns the underlying biochemical mechanisms by which foreign substances affect cellular activity. This argument holds that chemical interactions with biological systems can proceed only when a sufficiently large number of atoms or molecules of the foreign substance is present. Otherwise, no adverse effects will occur. Although there is not yet sufficient knowledge to build a model of these biochemical cellular interactions, Dinman[31] and others have argued that there will be little chance for biochemical activity if the concentration is less than 10,000 atoms or molecules per cell. Although the subject is still being debated, there is some experimental evidence to support the existence of a threshold level, especially for toxic substances such as lead.[32] Biological organisms often seem resistant to pollutants at very low concentrations, and no effects are observed. As the concentration increases, the resistance barrier is broken, and significant effects occur.

A damage function incorporating a threshold phenomenon would be horizontal at first, but the slope would rapidly increase when the threshold is reached (Figure 12). Such a function, with damage units plotted on the ordinate and pollutant concentration on the abscissa, has been described as resembling a hockey stick. Because of its importance in air pollution studies, the statistical application of this function has been investigated by Hasselblad, Creason and Nelson,[33] who give the following description of this curve:

> The establishment of criteria for air pollutants requires that a threshold level be established below which no adverse health effects are observed. Since standard dose-response curves, such as the logit or probit, assume an

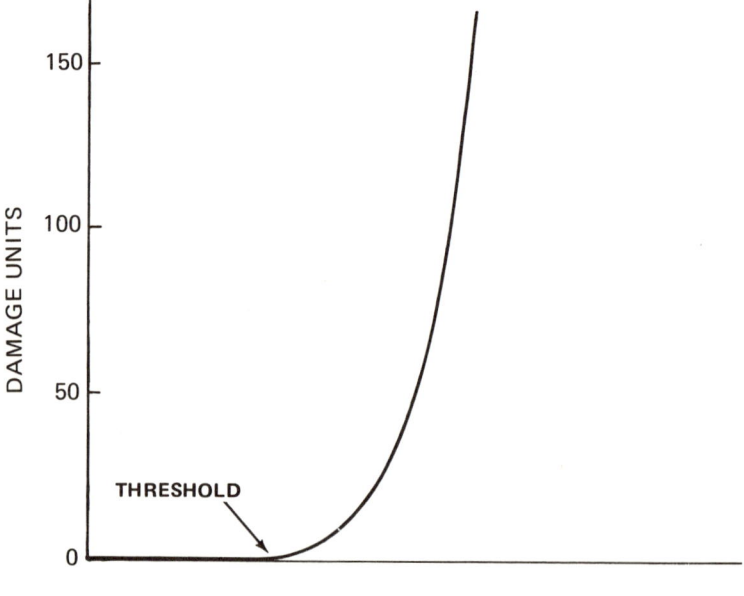

Figure 12. Hypothetical damage function showing characteristic "hockey stick" shape and threshold level.

effect at all levels, a segmented function was developed. This function has zero slope up to a point, and then increases monotonically from that point. Thus the name "hockey stick" function.

A general (hypothetical) damage function proposed by Hershaft, Morton and Shea [21] has a similar characteristic shape near the origin,* but these authors conclude that the overall damage function should also include a "saturation" level, giving a sigmoid, or s-shaped, curve (Figure 13):

> The ordinate may represent either the number of individuals affected or severity of effect. The abscissa indicates the dosage in terms of time at a given ambient concentration, or in terms of ambient concentration for a fixed period of time. The lower portion of the curve suggests that, up to a certain exposure value, known as a threshold level, no damage is observed, while the upper portion indicates that there exists a damage saturation

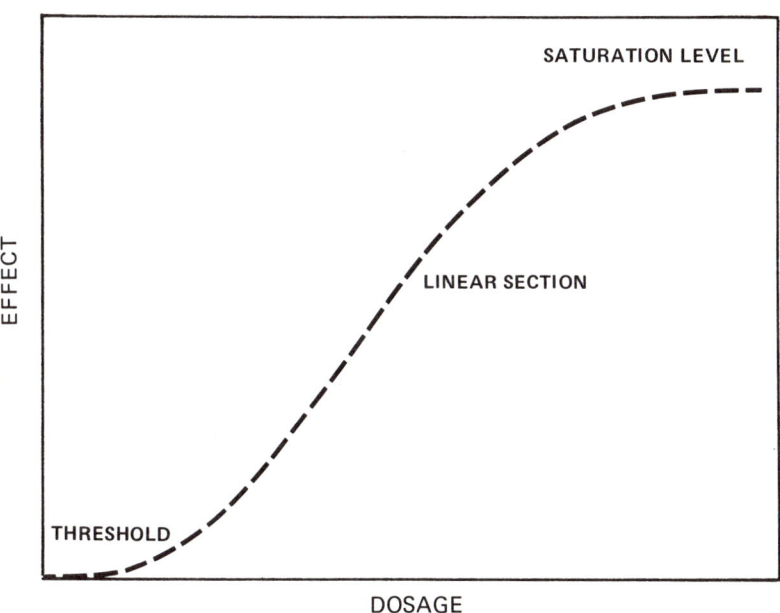

Figure 13. Hypothetical damage function proposed by Hershaft, Morton and Shea[21] showing characteristic sigmoid curve.

*The curvature of Figure 13 near the origin is more gradual than that of Figure 12, and the two curves may not, at first, seem similar. However, expanding the vertical scale of a curve tends to increase the sharpness of its curvature, and Figure 12 may be viewed as similar in shape to the threshold portion of the curve in Figure 13 when plotted on a different scale.

level (*e.g.*, death of the target population or total destruction of the crops), beyond which increased exposure levels do not produce additional damage. The middle, quasilinear portion is very useful in that any data points here can be readily interpolated, and the frequent assumption about linearity of a damage function is most valid in this sector.

Unfortunately, it has proved extremely difficult in actual research investigations to establish real damage functions like the theoretical curves presented here. Research studies on the effects of a given pollutant usually establish only that the two variables (for example, particulate matter and mortality) are "associated"; that is, that they have a statistically significant correlation. Such a result is too general to allow plotting of a specific dose-effect function. Further, such a finding does not establish a cause-effect relationship, which usually is implied in the theoretical damage functions. Actual research studies are further complicated by the fact that different persons will have different dose-effect functions and that synergistic effects occur when more than one pollutant act together, as is usually the case. Thus, it is often not possible to identify a dose-effect function which applies to an individual pollutant variable and properly covers all segments of the population.

Development of Dose-Effect Information

The underlying assumption behind development of damage functions—that a single cause-effect relationship can be established between pollutant variables and observed deleterious effects—is often regarded as highly simplistic and is not readily accepted by many researchers in the field. A major practical problem is that the scientist attempting to develop meaningful damage or dose-effect functions must contend with the complexity of controlling for the many extraneous factors which also have impact on observed effects. For example, studies examining correlations between mortality and air pollutant concentrations in different cities must control for differences in population density, income levels, racial composition, age distribution, type of industry, recreational habits, and other variables. Similarly, epidemiological studies undertaken within a single city must control for such factors as the effect of weather on mortality, delays between daily instances of high concentrations and increases in death rates, effects of holidays versus weekdays, effects of other pollutants, lack of spatial representativeness of the pollution data, and errors in the mortality data. Further, as suggested above, any observed correlation between a pollutant variable and an effect does not necessarily imply causation; the actual cause may be some other, unmeasured variable which also happens to be correlated with the pollutant variable. Although many problems exist with such research, some interesting correlation studies have been carried out, and it is useful to consider several examples to illustrate the complexity of the process. In these examples, the pollutant of interest is atmospheric sulfur dioxide.

Glasser and Greenburg[34] examined the relationship between day-to-day variations in mortality and sulfur dioxide concentrations in New York City during normal (nonepisodic) meteorological conditions over a 5-year period. Although their study was restricted to the 6 months between October and March, they attempted to reduce the effect of seasonal factors by transforming daily deaths into a 15-day moving average and by computing deviations from the "normal" death rate. In addition to sulfur dioxide, their regression analysis included such variables as wind speed, sky cover, rainfall, temperature deviations, day of the week, and Coefficient of Haze (COH).* Some of the air pollution data were missing due to instrument malfunction; sulfur dioxide data were available for 741 of the 910 days of the study (81% of the total), and COH data were available for 854 days (94% of the total). The findings of Glasser and Greenburg show that the mean number of daily deaths and the SO_2 level were associated, with daily mortality (expressed as the deviation from normal daily mortality) increasing as the SO_2 concentration increased (Figure 14). After taking into account the error in their analysis, the authors conclude that an SO_2 level of 0.4 ppm (parts per million) is associated with 10-20 more deaths per day than a level of 0.2 ppm:

> Grouping all days having less than 0.20 ppm of SO_2 compared with all days having mean SO_2 levels of 0.40 ppm or more, we find a difference of 15 deaths per day. The estimated standard error of this figure is 2.6 deaths. Thus, the 95 confidence interval for the difference in the mean number of daily deaths on days with mean SO_2 levels of 0.20 ppm or less compared with days having mean SO_2 levels of 0.40 ppm or more is approximately 10 to 20 deaths per day.[34]

In this study, a positive association also was found between particulates, as measured by COH, and mortality. However, the COH values also were found to be correlated with the SO_2 values. Regression analysis revealed that the relationship between SO_2 and mortality was much stronger than the relationship between COH and mortality, when each was adjusted for the other in the presence of weather variables. In summary, SO_2 appeared to be strongly related to mortality, even when all other variables (rainfall, temperature variation, day of the week, etc.) were taken into account. Thus, the authors concluded that "there is a positive relationship of SO_2 to mean daily mortality—a relationship which cannot be explained by associated weather factors."[34]

---

*Coefficient of Haze (COH) is a reading obtained by passing air through a paper tape and measuring the reduced light transfer (opacity) that results. Although widely employed throughout the United States, the tape sampler is not the reference method generally recommended by the Environmental Protection Agency for measuring particulate matter (see Chapter III). Because it is sensitive to the darkness of particles, the method sometimes is called a measure of Smoke Shade.

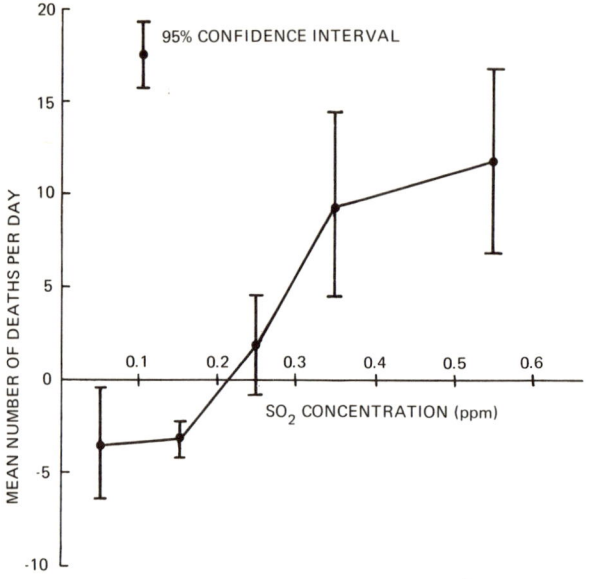

Figure 14. Relationship between mean number of deaths per day, expressed as deviations from normal, and daily mean SO_2 concentration, based on data from Glasser and Greenburg.[34]

A study by Schimmel and Greenburg[35] examined over one-half million death certificates in New York City and compared the number of deaths by various causes with daily measurements of SO_2, COH and weather variables, including average temperature. The data covered 2192 days over a 6-year study period from 1963 to 1968 and employed a variety of statistical approaches—correlation and partial correlation functions, autocorrelation functions, "cross-correlation" functions, and spectral analysis. Their study is interesting because it took into account the time delay between exposure to high concentrations and increased death rates. Using a cross-correlation analysis approach, the study examined mortality from different causes (for example, coronary heart disease, respiratory illness) by comparing death rates at various time lags after high concentrations occurred with death rates before high concentrations occurred. These comparisons showed higher correlations on days following high pollution levels than before, and the authors concluded, "the consistently higher values for positive days of lag with respect to pollution show that pollution tends to influence mortality on the same day and on subsequent days."[35]

Like Glasser and Greenburg's earlier study, the study by Schimmel and Greenburg also found evidence to support a threshold level, but there was no evidence of a saturation effect. When comparing the statistical contribution of SO_2 and COH to mortality, they concluded that approximately 80% of the excess deaths could be attributed to the COH data, while only 20% could be attributed to SO_2. Although this finding appears inconsistent with the Glasser and Greenburg finding that SO_2 appeared more dominant, the authors attributed the inconsistency to differences in methodology. Overall, they concluded that "Estimated Average Daily Excess Deaths" due to SO_2 and COH in New York City ranged from 18.2 to 36.74, with an intermediate estimate of 28.63. If projected to an annual basis (and if causation is assumed), this figure would represent ". . . about 10,000 deaths a year which would not have occurred at the time they did, if there had been no pollution on the day of death or on immediately preceding days. In percentage terms, this represents 12% of the over one-half million deaths which occurred during these six years."[35] When listed by cause of death (Table III), 41.6% of the deaths from respiratory disease were associated with daily air pollution levels. Because of the large number of deaths that ordinarily occur from coronary heart disease, coronary deaths associated with air pollution, although they represented only 12% of all the coronary deaths, were actually twice as large as respiratory deaths.

The two studies discussed above show evidence of a strong associative relationship between air pollutant variables and death rates, even though it is difficult precisely to quantify the relative contribution of each pollutant variable. It may seem tempting to assume that causation exists and to accept an SO_2 damage function such as that proposed by Glasser and Greenburg (Figure 14) as a valid relationship between SO_2 levels and resulting deaths. However, considerable regulatory action has occurred in New York City over

Table III. Average Daily Excess Deaths from Air Pollution Compared with All Deaths, by Cause, New York City, 1963-68[35]

Cause of Death	All Deaths	Excess Deaths	Percent
Tuberculosis	1.57	0.404	25.8
Respiratory Disease	12.71	5.277	41.6
Vascular Lesion	17.13	2.619	15.3
Coronary Heart Disease	85.54	10.373	12.1
Hypertensive Heart Disease	7.39	2.327	31.5
Other Circulatory Disease	16.90	3.205	19.0
Respiratory Cancer	8.74	0.476	5.4
Infant Diseases	6.84	a	a
All Other Diseases	86.40	3.888	4.5
Total Mortality	243.21	28.630	11.8

[a]Not significantly different from zero.

the past decade accompanied by a significant decline in measured SO_2 levels. On the basis of the above damage function, we would expect death rates to show a corresponding decline. However, a more recent study by Schimmel and Murawski[36] of mortality in New York City for the 10-year period from 1963 to 1972 showed no decline in pollution-related mortality:

> Within the limits of accuracy of the data, despite a 70% reduction in SO_2 levels for the latter period (1970-72) as compared to the earlier years, there has been no reduction in health effects associated with SO_2 measure. This tends to confirm that SO_2 is serving as a day-to-day indicator of air quality rather than functioning as a harmful pollutant.

Thus, Schimmel and Murawski attribute the seeming inconsistency in findings to the fact that SO_2 itself might not be responsible for mortality but happens to be correlated with some other, more injurious pollutant. Therefore, they conclude that controlling SO_2 has no effect on mortality. Because of the high cost of controlling sulfur oxides emissions from combustion of coal and oil and the scarcity of low-sulfur fuels for energy production, these authors argue that the existing NAAQS for SO_2 should be changed: "In the light of our immediate study and the available health information, it would appear that the SO_2 standards should be either greatly relaxed or abandoned entirely."[36]

Although the findings from these studies may seem somewhat confusing, they demonstrate the difficulty of establishing valid dose-effect relationships and damage functions based on existing knowledge. Under the present environmental legislation, the primary emphasis has concentrated more on establishing environmental standards than on developing damage functions. These standards are intended to reflect the best judgement of the legal and scientific communities as to what air quality levels should be attained to protect public health and welfare. Although the SO_2 NAAQS has been in effect for more than 5 years, there is a possibility that it will be revised in the future as additional research establishes a better understanding of the effect of different pollutant levels on health and welfare.

ENVIRONMENTAL MONITORING

In subsequent chapters, the structural characteristics of various environmental indices will be covered. Prior to this, it is useful to briefly describe the environmental monitoring activities which generate the data on which all index calculations are based. In the United States, most environmental monitoring is carried out by private firms, universities, and governmental agencies (local, state and federal). Although the data are often used immediately to assess the impact of control actions and to inform the public about local environmental conditions, these data represent one of the most important means available for evaluating the state-of-the-environment. Most of these data ultimately find their way into large data banks maintained by the federal government. In turn, these data are summarized in various official reports, such as the annual reports to Congress prepared by the President's Council on Environmental Quality.[37,38]

As currently constructed, most environmental indices utilize environmental quality data in their calculations rather than pollutant source data; therefore, our discussion emphasizes environmental quality monitoring. Because the largest existing environmental data banks contain air and water pollution data, we shall focus on air and water quality monitoring.

Air Quality Monitoring

Air quality monitoring provides a valuable tool for determining the effectiveness of a community's air pollution control effort. Because of the uniform NAAQS adopted under the Clean Air Act,[11] the raw data collected in any community can be readily compared with existing air quality standards, giving an indication of the severity of the air pollution problem and its health significance. Considering the data from many communities also permits an assessment of the Nation's progress in attaining clean air goals.

Measurements of pollutant concentrations in the atmosphere are routinely carried out at air monitoring stations in most urban areas throughout the United States. State and local air pollution control agencies have primary responsibility for constructing these networks and operating the monitoring stations. These agencies use the data to:

- assess compliance with the NAAQS
- examine long-term air quality trends
- assess critical air pollution episodes requiring emergency control actions
- conduct special-purpose surveys to evaluate particular air pollution problems
- report daily air quality levels to the public

The federal government provides technical assistance and other guidance to these agencies. Although it collects and stores much of the data, the federal government does not operate large-scale air monitoring networks, except as integral parts of specific research projects.

A typical modern air monitoring station generally consists of a building housing a number of complex instruments, each of which is designed to measure a particular air pollutant. For the gaseous pollutants, air usually enters through an intake probe mounted outside the building and is pumped to these monitoring instruments. To reduce the turbulent effect of air currents passing the building, the probe often is mounted several feet above the roof. A particular air monitoring station may measure six or more different air pollutants, and meteorological data may be collected as well (temperature, wind direction, and wind speed). A part-time operator usually is required at the station to maintain and calibrate the instruments.

Over the last 15 years, there has been increasing use of "continuous monitoring" instruments, which operate continuously 24 hours a day and produce readings which can be recorded in ink on a moving chart (that is, a "strip chart") or translated to digital form electronically. A great many continuous air monitoring instruments were operating in the United States in 1969 and were widely geographically distributed (Figure 15).[39] California, which entered the field of air pollution control at a relatively early date, operated more continuous monitoring instruments than any other state.

Subsequent to the 1970 Clean Air Act, the Environmental Protection Agency published guidance on the number of air monitoring stations required for each air pollutant as a function of a community's population.[19] Over the period from 1970 to 1975, there was considerable growth in the number of air monitoring stations, as reflected by the number of measuring instruments in operation for each pollutant (Table IV).[40] The annual rate of increase declined in 1974 and has stabilized at slightly less than 8000 measuring instruments currently in operation nationwide.

The most commonly measured air pollutant is total suspended particulates, partly because the instrument is relatively inexpensive and convenient to use. It operates like a vacuum cleaner, with air forced through a filter that

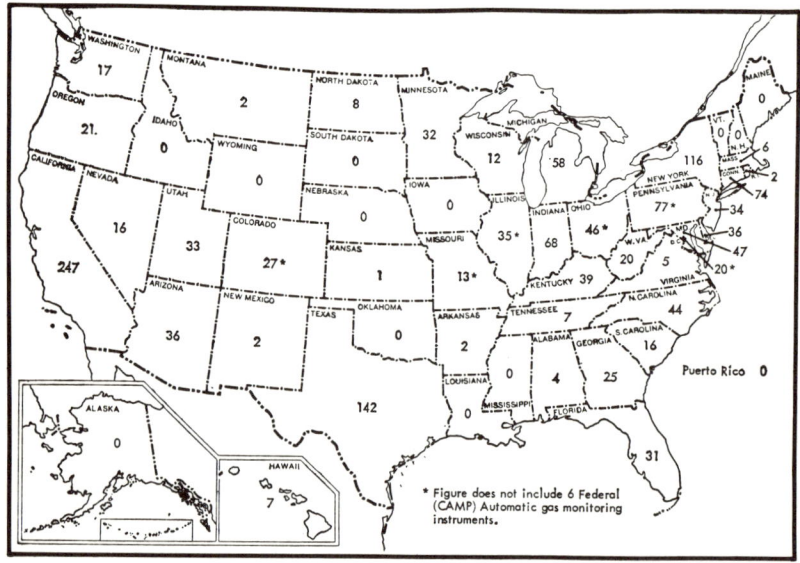

Figure 15. Distribution of continuous air monitoring instruments in the United States.[39]

INTRODUCTION 39

Table IV. Number of Air Pollution Measuring Systems Operating in the U.S., 1970-1975[40]

Pollutant	1970	1971	1972	1973	1974	1975
TSP	1280	2040	2980	3760	3790	3860
SO_2	403	730	1310	2000	2240	2400
CO	73	133	191	299	377	415
Oxidant	51	82	162	265	343	395
NO_2	28	32	47	67	582	700
Total	1835	3017	4690	6391	7332	7770

is weighed both before and after the sample is collected. The mass of the material accumulated on the filter and its chemical composition form the basis for the measurement. By this means, it is possible to determine the concentration of a variety of pollutants in the atmosphere (Table V). The TSP instrument is regarded as a "discrete" monitoring system rather than a continuous monitoring system, because it produces one value per averaging period (usually 24 hours). In 1975, the total number of TSP instruments in operation (3860) was roughly equal to the total number of instruments used to measure four major gaseous air pollutants (sulfur dioxide, carbon monoxide, photochemical oxidants, and nitrogen dioxide). The gases usually were measured with continuous monitoring systems.

Water Quality Monitoring

Like air quality monitoring, water quality monitoring provides one of the most important and direct yardsticks available for evaluating the effectiveness of the Nation's water quality management program. Interpreting water quality data is somewhat more complex than interpreting air quality data, however, because (1) more variables usually are measured, (2) water has a variety of different uses, and (3) nationally uniform ambient standards are not available for most use categories.

Table V. Pollutant Variables Often Reported in Air Pollution Measurements with the High-Volume Sampler

Benzene-Soluble Organics	Beryllium	Lead
Nitrates	Bismuth	Manganese
Sulfates	Cadmium	Molybdenum
Ammonium	Chromium	Nickel
Antimony	Cobalt	Tin
Arsenic	Copper	Titanium
	Iron	Vanadium
		Zinc

Although more variables are routinely measured in water than in air, there is less use of continuous monitoring techniques in water monitoring, and it is often necessary to rely on a few samples collected at some arbitrary point on a stream. The greater reliance on discrete measurements in water monitoring creates somewhat more difficult statistical sampling problems in water than in air.

The principal uses of water include:

- supply of drinking water for humans, livestock and other animals
- supply of water for industrial processes, cooling and other commercial uses
- irrigation of crops
- maintenance of a suitable fishery and wildlife habitat
- recreation (principally, boating and swimming) and aesthetics

Nationally uniform water quality standards have been adopted (on an interim basis) for one use category, drinking water supplies.[41] The lack of uniform national ambient water quality standards for the other use categories can be attributed primarily to variation in natural water quality characteristics from one geographical area to another. For example, high turbidity levels may be acceptable in some waters used for boating and similar recreational activities, such as in the Missouri River in the mid-western United States, while high turbidity may be unacceptable in streams elsewhere in the country where natural turbidity levels are lower. In response to existing legislation, states have established standards which take into consideration natural variations in water quality within their boundaries. Under the Federal Water Pollution Control Act,[9] all states also have complied with a requirement to promulgate standards for interstate waters and to submit these to the Environmental Protection Agency for acceptance. Thus, although ambient standards exist for surface waters throughout the United States, the standards vary greatly from location to location.

Monitoring the receiving waters to evaluate and enforce water quality standards is a primary reponsibility of state and local water pollution control agencies. The federal government also has an important role in the monitoring area, because it may be required to enforce water quality standards when a state fails to exercise its responsibilities.

At the local level, most water treatment facilities supplying public drinking water monitor raw water quality on a daily basis. They also monitor the finished drinking water after it is treated. There are over 6000 communities in the United States with water supplies served by surface water sources. Monitoring activities also are carried out by operators of local wastewater treatment facilities designed to treat sewage and industrial wastes. Monitoring the influents and effluents of individual wastewater treatment plants is necessary to evaluate each plant's efficiency and to assess the effects of the waste source on the waters into which the plant discharges. Many municipal wastewater treatment programs and county agencies also routinely monitor receiving waters at various distances upstream and downstream from treatment plant discharges. Thus, considerable water quality data are being gathered in the United States by operators of water and wastewater treatment facilities.

Most state water pollution control agencies also have monitoring programs for assessing surface water quality. These programs vary in scope among the states from near-minimal coverage to comprehensive monitoring systems. Other water-oriented state agencies, such as state conservation and geology departments, also collect various kinds of water data.

More than a dozen federal agencies also collect water quality data. Among these are the U.S. Geological Survey, the Environmental Protection Agency, the Tennessee Valley Authority, the Army Corps of Engineers, the U.S. Forest Service, and the Bureau of Reclamation. The federal water monitoring activities are coordinated by the U.S. Geological Survey, consistent with an Office of Management and Budget requirement for interagency coordination to avoid duplication of effort.

The diverse monitoring activities of municipal, state and federal agencies can be viewed as a national network of water monitoring stations, covering the quantity and quality of water that flows out of 306 hydrologic basins which cover the conterminous United States. As indicated by Sayers,[42] this network serves to identify:

- compliance and noncompliance with water quality standards
- water quality baselines and trends
- improvement in water quality produced by abatement and control efforts (for example, construction of waste treatment facilities)
- new or emerging water quality problems in sufficient time to effect adequate preventive measures

From an initial evaluation of the local, state and federal long-term water quality monitoring effort required to provide adequate coverage of the nation's water resources, Sayers estimated that at least 10,000 monitoring stations would be needed on both interstate and intrastate waters. Because of the considerable length of all the streams, impoundments and shorelines in the United States, the number of stations per unit of distance still would be relatively small. For example, these 10,000 stations, if equally spaced, would be located at 350-mile intervals. Of course, the stations would not be equally spaced, but would be more dense in highly developed areas and less dense in less developed areas. Locations of prime concern from a federal standpoint would be along state and international boundaries, principal estuaries, mouths of major tributaries, large metropolitan complexes, and major water resource projects. There are approximately 900 stream locations and 1500 open-water locations in this category.

As shown by Sayers,[42] the actual locations of the local, state and federal water monitoring stations in 1968, when plotted on a map of the United States (Figure 16), showed considerable geographical variation, giving a "shotgun" pattern. Because of the growth of water monitoring activities, the same map, if plotted with today's monitoring station locations, would be so dense in some regions that parts of the map would appear completely black. The sampling frequencies of the various U.S. monitoring stations in 1968 varied greatly (Table VI). Although some variables were measured continuously, there was a preference for sampling less frequently than on a continuous or daily basis, and many stations collected samples on a monthly, quarterly or annual basis.

The Environmental Protection Agency's STORET (STOrage and RETrieval of water quality data) system, operational since 1964, utilizes many computer terminals throughout the United States linked to a central computer. Data from federal, state and local agencies are stored routinely in this system. To date, more than 200,000 observations from locations on nearly all of the nation's rivers, lakes, streams and other waterways are stored in the central computer.[43] Each observation represents a measurement of a single variable at a specific location or station. Although more than 1800 unique water quality variables are defined within the STORET system, a nucleus of about 200 variables are most commonly monitored and constitute the bulk of the data. Approximately 80% of the 40 million individual observations within the system pertain to these 200 variables. The 200 most common variables can be grouped into 14 general categories (Figure 17). As can be seen, the system has undergone considerable growth since its original development, with the quantity of data stored doubling in the 5-year period from 1966 to 1970 and again nearly doubling in the 5-year period from 1971 to 1975.

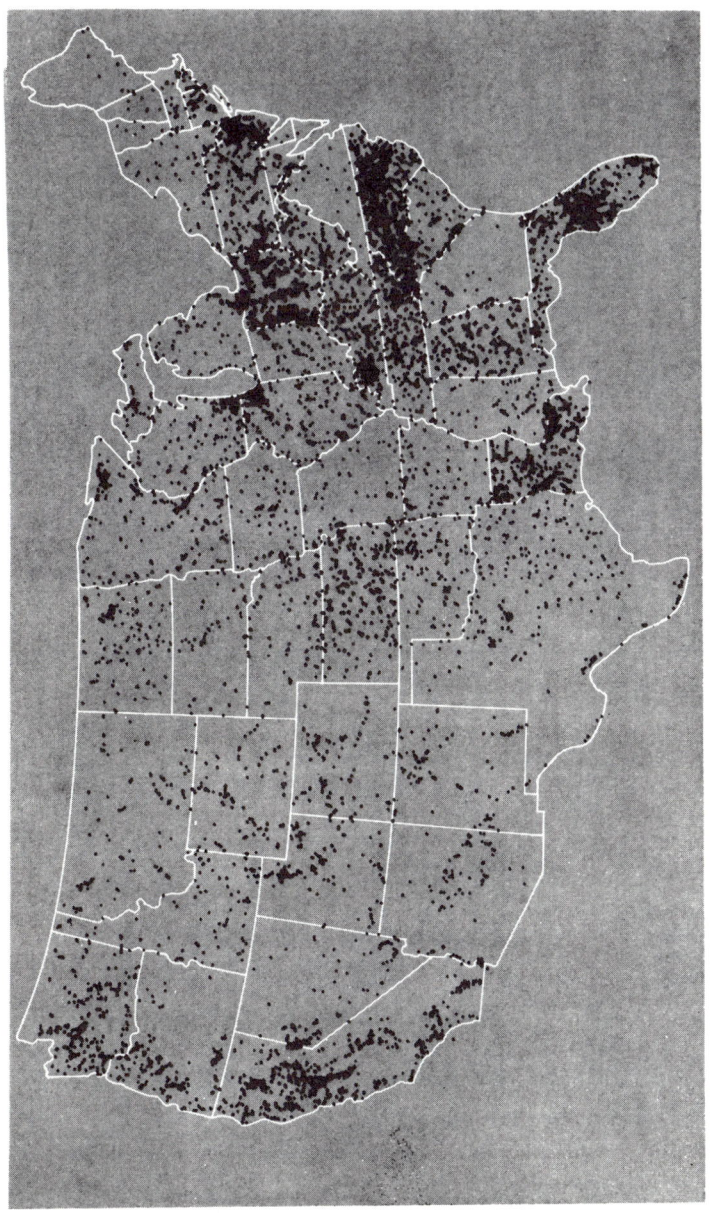

Figure 16. Locations of water quality monitoring stations in the United States, 1968. Data for AL, MA, ME, NH, RI, VA, and VT show federal stations only; data for other states show federal and nonfederal stations. (Reprinted with permission from William T. Sayers, copyright by the American Chemical Society[42]).

44 ENVIRONMENTAL INDICES

Table VI. Variables and Sampling Frequencies of U.S. Water Quality Monitoring Stations[a]

Variable	Continuous	Daily	Weekly	Monthly	Quarterly	Annually	Other	Total
Temperature	871	822	529	2270	832	1064	1276	7664
Specific Conductance	241	432	97	1636	491	1287	1045	5229
Turbidity	31	285	332	1053	615	57	623	2996
Color	14	87	205	939	611	1122	747	3725
Odor	9	32	14	389	406	13	248	1111
pH (Field)	77	55	156	745	372	122	721	2251
pH (Lab)	16	451	268	1914	749	1398	1111	5907
EH	18	0	0	7	0	0	1	26
Suspended Solids	0	2	48	328	213	2	45	594
Other Physical Analyses	12	19	73	146	179	41	193	663
Dissolved Solids	42	188	232	1495	624	1187	1063	4831
Chloride	10	121	118	1196	354	552	931	3282
Nutrients (Nitrogen)	14	32	243	1266	507	896	775	3733
Nutrients (Phosphorus)	23	58	135	853	609	889	918	3485
Common Ions	20	412	119	1775	555	1357	1288	5526
Hardness	6	362	225	1490	674	1310	907	4974
Radiochemical	21	3	37	339	448	31	265	1144
Dissolved Oxygen	85	45	345	1379	500	913	681	3948
Other Gases	10	86	13	34	0	5	18	166
Minor Elements	0	12	3	164	45	149	175	548
Pesticides	1	1	7	69	11	24	80	193
Detergents	0	1	85	363	463	33	114	974
Biochemical Oxygen Demand	0	12	239	866	460	30	176	1783
Carbon (Tot., Diss., etc.)	0	4	4	22	0	31	4	65
Coliforms	13	240	423	1261	637	160	345	3079
Other Microorganisms	2	13	185	252	128	4	113	697
Biologic	6	3	102	127	60	13	76	387
Sediment Conc. (Susp.)	25	291	96	269	41	9	328	1059
Particle Size (Susp.)	5	56	7	134	12	51	315	580
Particle Size (Bed Load Mat.)	0	0	0	10	2	6	33	42
Other Sediment	1	8	5	97	2	6	71	190

[a] Reprinted with permission, copyright by the American Chemical Society.[42]

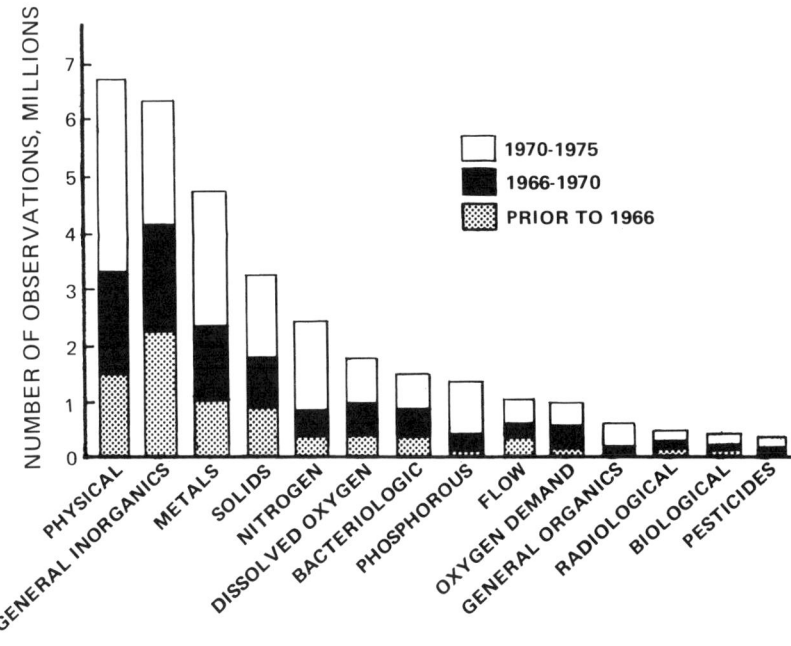

Figure 17. Number of observations in EPA's STORET water quality data bank for the 200 most common variables, grouped into 14 general categories.[43]

PROBLEMS FOR STUDY

1. In your own language, discuss the difference between an environmental indicator and an environmental index.

2. List five examples of environmental indicators. What difficulties would you have in attempting to combine these five indicators into an overall index?

3. In your own words, describe the classic dichotomy of views toward environmental indices. (a) Do any other fields have the same problems? (b) Why or why not?

4. In your own words, discuss the difference between a source-related pollutant variable and an environmental quality pollutant variable. List two examples of each.

5. Describe the difference between a point source and a nonpoint source of water pollution.

6. In your own words, discuss the ranking paradox that is characteristic of environmental quality measurements. Can you think of any other fields where this same paradox may occur?

7. Describe the difference between continuous and discrete monitoring data. Why do you think continuous monitoring data is more often collected for air pollutants than for water pollutants?
8. Discuss the difference between an economic damage function and a dose-effect function. What difficulties would exist in creating an economic damage function that reflected the economic impact of adverse health effects if a valid dose-effect function were available?
9. Discuss the reasons that have been given for the existence of a threshold limiting value (TLV) in environmental damage functions.
10. The data from Schimmel and Greenburg,[35] when extrapolated to the year, give 10,000 deaths in New York City due to air pollution (sulfur dioxide). In view of the findings of earlier and later studies, discuss why you would either (1) accept or (2) challenge this estimate.

REFERENCES

1. Train, Russell E. "Management for the Future," presented at the National Conference on Managing the Environment, Washington, DC, May 1973.
2. Train, Russell E. "The Quest for Environmental Indices," *Science* 178(4057):121 (October 13, 1972).
3. "Planning for Environmental Indices," a report of the Planning Committee on Environmental Indices, National Academy of Sciences, Washington, DC, February 1975.
4. "National Environmental Policy Act of 1969," *Public Law 91-190.*
5. Coate, Edwin L. and Anthony K. Mason. "Some Practical Problems in Developing and Presenting Environmental Quality Indices to the Public," proceedings of the International Conference on Environmental Sensing and Assessment, Las Vegas, NV, IEEE #75-CH 1004-1 ICESA, September 1975, p. 31-32.
6. Thom, Gary C. and Wayne R. Ott. *Air Pollution Indices: a Compendium and Assessment of Indices Used in the United States and Canada* (Ann Arbor, Michigan: Ann Arbor Science Publishers, Inc., 1976).
7. Inhaber, Herbert. *Environmental Indices* (New York: John Wiley and Sons, 1976).
8. "Environmental Quality Profile 1976," U.S. Environmental Protection Agency, Region X, Seattle, Washington.
9. "Federal Water Pollution Control Act Amendments of 1972," *Public Law 92-500.*
10. "Water Quality Criteria 1972," a report of the Committee on Water Quality Criteria, National Academy of Sciences, Washington, DC, 1972.
11. "Clean Air Act Amendments of 1970," *Public Law 91-604.*
12. "National Primary and Secondary Ambient Air Quality Standards," *Federal Register* 36(84):8187, Part II (April 30, 1971) Washington, DC.
13. "Air Quality Criteria for Carbon Monoxide," U.S. Department of Health, Education, and Welfare, Washington, DC, NAPCA Publication No. AP-62, March 1970.

INTRODUCTION 47

14. "Air Quality Criteria for Nitrogen Oxides," U.S. Environmental Protection Agency, Washington, DC, Publication No. AP-84, January 1971.
15. "Air Quality Criteria for Hydrocarbons," U.S. Department of Health, Education, and Welfare, Washington, DC, NAPCA Publication No. AP-64, March 1970.
16. "Air Quality Criteria for Photochemical Oxidants," U.S. Department of Health, Education, and Welfare, Washington, DC, NAPCA Publication No. AP-63, March 1970.
17. "Air Quality Criteria for Particulate Matter," U.S. Department of Health, Education, and Welfare, Washington, DC, NAPCA Publication No. AP-49, January 1969.
18. "Air Quality Criteria for Sulfur Oxides," U.S. Department of Health, Education, and Welfare," Washington, DC, NAPCA Publication No. AP-50, January 1969.
19. "Requirements for Preparation, Adoption, and Submittal of Implementation Plans," *Federal Register* 36(158):15486-15506, Part II (August 14, 1971) Washington, DC.
20. "Noise Control Act of 1972," *Public Law 92-574.*
21. Hershaft, A., J. Morton and G. Shea. "Critical Review of Air Pollution Dose-Effect Functions," Enviro Control, Inc., prepared for the Council on Environmental Quality and the Environmental Protection Agency, March 1976.
22. Haynie, F. H. and J. B. Upham. "Effects of Atmospheric Sulfur Dioxide on the Corrosion of Zinc," *Materials Protection and Performance*, 9(8):35-39 (August 1970).
23. Heck, W. W., *et al.* "Ozone; Non-Linear Relation of Dose and Injury in Plants," *Science*, 151:577-578 (February 4, 1966).
24. Dow Chemical Company. "An Economic Analysis of Erosion and Sediment Control Methods for Watersheds Undergoing Urbanization," February 1972.
25. Tihansky, Dennis P. "Economic Damages From Residential Use of Mineralized Water Supply," *Water Resources Res.* 10(2):145-154 (April 1974).
26. Liu, Ben-chieh and Eden Siu-hung Yu. "Physical and Economic Damage Functions for Air Pollutants by Receptors," U.S. Environmental Protection Agency, Corvallis, OR, EPA-600/5-76-011, September 1976.
27. Hunt, William F., William M. Cox, Wayne R. Ott and Gary C. Thom. "A Common Air Quality Reporting Format, Precursor to an Air Quality Index," Proceedings of the Fifth Annual Environmental Engineering and Science Conference, Louisville, Kentucky, March 3-4, 1975, pp. 99-121.
28. "Quality Criteria for Water," U.S. Environmental Protection Agency, Washington, DC, EPA-440/9-76-023, 1976.
29. Stokinger, Herbert E. "Concepts of Thresholds in Standards Setting," *Archives of Environmental Health* 25:153-157 (September 1972).
30. Robinson, E. and R. C. Robbins. "Sources, Abundance, and Fate of Gaseous Atmospheric Pollutants," Stanford Research Institute, Menlo Park, California, June 1969.
31. Dinman, Bertram. "'Non-Concept' of 'No-Threshold': Chemicals in the Environment," *Science* 175:495-497 (February 4, 1972).

32. Waldron, Harry A. "The Blood Lead Threshold," *Archives of Environmental Health* 29:271-273 (November 1974).
33. Hasselblad, Victor, John P. Creason and William C. Nelson. "Regression Using 'Hockey Stick' Functions," U.S. Environmental Protection Agency, Research Triangle Park, North Carolina EPA-600/1-76-024, June 1976.
34. Glasser, Marvin and Leonard Greenburg. "Air Pollution, Mortality, and Weather," *Archives of Environmental Health* 22:334-343 (March 1971).
35. Schimmel, Herbert and Leonard Greenburg. "A Study of the Relation of Pollution to Mortality," *J. Air Poll. Control Assoc.* 22(8):607-616 (August 1972).
36. Schimmel, Herbert and Thadeus J. Murawski. "SO_2—Harmful Pollutant or Air Quality Indicator?," *J. Air Poll. Control Assoc.* 25(7): 739-740 (July 1975).
37. "Environmental Quality: the Sixth Annual Report of the Council on Environmental Quality," U.S. Government Printing Office, Washington, DC, December 1975.
38. "Environmental Quality: the Seventh Annual Report of the Council on Environmental Quality," U.S. Government Printing Office, Washington, DC, September 1976.
39. "Progress in the Prevention and Control of Air Pollution," third report of the Secretary of Health, Education, and Welfare to the Congress of the United States in compliance with Public Law 90-148, U.S. Government Printing Office, Washington, DC, March 1970.
40. Hoffman, Alan J. "EPA Siting Criteria for State Implementation Plan Monitoring Networks," presented at the Environmental Protection Agency Air Monitoring Siting Workshop, Las Vegas, NV, July 1976.
41. "Interim Primary Drinking Water Standards," *Federal Register* 40(51): 11990-11998, Part II (March 14, 1975), Washington, DC.
42. Sayers, William T. "Water Quality Surveillance: the Federal-State Network," *Environ. Sci. Technol.* 5:114-119 (February 1971).
43. "STORET: EPA's Computerized Water Quality Data Base," U.S. Environmental Protection Agency, Office of Water and Hazardous Materials, Washington, DC, 1976.

CHAPTER II

STRUCTURE OF ENVIRONMENTAL INDICES

The purpose of an index, as discussed in Chapter I, is to *simplify*. The simplification process strives toward parsimony—presentation of the least amount of information possible that will convey necessary meaning. Through mathematical manipulation, an environmental index seeks to reduce measurements of two or more environmental variables to a single number (or a set of numbers, words or symbols) that retains meaning. Although the environmental indices which have been developed show great variety and striking differences, it is possible to construct a general mathematical framework which accommodates most existing environmental indices. This structure is intended to provide a conceptual tool for understanding and comparing environmental indices.

Before discussing this framework, it is important to distinguish between two general environmental index forms: (1) those in which the index numbers increase with increasing environmental pollution, and (2) those in which the index numbers decrease with increasing environmental pollution. Some specialists in the field refer to the former as "environmental pollution" indices and the latter as "environmental quality" indices. Using this terminology, an index in which $I = 0$ corresponds to pristine water and $I = 100$ corresponds to severe water pollution would be called a "water pollution" index. Conversely, an index in which $I = 0$ means poor water quality and $I = 100$ represents good water quality would be called a "water quality" index. These terms are not universally accepted, however. To avoid confusion, we shall describe each index as either (1) an "increasing scale" form or (2) a "decreasing scale" form. In an increasing scale index, the index value increases with increasing pollution; in a decreasing scale index, the index value decreases with increasing pollution. As will be shown in Chapter III, air pollution indices generally are of the increasing scale form, while most water pollution (or water quality) indices are of the decreasing scale form.

These differences apparently stem from the limited amount of communication which occurred historically between index developers in the water pollution field and those in the air pollution field.

MATHEMATICAL STRUCTURE

In this framework, calculation of an environmental index is viewed as consisting of two fundamental steps: (1) calculation of the subindices for the pollutant variables used in the index and (2) aggregation of the subindices into the overall index. Suppose we are considering a set of observations for n pollutant variables, in which X_1 denotes the observed value for the first pollutant variable, X_2 denotes the observed value for the second pollutant variable, and X_i denotes the value of the i*th* pollutant variable. Then the set of observations is denoted as $(X_1, X_2, \ldots, X_i, \ldots, X_n)$. For each pollutant variable X_i, a subindex I_i is computed using subindex function $f_i(X_i)$:

$$I_i = f_i(X_i) \tag{1}$$

In most environmental indices, a different mathematical function is used to compute each pollutant variable, giving the subindex functions $f_1(X_1)$, $f_2(X_2), \ldots, f_n(X_n)$. Each subindex function is intended to represent the environmental characteristics of the particular pollutant variable. It may consist of a simple multiplier, or the pollutant variable raised to a power, or some other functional relationship.

Once the subindices are calculated, they usually are aggregated together in a second mathematical step to form the final index:

$$I = g(I_1, I_2, \ldots, I_n) \tag{2}$$

The aggregation function, Equation 2, usually consists either of a *summation* operation, in which individual subindices are added together, or a *multiplication* operation, in which a product is formed of some or all of the subindices, or a *maximum* operation, in which just the maximum subindex is reported.

The overall process—calculation of subindices and aggregation of subindices to form the index—can be illustrated in a flow diagram (Figure 1). In this process, the "information" contained in the raw data (environmental measurements) flows from left to right and is reduced to a more parsimonious form. Some information may be lost; however, in a properly designed index, the information loss should be of such a nature that it does not cause the results to be distorted or ultimately misinterpreted.

Environmental indices sometimes appear more complex in structure than the framework outlined here, with subindices calculated at intermediate steps in the process and many variations in approach. Occasionally, for example, some of the subindex values are reported along with the final index. In other

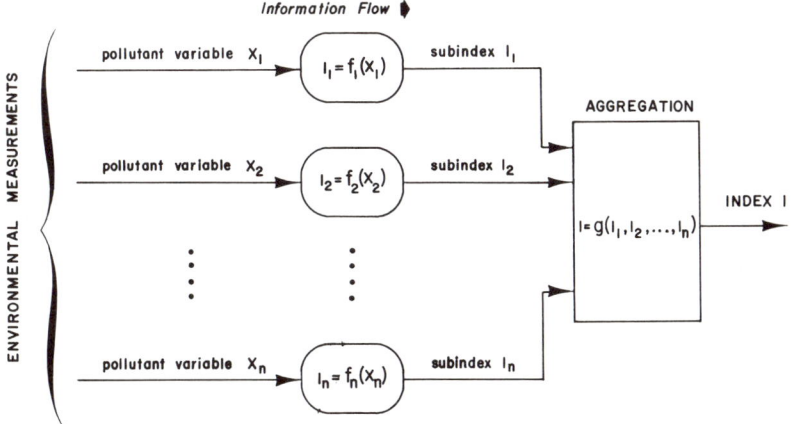

Figure 1. Information flow process in an environmental index.

instances, a subindex depends on the frequency of occurrence of some environmental event, such as the number of times that an environmental standard is exceeded in a given season or year. However, such structures can, with some limitations, be accommodated within this general framework.

Although this framework is suited to a large variety of indices, there are important exceptions. In the above index structure, a particular observation always produces the same subindex value, regardless of other observations comprising the data set. Because the relationship of the pollutant variable to the index is fixed, we call this an *absolute* index. Some structures, however, are designed for the purpose of ranking different observations. In such cases, the index depends not only on a given observation but on all other observations comprising the data set. We call this form a *relative* index. Although such indices can be, with difficulty, incorporated into the general framework presented here, they usually are better treated as special cases.

SUBINDICES

A large number of different functions for Equation 1—relationships between subindex I and environmental pollutant variable X—are possible.

Linear Function

The simplest subindex function (Figure 2) is the linear equation:

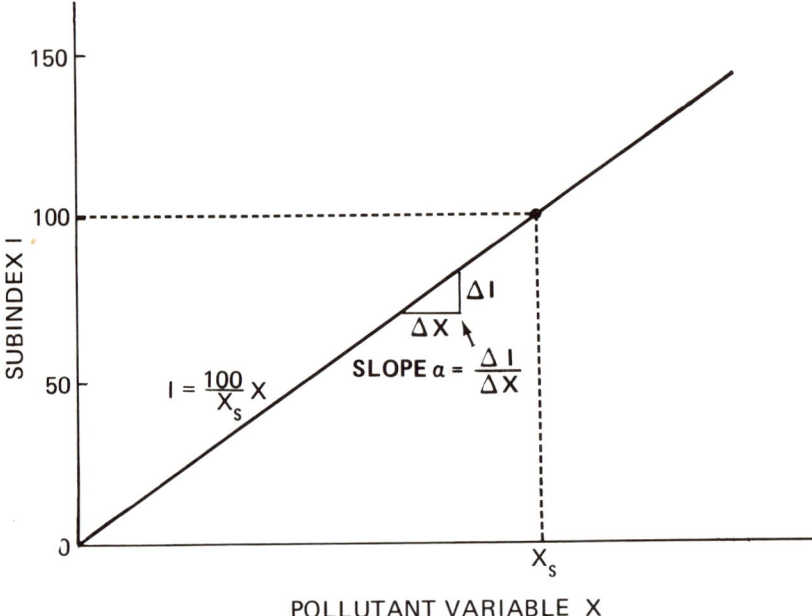

Figure 2. Monotonically increasing linear subindex function.

$$I = \alpha X \tag{3}$$

where I = subindex
       X = pollutant variable
       α = constant

With this function, a direct proportion exists between the subindex and the pollutant variable. That is, a doubling of pollutant variable X results in a doubling of subindex I. The slope* of the curve is α, a constant, implying that, for each absolute change ΔX, the subindex changes by $\Delta I = \alpha \Delta X$ units. If, for example, α were chosen so that I = 100 corresponds to an environmental standard, $X = X_s$, then $\alpha = 100/X_s$ and the subindex I could be interpreted as the "percentage of the standard."

In the above example, the straight line passes through the origin, giving I = 0 for X = 0. This simple linear subindex function can be offset from the origin by adding a constant, β, to Equation 3:

$$I = \alpha X + \beta \tag{4}$$

The constant β corresponds to the intercept of the I-axis, giving I = β for X = 0 (Figure 3). In the example, α = 125 and β = 75.

─────────
*For any subindex curve, the slope at some point $X = X_o$ is defined as the ratio of the change in I, ΔI, to the change in X, ΔX. If ΔX is allowed, in the limit, to approach zero, this ratio is defined as the first derivative.

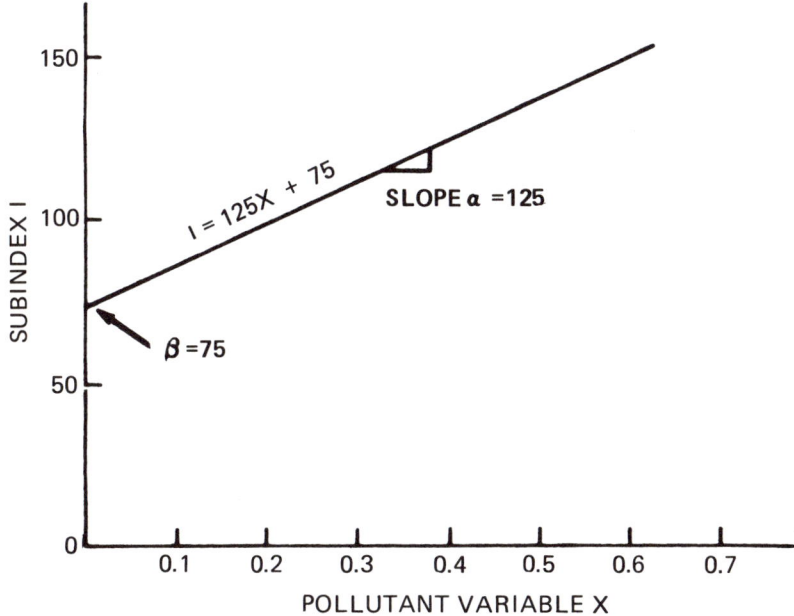

Figure 3. Simple linear (increasing scale) subindex function which does not pass through the origin.

If α is greater than zero, Equations 3 and 4 give increasing scale subindex functions of X. That is, for each upward increase of X, I also increases upward. If α is less than zero, the subindex function has a decreasing scale (it decreases monotonically). If we choose $\alpha = -100/X_s$ and $\beta = 100$, a decreasing scale subindex function will result which gives $I = 100$ for $X = 0$ and $I = 0$ for $X = X_s$ (Figure 4):

$$I = -\frac{100}{X_s} X + 100 \qquad (5)$$

Equation 5 may be viewed as the decreasing scale version of the "percentage of the standard" index discussed above. Here, 100 could correspond to 100% quality (zero pollution), and 0 could correspond to 0% quality (reaching the standard). For illustrative purposes, we assume that I cannot be negative.

An example of an environmental index using a simple linear function is the "common air quality reporting format" proposed by Hunt et al.[1] In this increasing scale system, $I = 0$ corresponds to zero air pollutant concentration, and $I = 1.0$ corresponds to the National Ambient Air Quality Standard (NAAQS). Thus, an atmospheric total suspended particulate measurement of

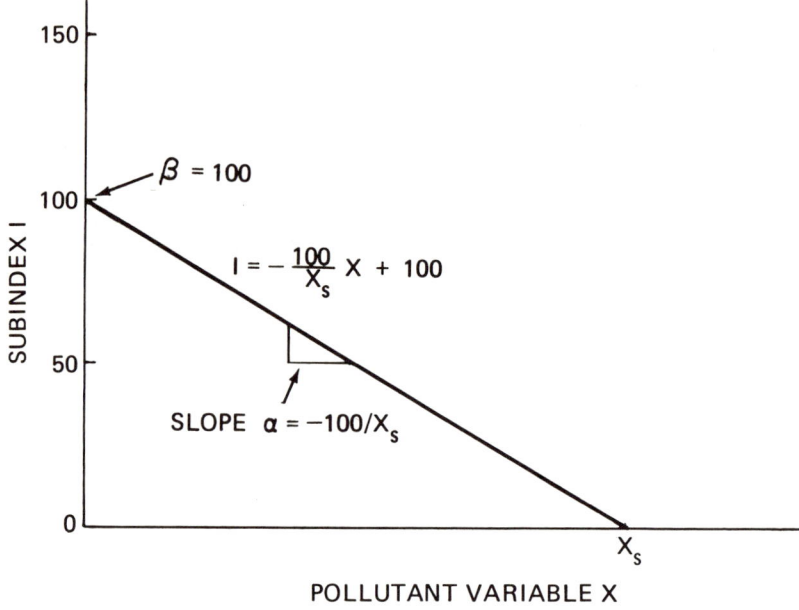

Figure 4. Linear subindex function with decreasing scale.

350 micrograms per cubic meter ($\mu g/m^3$) would be compared with the NAAQS of 260 $\mu g/m^3$ (24-hr average) and reported as I = 350/260 = 1.35. In the (increasing scale) percentage type of index, the result would be reported as 135%.

The linear indices have the advantages that they are simple to compute and easy to understand. The linear concept probably is familiar because so many variables in our daily lives are related linearly. Examples include weight (conversion from pounds to kilograms), temperature (conversion from Fahrenheit to Celsius), and motion (relationship of acceleration of an object to the force acting on it, Newton's second law).

Segmented Linear Function

As indicated in the earlier discussion of damage functions (Chapter I), the possible existence of a "threshold level," below which no effects occur, gives rise to the use of "hockey stick" functions. Suppose, in a hypothetical situation, that no adverse environmental effects occur at concentrations below some recommended administrative limit $X = X_s$, while extremely serious detrimental effects occurred at concentrations $X > X_s$. Then a suitable subindex function might consist of two straight lines joined approximately at a right angle at $X = X_s$ (Figure 5). This particular function may be viewed as an extreme case of the curved hockey stick function discussed earlier (Figure 12, Chapter I). In this extreme case, the subindex is zero for all concentrations below the recommended limit, but the subindex becomes a very large number as soon as the limit is exceeded. Because such a function consists of two straight line segments joined together at a point (a "breakpoint"), we call it a "segmented linear" function.

If an index is to be based on recommended administrative limits, such as legally promulgated standards, one often finds that two or more recommended limits must be incorporated into each subindex. In the air pollution field, for example, the federal government has established a "Significant Harm" level for each air pollutant. The Significant Harm level is much higher than the NAAQS and represents a concentration that should never be reached if public health is to be protected adequately. Suppose one wishes to design a sulfur dioxide subindex (for 24-hr averaging periods) which includes three points as follows:

A: $I = 0$ corresponds to zero ($X = 0$ ppm SO_2)
B: $I = 100$ corresponds to the NAAQS ($X = 0.14$ ppm SO_2)
C: $I = 500$ corresponds to the Significant Harm level ($X = 1.0$ ppm SO_2)

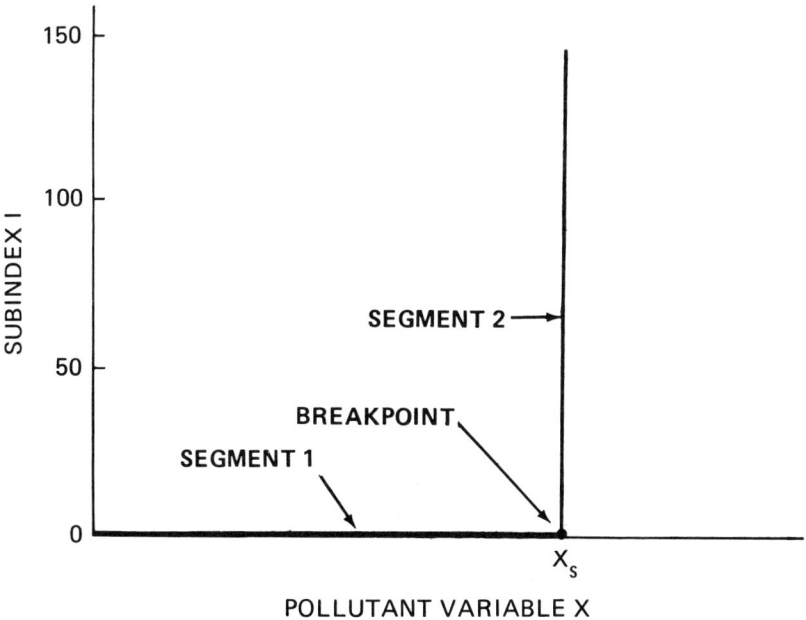

Figure 5. Example of a segmented linear (hockey stick) function.

56 ENVIRONMENTAL INDICES

If these values are plotted on the (X, I)-plane, the coordinates, (0,0), (0.14, 100) and (1.0,500), do not all lie on a straight line (Figure 6). How are the intermediate values of the index—those between adjacent points—to be calculated? Obviously, a variety of arbitrary curves, such as the dotted line (line a), can be drawn through the three points in Figure 6. If, however, there is limited knowledge about the damage associated with intermediate concentrations of the pollutant, it may be impossible to say that one curve is more valid than another. Another approach is to make the simplest possible assumption—that adjacent points are connected by straight lines, such as line b. This result gives another important example of a segmented linear function.

In general, a segmented linear function consists of two or more straight line segments, usually with different slopes, joined at successive breakpoints (Figure 7). If the X and I coordinates of the breakpoints are represented by $(a_1, b_1), (a_2, b_2), \ldots, (a_j, b_j)$, any segmented linear function with m segments can be presented by the following general equation:

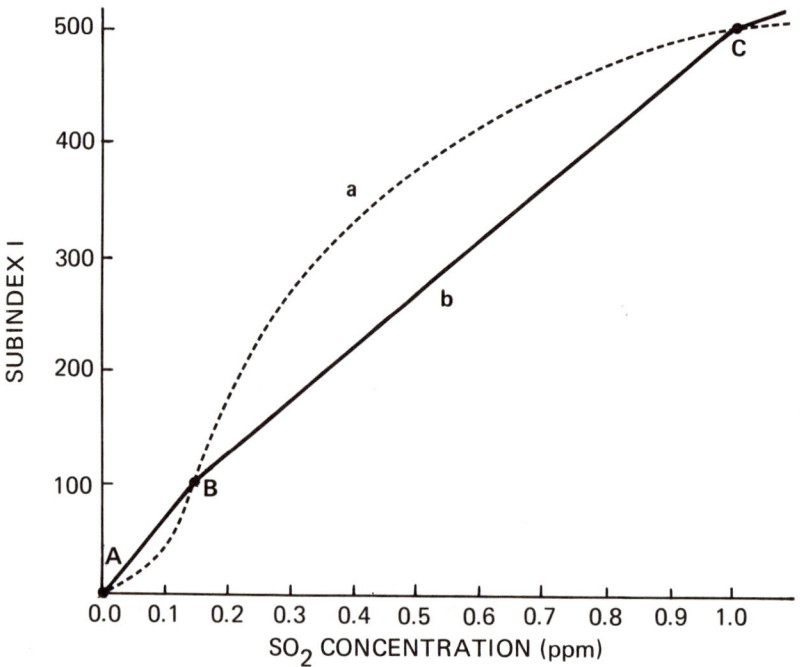

Figure 6. Example of subindex function which incorporates recommended limits, showing (a) arbitrary curve and (b) segmented linear function.

STRUCTURE OF ENVIRONMENTAL INDICES 57

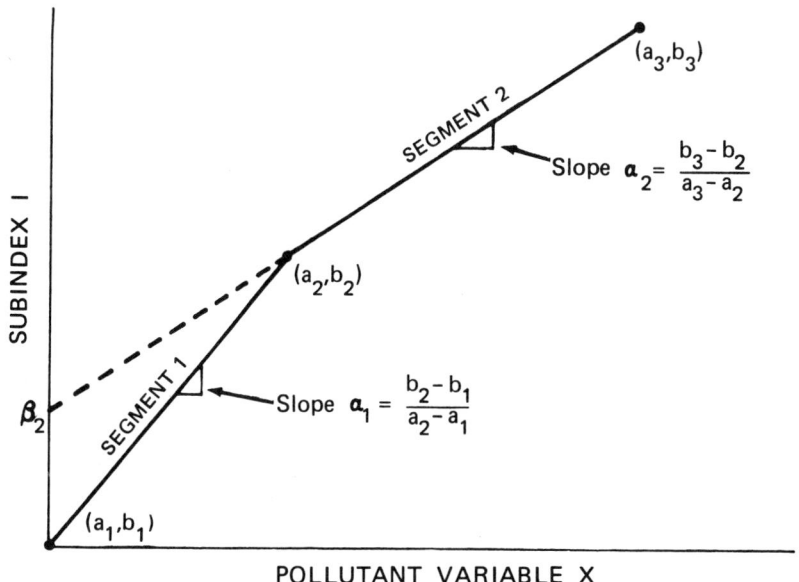

Figure 7. General form of segmented linear function.

$$I = \frac{b_{j+1} - b_j}{a_{j+1} - a_j}(X - a_j) + b_j \quad (6)$$

$$\text{for } a_j \leqslant X \leqslant a_{j+1}$$

where $j = 1, 2, 3, \ldots, m$.

In most instances, a_1 and b_1 are zero, and the first segment passes through the origin as in Figure 7, but this need not be the case.

Using Equation 6, the first segment is represented as follows:

$$I = \frac{b_2 - b_1}{a_2 - a_1}(X - a_1) + b_1 \quad (7)$$

$$\text{for } a_1 \leqslant X \leqslant a_2$$

This result is easily checked by substituting $X = a_1$ into Equation 7, giving b_1 as expected:

$$I = \frac{b_2 - b_1}{a_2 - a_1}(a_1 - a_1) + b_1 = 0 + b_1 = b_1 \quad (8)$$

Next, $X = a_2$ is substituted into Equation 7, giving b_2 as expected:

$$I = \frac{b_2 - b_1}{a_2 - a_1}(a_2 - a_1) + b_1 = b_2 - b_1 + b_1 = b_2 \tag{9}$$

Thus, Equation 7 gives the proper coordinates, (a_1, b_1) and (a_2, b_2), for both ends of the first segment.

An equation for the second segment is obtained from Equation 6 in a similar manner:

$$I = \frac{b_3 - b_2}{a_3 - a_2}(X - a_2) + b_2 \tag{10}$$

$$\text{for } a_2 \leq X \leq a_3$$

Equation 7 must give the same value for I at the end of the first segment as is obtained from Equation 10 for the second segment, because the two segments are joined at the breakpoint. Of course, certain applications may require that the inequalities be specified differently from those shown here.

By multiplying the product terms to the right of the equal sign in Equation 6 and rearranging, Equation 6 may be written in a form similar to Equation 4:

$$I = \alpha_j X + \beta_j \tag{11}$$

$$\text{for } a_j \leq X \leq a_{j+1}$$

where

$$\alpha_j = \frac{b_{j+1} - b_j}{a_{j+1} - a_j} \tag{11a}$$

$$\beta_j = \frac{b_j a_{j+1} - a_j b_{j+1}}{a_{j+1} - a_j} \tag{11b}$$

Here, α_j represents the slope of each line segment, and β_j represents the I-intercept if the line is extended as shown in Figure 7 (dotted line).

In the example of a segmented linear function discussed above and shown in Figure 6, the equation for the first segment is calculated as follows using Equation 7:

$$I = \frac{100 - 0}{0.14 - 0.0}(X - 0.0) + 0 = 714.29X$$

$$\text{for } 0.0 \leq X \leq 0.14$$

Using this equation, $I = 0$ for $X = 0.0$ ppm SO_2, and $I = 100$ for $X = 0.14$ ppm SO_2. The equation for the second segment is calculated in a similar fashion:

$$I = \frac{500 - 100}{1.0 - 0.14}(X - 0.14) + 100 = 465.12X + 34.88$$

$$\text{for } 0.14 \leq X \leq 1.0$$

Using this equation, I = 100 for X = 0.14 ppm SO_2, and I = 500 for X = 1.0 ppm SO_2.

An extreme case of the segmented linear function is the "step function." Here, successive segments alternate between horizontal and vertical directions, giving slopes that vary between zero and infinity. The simplest form is the "dichotomous," or two-state, step function, in which I takes on just two values. In the example shown (Figure 8), I = 0 when pollutant variable X is below the recommended administrative limit X_s, and I = 1 when X exceeds X_s. Step functions often consist of several successive steps, or multiple states, with particular subindex values corresponding to various ranges of the pollutant variable. For example, Horton's index,[2] published in 1965, uses subindex functions containing three, four, and five steps (Chapter IV). In Horton's dissolved oxygen subindex, I = 0 for X less than 10% saturation, while I = 30 for X between 10% and 30% saturation, and I = 100 for X above 70% saturation. The entire function, when plotted, consists of a "staircase" of steps (Figure 9).

Nonlinear Function

Although segmented linear functions are flexible, they are not ideally suited to some situations, particularly those in which the slope (rate of change) changes very *gradually* with increasing levels of environmental pollution. In these instances, a nonlinear function usually is more appropriate. A nonlinear function is any relationship which, when plotted, does not give a straight line. We shall consider two basic types of nonlinear functions: (1) an implicit function, which can be plotted on a graph but for which no equation is given, and (2) an explicit function, for which a mathematical equation is given.

Implicit functions usually arise when some empirical curve has been obtained from a process under study, but an exact equation is unknown for this curve. For example, Brown *et al.*[3] sent a questionnaire to water quality experts throughout the United States asking them to rate "water quality" on a scale from 1 to 100 as a function of the level of certain pollutant variables. The curves drawn by these respondents were averaged to give composite curves (Chapter IV). An example is the bell-shaped pH subindex curve[4] (Figure 10). No mathematical equation is available for this relationship, although explicit equations probably could be developed to approximate the shape of the curve. For a given pH measurement, the index user reads the appropriate subindex value from the graph.

In explicit nonlinear functions, curvature is achieved mathematically. An important general nonlinear function is one in which the pollutant variable is raised to a power other than one, the "power" subindex function:

60 ENVIRONMENTAL INDICES

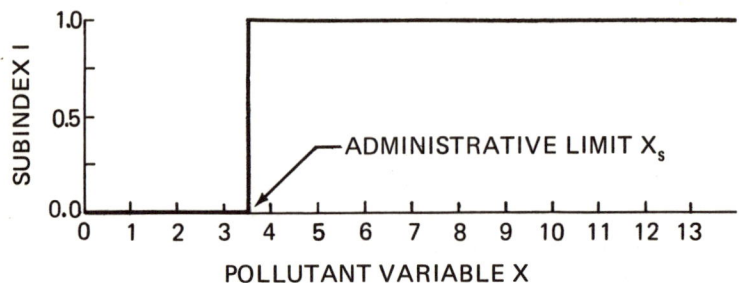

Figure 8. Example of a dichotomous step function.

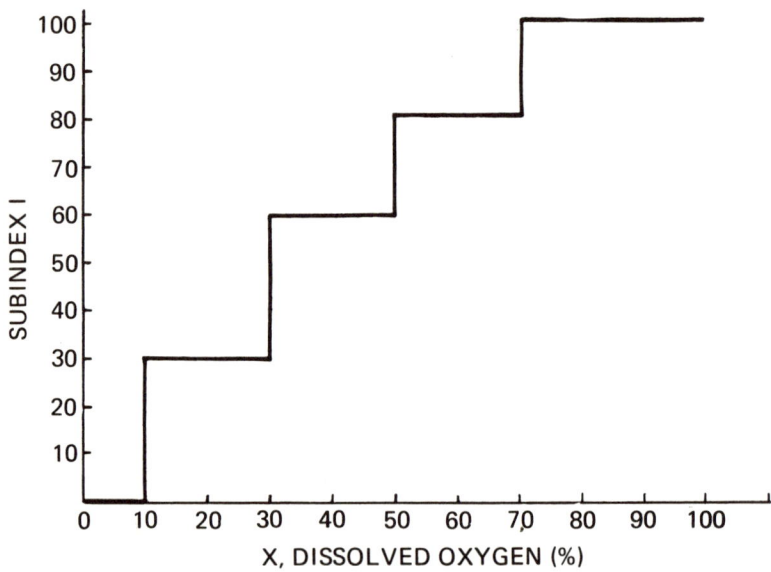

Figure 9. Staircase step function for dissolved oxygen from a water quality index proposed by Horton.[2]

$$I = X^c \qquad (12)$$

where $c \neq 1$.

If $c = 2$, the parabola $I = X^2$ results, which is a curve with rapidly increasing magnitude of slope (Figure 11). Actually, the slope is always twice the value

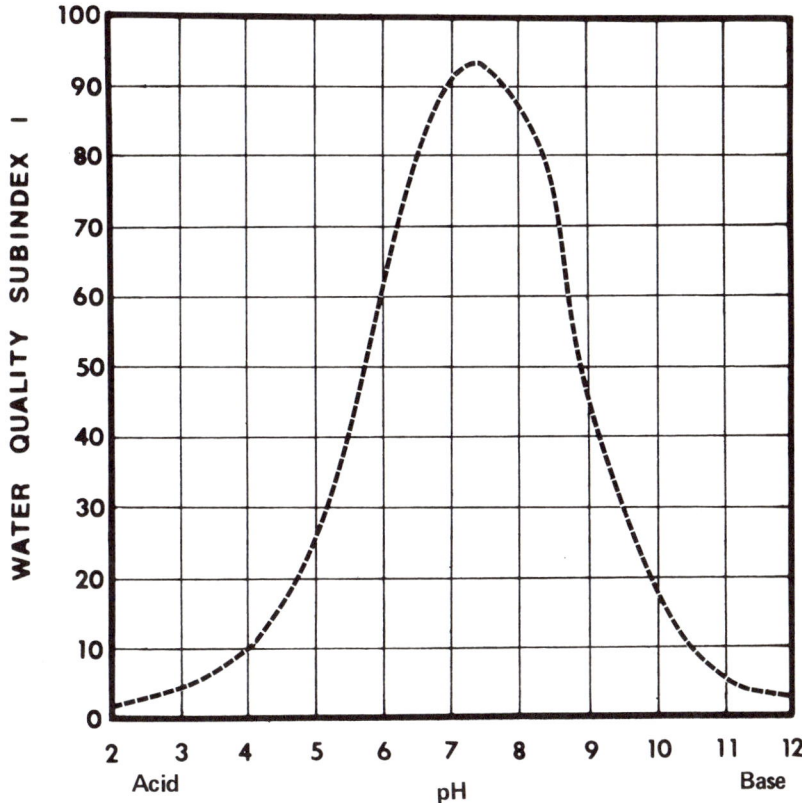

Figure 10. Example of implicit nonlinear subindex function for pH from a water quality index proposed by Brown et al.[3] and McClelland.[4]

of X, and a doubling of X results in a fourfold increase in the magnitude of I. In actual applications, the parabola often is translated from the origin and inverted. In a slightly more general parabolic form (Figure 12), I = b, a maximum, when X = a, and I = 0 when either X = 0 or X = 2a:

$$I = -\frac{b}{a^2}(X - a)^2 + b \qquad (13)$$

for $0 \leqslant X \leqslant 2a$

An index proposed by Walski and Parker[5] (Chapter IV) uses Equation 13 in its subindices for temperature and pH. Because this subindex has a single mode, or maximum, we say that it is "unimodal." We shall use the term unimodal to denote any function that has just one maximum or minimum.

62 ENVIRONMENTAL INDICES

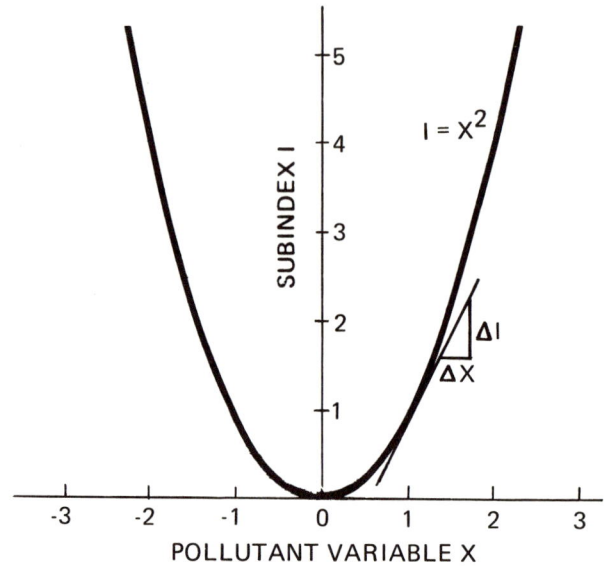

Figure 11. Example of explicit nonlinear subindex function, the parabola $I = X^2$.

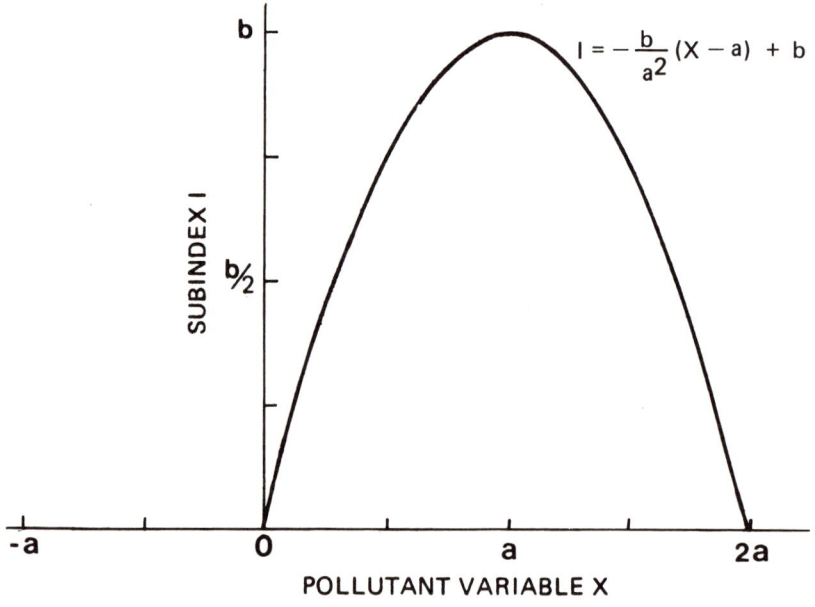

Figure 12. Example of a parabolic subindex function which was translated from the origin and inverted, based on a water quality index by Walski and Parker.[5]

Of course, pollutant variable X in Equation 12 can be raised to powers other than 2, creating a variety of other nonlinear shapes (Figure 13). All of these curves pass through the point (1,1) because 1 raised to any power is 1. The curve for c = 2.0 is the same parabolic function that was plotted in Figure 11, except that X is not allowed to be negative, giving only the right half of the parabola. Note that the curve for c = 0.5 is really a parabola lying on its side; that is, pollutant variable X is a parabolic function of I, because $I = X^{1/2}$, or $X = I^2$.

Another common nonlinear function is the exponential function, in which pollutant variable X is the exponent of a constant:

$$I = c^X \tag{14}$$

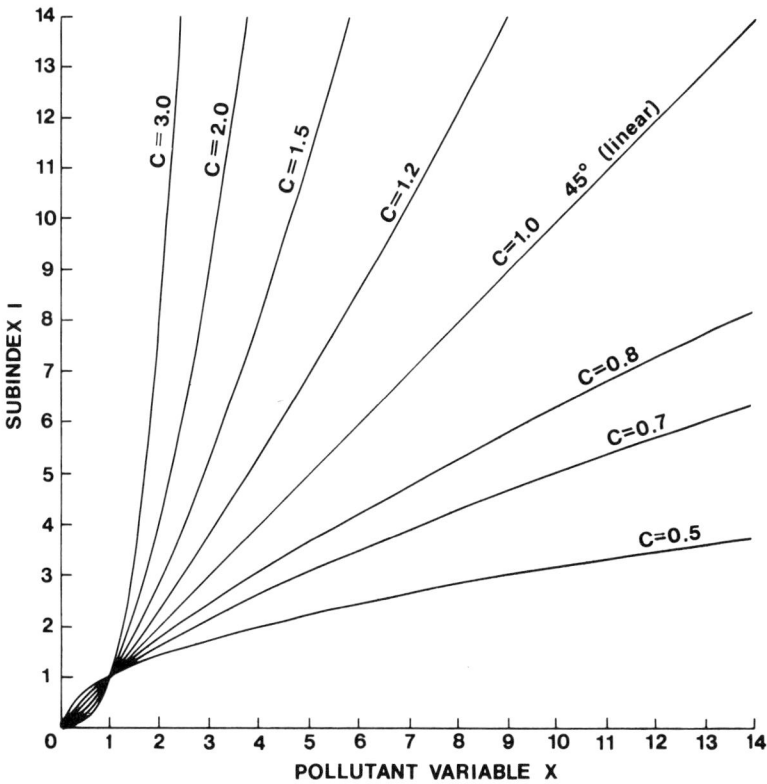

Figure 13. Plot of the power subindex function $I = X^c$ for selected values of c.

64 ENVIRONMENTAL INDICES

The constant usually selected is either 10 or e, the base of the natural logarithm (e = 2.71828128 . . .). If a and b are constants, the general form of an exponential function is written as follows:

$$I = ae^{bX} \qquad (15)$$

The exponential curve has the property that, for equal increments of X, I changes by a constant proportion. For the example given in Figure 14, I = 100 for X = 0. For X = 1, I = 85.2, which is 85.2% of 100. For X = 2, I = 72.6, which is 85.2% of the previous value (that is, 0.852 x 85.2 = 72.6). For X = 3, I = 61.9, which is also 85.2% of the previous value (that is, 0.852 x 72.6 = 61.9). Thus, in Figure 14, equal increments of size ΔX = 1.0 always produce values which are the same proportion, 85.2%, of the previous value.

Another common nonlinear function is the logarithmic function. In this book, ln X will denote the logarithm taken to the base e, or $\log_e X$, and log X will denote the logarithm taken to the base 10, or $\log_{10} X$.

Segmented Nonlinear Function

Although increasingly complex nonlinear equations can be used to generate curves of a variety of different shapes, the equations sometimes become unwieldy. A more flexible approach is to divide the curve into particular ranges and to represent each range by a nonlinear equation. The

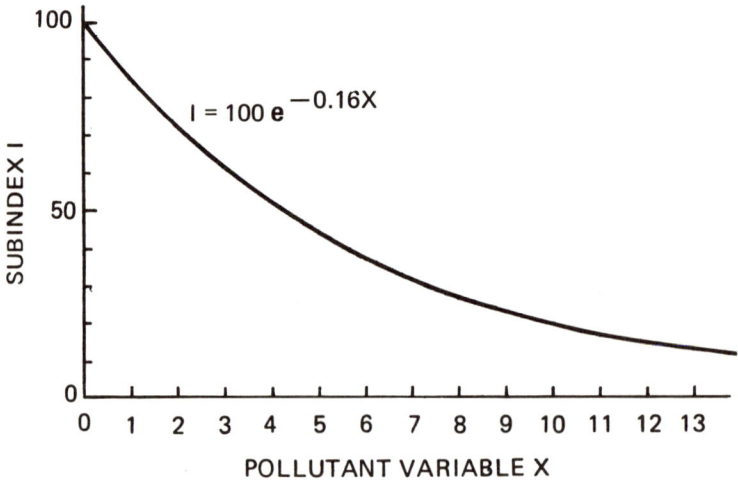

Figure 14. Example of exponential subindex function $I = ae^{bX}$ for a = 100 and b = -0.16.

result is a "segmented nonlinear function," which is used in a number of water quality indices. Like the segmented linear function, the line segments are joined at breakpoints. We shall describe a subindex function as segmented nonlinear if at least *one* of its segments is curved (nonlinear).

A segmented nonlinear function is used for the pH subindex in a water quality index proposed by Prati, Pavanello and Pesarin[6] (Chapter IV). The function (Figure 15) contains four segments:

Segment 1 (AB)	$0 \leqslant X \leqslant 5$	$I = -0.4X^2 + 14$
Segment 2 (BC)	$5 \leqslant X \leqslant 7$	$I = -2X + 14$
Segment 3 (CD)	$7 \leqslant X \leqslant 9$	$I = X^2 - 14X + 49$
Segment 4 (DE)	$9 \leqslant X \leqslant 14$	$I = -0.4X^2 + 11.2X - 64.4$

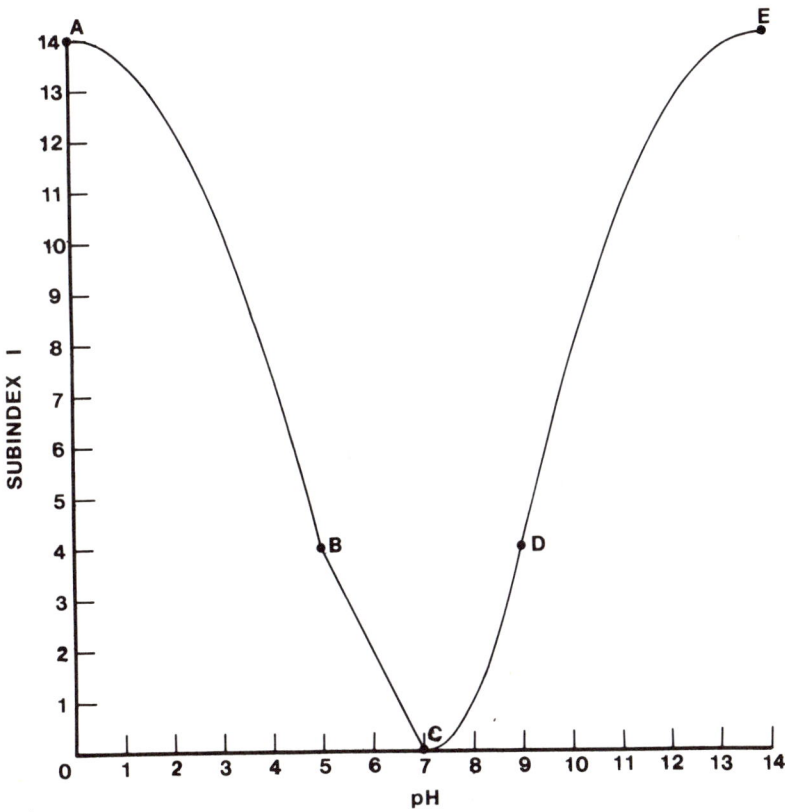

Figure 15. Example of a segmented nonlinear function for pH, from the water quality index of Prati, Paranello and Pesarin.[6]

Line segment 1 (between breakpoints A and B) is a parabola, and line segment 2 (between breakpoints B and C) is a straight line. As with the segmented linear function described above, the two equations give the same value at each coincident breakpoint. For example, at $X = 5$ the first equation gives $I = -0.4[5]^2 + 14 = -10 + 14 = 4$, and the second equation gives $I = -2[5] + 14 = 4$. Note that this pH subindex cannot be compared directly with the decreasing scale pH subindex shown in Figure 10, because the subindex in Figure 15 has an increasing scale and ranges from 0 to 14.

AGGREGATION OF SUBINDICES

The aggregation process is one of the most important steps in calculating any environmental index. Here is where most of the simplification (reduction of information) takes place, and here is where most of the distortion is likely to be introduced. In the following sections, we shall list the general forms of each of the more common aggregation functions. We shall illustrate these general forms by means of a graphical technique involving just two subindices. This technique allows us to examine the behavior and limitations of each aggregation function, and the conclusions can be expected to apply also to the more general case in which more than two pollutant variables are involved.

Additive Forms

The simplest aggregation functions are the additive forms. We shall call the addition of unweighted subindices, in which no subindex is raised to a power other than 1, the "linear sum":

$$I = \sum_{i=1}^{n} I_i \tag{16}$$

where I_i = subindex for pollutant variable i
       n = number of pollutant variables

Suppose that a linear sum air pollution index is formed consisting of just two subindices, I_1 and I_2:

$$I = I_1 + I_2 \tag{17}$$

In this simple index, we shall assume that I_1 and I_2 are dichotomous subindices in which $I_1 = 0$ and $I_2 = 0$ represent zero air pollutant concentrations for pollutant variables X_1 and X_2, and $I_1 \geq 100$ or $I_2 \geq 100$ represent concentrations at or above the NAAQS. Does Equation 17 combine the two subindices in a manner that properly reflects violation of the NAAQS?

STRUCTURE OF ENVIRONMENTAL INDICES 67

If either subindex in Equation 17 exceeds 100, then the overall index will exceed 100. That is, if either $I_1 \geq 100$ or $I_2 \geq 100$, then $I \geq 100$. This system aggregates subindices in a reasonable fashion if the magnitude of either subindex is large (extremely severe air pollution). However, most users will expect I above 100 to mean unequivocally that a NAAQS is violated for at least one subindex, and it is unfortunately possible for I to exceed 100 without a NAAQS being violated. For example, if moderate pollution levels occur for both pollutant variables, giving, say, $I_1 = 50$ and $I_2 = 50$, then $I = 100$. Similarly, if $I_1 = 60$ and $I_2 = 70$, then $I = 130$. The index conveys the impression that a NAAQS has been violated when it has not been, giving an exaggerated and ambiguous reading. A variety of other combinations of subindex values (I_1, I_2), will give an ambiguous reading for $I = 100$; for example: (5,95), (20,80), (35,65), (60,40), (75,25), and (99,1). These combinations all satisfy the equation $I_2 = 100 - I_1$, and this function is readily plotted on a graph of I_2 versus I_1 (Figure 16). The result is a straight line

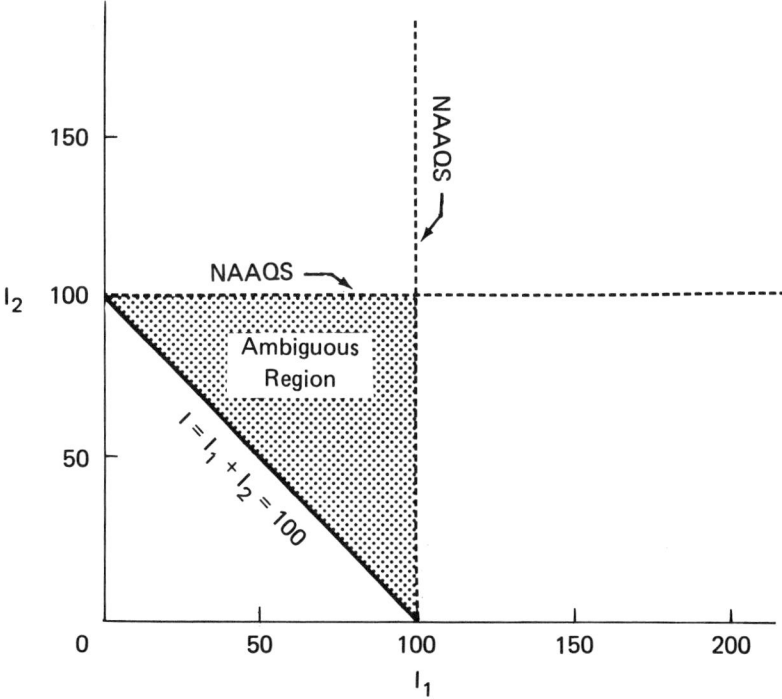

Figure 16. Plot of the linear sum $I_1 + I_2 = 100$ showing ambiguous region for which I exceeds 100 without either subindex exceeding 100.

denoting all possible combinations of I_1 and I_2 that give I = 100. On this same figure, it is possible to denote values $I_2 \geq 100$ as one region (area above the horizontal dotted line), and values $I_1 \geq 100$ as another region (area to the right of the vertical dotted line). In either of these regions, the index reports violations of the NAAQS in an unambiguous fashion. However, in the triangular area above the straight line but bounded by the two dotted lines (shaded area), the index exceeds 100 *without* either subindex exceeding 100. We shall define this area as the "ambiguous" region. In this area, the linear sum exaggerates the severity of the air pollution problem.

It is conceivable, of course, that the ambiguity might not be sufficiently serious to restrict the use of the index in certain applications; however, as more pollutant variables are added, the ambiguity tends to increase. For example, if 10 subindices are aggregated, the index reaches I = 100 when each subindex is only 10. In this system, the index ordinarily would exceed 100 so much of the time that real violations of the NAAQS would be difficult to detect. This analysis illustrates why the linear sum is not well-suited to aggregating dichotomous subindices.

If three variables—I_1, I_2, and I—are to be plotted in two dimensions, the usual custom is to choose I as the dependent variable and to plot it as a function of one of the other two variables, say I_1, while holding the third variable, I_2, constant. However, any of the three variables may be held constant, and it is more instructive, for analysis purposes, to plot I_2 as a function of I_1 while holding I constant. Figure 17, for example, shows a plot of $I_2 = I - I_1$ for I = 50, 100, 150, and 200. The result is a series of parallel straight lines. Because I may be any positive real number, the lines shown are part of a surface. We shall refer to this graph of I_2 plotted against I_1, with aggregation function $g(I_1, I_2)$ = constant, as the "(I_1, I_2)-plane." The (I_1, I_2)-plane is intended as a mathematical system introduced here to facilitate comparison of environmental indices.

Can the linear sum be modified so that the ambiguous region is eliminated? By multiplying each subindex by an appropriate coefficient, or "weight," we can modify the aggregation function so that the exaggeration problem does not occur. Usually, the weights are selected so that their sum is unity. The weighted linear sum has the following general form:

$$I = \sum_{i=1}^{n} w_i I_i \qquad (18)$$

where

$$\sum_{i=1}^{n} w_i = 1 \qquad (18a)$$

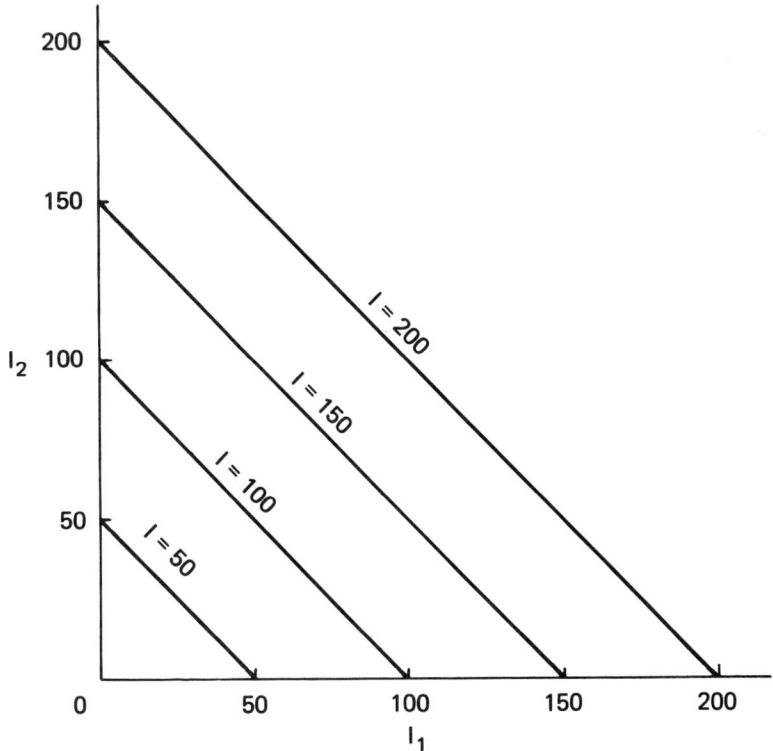

Figure 17. Plot of $I = I_1 + I_2$ in the (I_1, I_2)-plane for selected values of I.

For the two-variable case discussed above,

$$I = w_1 I_1 + w_2 I_2 \qquad (19)$$

where
$$w_1 + w_2 = 1 \qquad (19a)$$

In Equation 19, zero pollution is reported properly, because $I = 0$ when $I_1 = 0$ and $I_2 = 0$. Likewise, if both subindices are 100, then $I = 100$. This is shown by substituting $w_2 = 1 - w_1$ from Equation 19a into Equation 19 and setting both $I_1 = 100$ and $I_2 = 100$:

$$I = w_1(100) + (1 - w_1)(100) = 100 \qquad (20)$$

If both of the subindices are less than 100, then I will be less than 100. Thus, it is no longer possible for I to be 100 (or more) without *at least one* subindex being 100 (or more). Mathematically, if $I \geqslant 100$, then $I_1 \geqslant 100$ or $I_2 \geqslant 100$. Therefore, if I exceeds 100 in the weighted linear sum, we can unequivocally conclude that at least one NAAQS is violated.

The result of including the weights, w_1 and w_2, can be examined in the (I_1, I_2)-plane. Assume, for example, that $w_1 = w_2 = 0.5$, giving the following equation:

$$I = 0.5 I_1 + 0.5 I_2 \qquad (21)$$

This equation is equivalent to calculating the arithmetic mean of the two subindices. Inclusion of these weights has the effect of moving the line for $I = 100$ in Figure 16 to a position twice as far from the origin without changing its slope (Figure 18). The line for $I = 100$ in Figure 18 now occupies the same position as the line for $I = 200$ in Figure 17. The ambiguous region (shaded area in Figure 16) is now eliminated.

Although the weighted linear sum does not have an ambiguous region, a more serious problem is introduced. This problem is called "eclipsing," and it reflects an underestimation rather than an exaggeration of pollution. Eclipsing is said to occur when extremely poor environmental quality exists

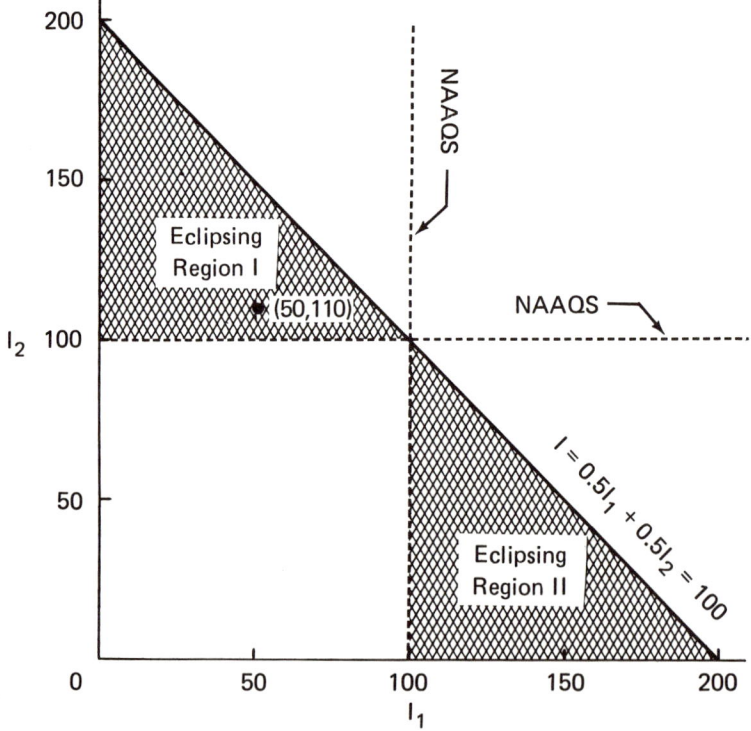

Figure 18. Plot of the weighted linear sum $0.5 I_1 + 0.5 I_2 = 100$ showing eclipsing regions for which a subindex exceeds 100 without the index exceeding 100.

for at least one pollutant variable, but the overall index does not reflect this fact. In Equation 21, suppose that $I_1 = 50$ and $I_2 = 110$, indicating violation of the NAAQS for pollutant variable X_2. This gives $I = 80$. Because the overall index is less than 100, violation of the NAAQS is eclipsed. This point is shown in Figure 18. We say that the poor air quality reflected in subindex I_2 has been eclipsed in the overall index.

When Equation 21 is plotted in the (I_1, I_2)-plane, this function reveals two eclipsing regions (crosshatched areas in Figure 18). In the upper region (Region I), $I_2 \geq 100$ without I exceeding 100. In the lower eclipsing region (Region II), $I_1 \geq 100$ without I exceeding 100. In the other areas in this figure, such as the large square formed between the origin and the two dotted lines, the aggregation function performs satisfactorily. Here, I is less than 100 and both subindices are less than 100.

In the example given in Equation 21, the two weights are equal, giving a straight line in the (I_1, I_2)-plane inclined at 135° relative to the horizontal axis (slope = -1). To facilitate plotting of the general form of the weighted linear sum in the (I_1, I_2)-plane, Equation 19 may be written in a manner similar to Equation 4:

$$I_2 = \alpha I_1 + \beta \qquad (22)$$

where
$$\alpha = -w_1/w_2$$
$$\beta = I/w_2$$

For different values of I, this equation gives parallel straight lines of slope $\alpha = -w_1/w_2$ and I_2-intercept $\beta = I/w_2$ (Figure 19). For example, if $w_1 = 2/3$ and $w_2 = 1/3$, the (I_1, I_2)-plane will contain parallel lines of slope $\alpha = -(2/3)/(1/3) = -2$, as shown in Figure 20. In the weighted linear sum, the line for $I = 100$ will always pass through the point (100,100), regardless of the slope. Therefore, regardless of the magnitudes of the two weights, the weighted linear sum will always contain two eclipsing regions.

This analysis shows that the unweighted linear sum exhibits ambiguity, which tends to exaggerate pollution, while the weighted linear sum exhibits eclipsing, which tends to underestimate pollution. These problems are characteristic of the linear additive forms and demonstrate the unsuitability of this aggregation function for representing dichotomous subindices. Is there another function available which would eliminate both the ambiguity and eclipsing problems?

Root-Sum-Power. To deal with these problems, a somewhat more complex additive form is available. The root-sum-power is a nonlinear aggregation function of the following form:

$$I = \left[\sum_{i=1}^{n} I_i^p \right]^{1/p} \qquad (23)$$

72 ENVIRONMENTAL INDICES

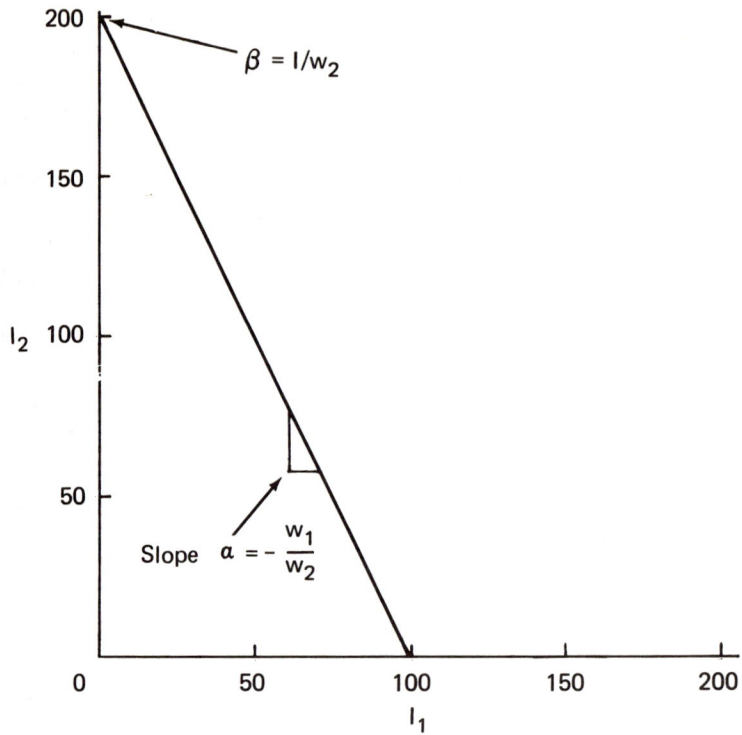

Figure 19. General form of the weighted linear sum $I = w_1 I_1 + w_2 I_2$ plotted in the (I_1, I_2)-plane.

The subindices are each raised to power p, added, and the p*th* root is taken. Generally, p is a positive real number greater than 1. For the two-variable case,

$$I = [I_1^p + I_2^p]^{1/p} \qquad (24)$$

If p = 2, the aggregation function is called the root-sum-square:

$$I = \sqrt{(I_1)^2 + (I_2)^2} \qquad (25)$$

In the (I_1, I_2)-plane, Equation 25 plots as a circle of radius I, with its center at the origin. The problem of ambiguity is less serious for the root-sum-square than it was for the linear sum. This can be seen by plotting Equation 25 in the (I_1, I_2)-plane for I = 100 (Figure 21):

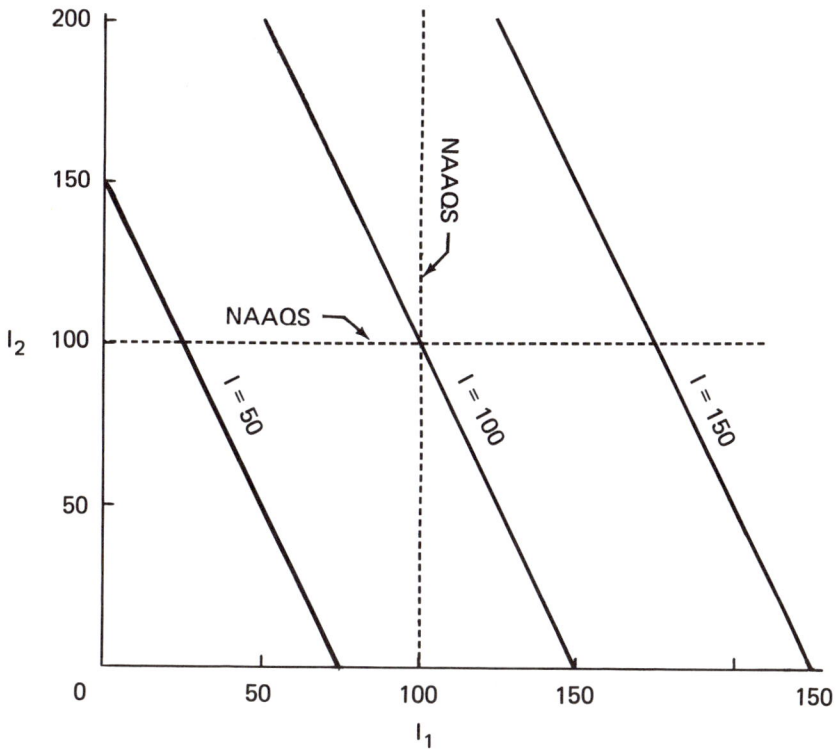

Figure 20. Plot of $I = (2/3)I_1 + (1/3)I_2$ in the (I_1, I_2)-plane for selected values of I.

$$100 = \sqrt{(I_1)^2 + (I_2)^2} \qquad (26)$$

In the linear sum plotted in Figure 16, the ambiguous region consists of 50% of the large square formed between the two axes and the dotted lines representing the NAAQS. In Figure 21, by comparison, the ambiguous region (shaded area) occupies only 21.5% of the total square. The root-sum-square aggregation function therefore reduces the tendency for ambiguity to occur.

As p in Equation 24 becomes larger, the shaded ambiguous region in Figure 21 becomes increasingly smaller. As can be seen from Figure 22, the root-sum-power can be viewed as a general form, of which several other aggregation functions are special cases, depending on the value of p. As p increases from 1 to infinity, the root-sum-power exhibits a progression of shapes. For example, if p = 1, a straight line results, which is the (unweighted) linear sum aggregation function discussed earlier (Figure 16). If p = 2, the root-sum-square results (Figure 21). If p = 3, the curvature becomes sharper, and, for

74 ENVIRONMENTAL INDICES

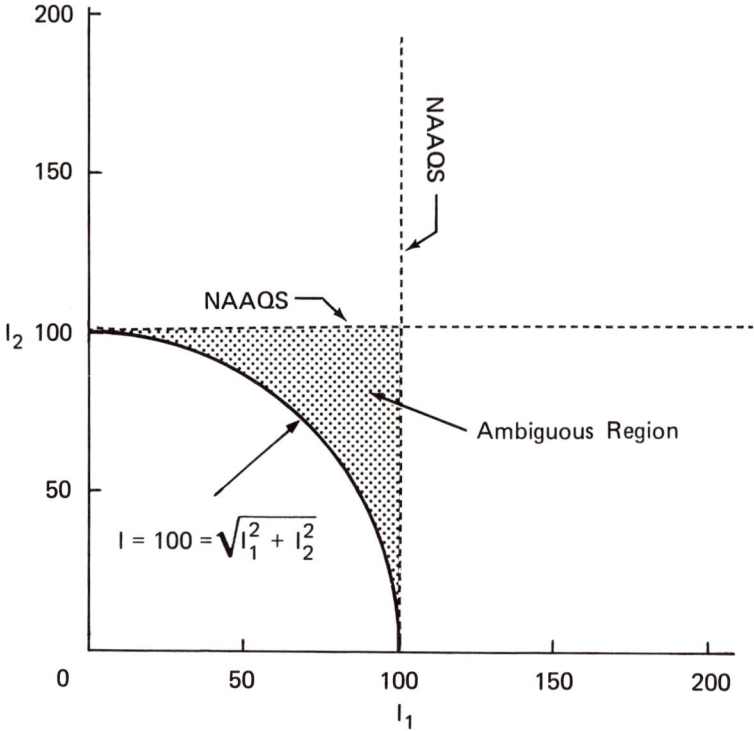

Figure 21. Plot of the root-sum-square aggregation function in the (I_1, I_2)-plane.

p = 5, the curvature becomes increasingly abrupt (Figure 22). For values of p greater than 10, the curve rapidly approaches the shape of the large square bounded by the NAAQS lines in Figure 22, and the shaded area becomes extremely small. Thus, for large values of p, the ambiguous region is almost entirely eliminated. In the limit, as p approaches infinity, the root-sum-power asymptotically approaches the lines $I_1 = 100$ and $I_2 = 100$.

For the limiting case in which p approaches infinity, the root-sum-power has desirable properties for aggregating subindices. It possesses neither an eclipsing region nor an ambiguous region. However, because it is a limiting function, it is somewhat unwieldy to write and use. A simpler function with the same properties is the maximum operator. Actually, the maximum operator can be viewed as the limiting case of the root-sum-power as p approaches infinity:

STRUCTURE OF ENVIRONMENTAL INDICES 75

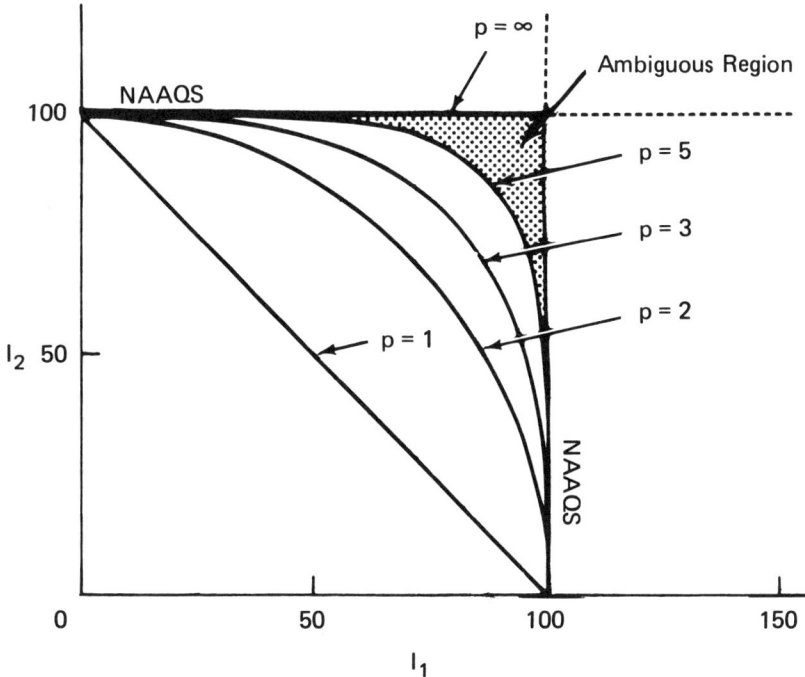

Figure 22. Plot of $I = (I_1^p + I_2^p)^{1/p}$ for selected values of p.

$$\lim_{p \to \infty} \left\{ \left[I_1^p + I_2^p \right]^{1/p} \right\} = \max \{I_1, I_2\} \qquad (27)$$

Root-Mean-Square. Another example of an additive form is the root-mean-square. This aggregation function is similar to the root-sum-square, except that the arithmetic mean of the square of the subindices is calculated before the square root is taken. For the case involving just two pollutant variables, the root-mean-square is calculated as follows:

$$I = \sqrt{(1/2)(I_1^2 + I_2^2)} \qquad (28)$$

In the (I_1, I_2)-plane, this function plots as a circle of radius $\sqrt{2}\,I$ (Figure 23). Like the weighted linear sum (Figure 18), the root-mean-square exhibits two eclipsing regions. For example, if $I_1 = 50$ and $I_2 = 120$, then $I = 91.2$, and violation of the NAAQS for pollutant variable X_2 is eclipsed.

76 ENVIRONMENTAL INDICES

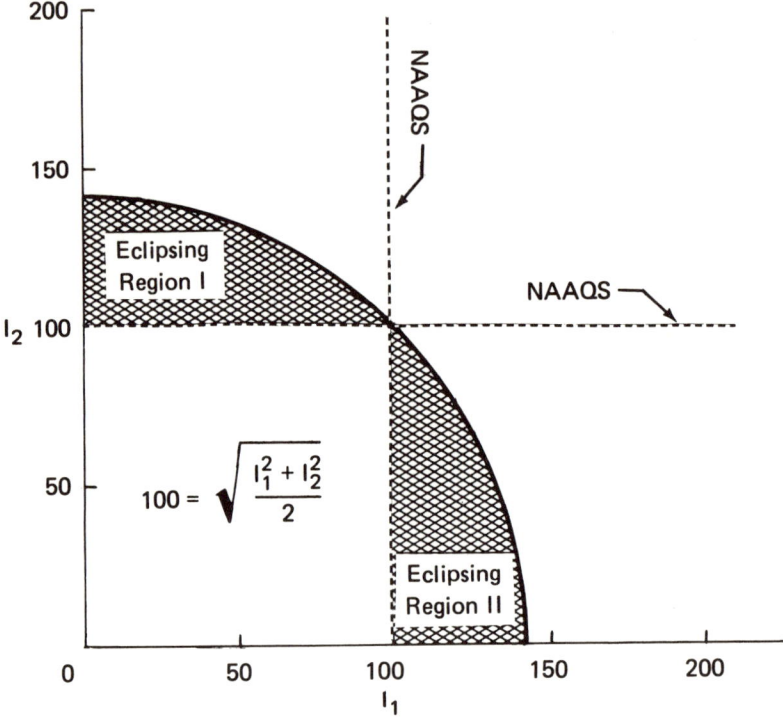

Figure 23. Plot of the root-mean-square aggregation function in the (I_1, I_2)-plane.

Maximum Operator

In 1976, Thom and Ott[7] conducted a study which defined the need for a nationally uniform air pollution index. They recommended that such an index incorporate the maximum operator, and they proposed a candidate index structure. The Pollutant Standards Index, which has since been adopted by the federal government, is based on this original structure and uses the maximum operator for its aggregation function (Chapter III).

The general form of the maximum operator is as follows:

$$I = \max \{I_1, I_2, \ldots, I_i, \ldots, I_n\} \qquad (29)$$

In the maximum operator, I takes on the value of the largest of any of the subindices, and $I = 0$ if and only if $I_i = 0$ for all i.

As with the other aggregation functions, it is instructive to examine the properties of the maximum operator for the simple case involving just two subindices:

$$I = \max\{I_1, I_2\} \tag{30}$$

To plot Equation 29 in the (I_1, I_2)-plane, we consider the set of all possible values which give $I = 100$:

$$100 = \max\{I_1, I_2\} \tag{31}$$

If $I_2 = 100$, Equation 31 is satisfied only if I_2 is greater than I_1, which can occur only if $0 \leq I_1 < 100$. This plots as the horizontal line segment AB in Figure 24. Conversely, if $I_1 = 100$, Equation 31 is satisfied only if I_1 is greater than I_2, which can occur only if $0 \leq I_2 < 100$. This plots as the vertical line segment BC in Figure 24. Thus, Equation 31 plots in the

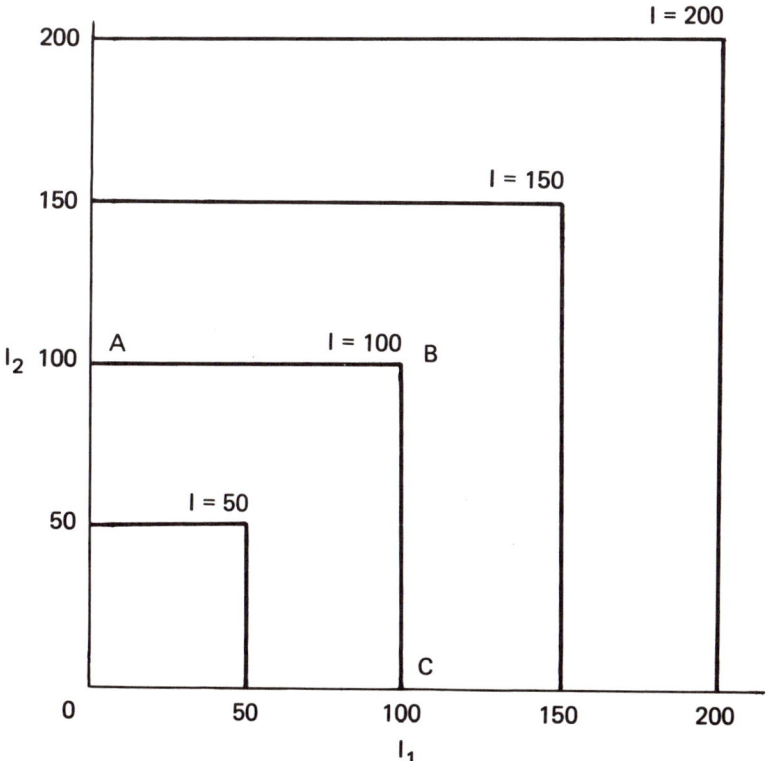

Figure 24. Plot of $I = \max\{I_1, I_2\}$ in the (I_1, I_2)-plane for selected values of I.

(I_1, I_2)-plane as two straight lines joined at right angles, giving the appearance of a square box.

With the maximum operator, there is no eclipsing region. Thus, if one subindex exhibits poor environmental quality ($I_i \geq 100$), then the overall index exhibits poor environmental quality ($I \geq 100$). There also is no ambiguous region, because, if the overall index exhibits poor environmental quality ($I \geq 100$), then at least one subindex must exhibit poor environmental quality ($I_i \geq 100$ for some i). Consequently, the maximum operator is particularly well suited for combining dichotomous subindices.

Although the above discussion applies mainly to dichotomous, or two-state, subindices, some of these arguments can be extended to subindices that include more than two states. Suppose, for example, that an air pollution index is constructed which reports three states. Index values less than I = 100 may denote that no air pollution problem exists, while those above I = 100 may denote that the NAAQS has been violated. Values above I = 200 may denote a more serious stage of air pollution, such as an air pollution "Alert." Here, the maximum operator will report all three states—zero air pollution, violation of the NAAQS, and the Alert level—without ambiguity or eclipsing. This is possible because each state corresponds to a dichotomous event (yes, an Alert has occurred, or no, it has not). Thus, within limits, the advantages of the maximum operator can be extended beyond the two-state case to the multiple-state case.

The limitations of the maximum operator become apparent when fine gradations of environmental quality, rather than discrete events, are to be reported and a number of subindices are to be aggregated. Consider, for example, an air pollution index consisting of four subindices and using the maximum operator. Suppose this index is used to report air quality for two different cases, based on observations on two successive dates. On the first date (Case 1), the subindices were as follows: $I_1 = 98$, $I_2 = 110$, $I_3 = 80$, and $I_4 = 90$; on the second date (Case 2), $I_1 = 0$, $I_2 = 110$, $I_3 = 5$, and $I_4 = 0$. The maximum operator gives the same value, I = 110, for both cases:

Case 1
$$I = \max\{98, 110, 80, 90\} = 110$$

Case 2
$$I = \max\{0, 110, 5, 0\} = 110$$

Some observers will be skeptical of this result because they will feel that the generally larger numbers in Case 1 denote more severe air pollution than those in Case 2. Note that an additive aggregation function, such as the arithmetic mean, gives I = 94.5 for Case 1 and I = 57.5 for Case 2. If fine gradations of air quality are important over the entire range of each variable, then the maximum operator would be unsuitable, and the arithmetic mean might be more appropriate.

However, if it is important to report violations of the NAAQS, as is usually the case for an air pollution index, the maximum operator is a more suitable aggregation function. Generally, the NAAQS have been interpreted as threshold values below which no adverse effects on health and welfare are observed. In other words, the NAAQS are assumed to be based on hockey stick damage functions (Chapter I). If this interpretation is correct, Case 1 and Case 2 are essentially equivalent in terms of the effects of air pollution on health and welfare. There is no important difference between $I_3 = 80$ in Case 1 and $I_3 = 5$ in Case 2, because I_3 is less than 100 in both instances. Thus, the maximum operator correctly reflects air pollution effects for these two cases. The arithmetic mean, in contrast, eclipses violation of the NAAQS for subindex I_2 in both cases.

We should note that air pollutants measured at a given monitoring station usually are all affected by the same meteorological conditions (sunlight, winds, inversion heights), causing concentrations of different pollutants to be correlated with each other. In Case 1, all the subindex values are high; in Case 2, one subindex is high and all the others are very low. These two cases taken together suggest a lack of correlation and therefore are unrealistic. If the correlations among pollutants are all positive and high, radically different combinations of values occur with low probability, and the maximum operator is an appropriate aggregation function.

The maximum operator is ideally suited to applications in which an index must report if at least one recommended limit (*e.g.*, the NAAQS) is violated and by how much. Of course, if several subindices violate a recommended limit, the maximum operator will report *only* the worst subindex. To provide more complete information, the Pollutant Standards Index adopted by the federal government reports not only the maximum subindex but also the other subindices which exceed 100 (Chapter III). The suitability of the maximum operator for use in water pollution indices has not been investigated, however, and none of the published water quality indices have employed this aggregation function (Chapter IV).

Multiplicative Forms

All previous examples of indices in this chapter have employed increasing scales. The multiplicative forms have found use primarily in indices that have decreasing scales. Most of the decreasing scale indices are water quality indices. The water quality index proposed in 1970 by Brown *et al.*[3] originally used an additive aggregation function, the weighted linear sum. Later, Landwehr[8] evaluated multiplicative aggregation functions that could be substituted for the additive form, and the multiplicative form has become the most popular version of this index (see Chapter IV).

Like increasing scale indices, decreasing scale indices have important characteristics which can be examined in the (I_1, I_2)-plane. To use the (I_1, I_2)-plane for this purpose, we must readjust our thinking somewhat. Now, points close to the origin will represent poor environmental quality, while points far away from the origin will represent good environmental quality. In most of the following examples, we shall assume that each subindex ranges between zero (poor environmental quality) and 100 (good environmental quality).

Like increasing scale indices, many decreasing scale indices exhibit serious eclipsing problems. In the decreasing scale system, eclipsing occurs if one subindex is zero but the overall index is greater than zero; that is, eclipsing occurs if $I_i = 0$ for any subindex i but $I > 0$. Because the scale is reversed, eclipsing problems show up somewhat differently in the (I_1, I_2)-plane.

To understand how the (I_1, I_2)-plane displays the characteristics of decreasing scale aggregation functions, it is first instructive to consider a very simple example, the weighted linear sum (previously, Equation 21):

$$I = 0.5I_1 + 0.5I_2 \qquad (32)$$

Here, I_1 and I_2 are decreasing scale subindices such that $0 \leqslant I_1 \leqslant 100$ and $0 \leqslant I_2 \leqslant 100$. As in Figure 18, Equation 32 plots in the (I_1, I_2)-plane as a straight line of slope -1 (Figure 25). As I varies from 0 to 100, the straight line moves from the origin to locations relatively far from the origin, arriving finally at the point (100,100).

Eclipsing occurs in Equation 32 for all values (I_1, I_2) for which $I_1 = 0$ with $I > 0$, or $I_2 = 0$ with $I > 0$. To find the eclipsing region for subindex I_1, for example, we set $I_1 = 0$ in Equation 32 and examine the range of values of I and I_2:

$$I = 0.5I_2 \qquad (33)$$

If I_2 is allowed to vary over the range $0 < I_2 \leqslant 100$, then I in Equation 33 varies over the range $0 < I \leqslant 50$. Thus, eclipsing of I_1 can occur for nearly all values of I_2, which correspond to values of I ranging from slightly above zero to 50. Graphically, the eclipsing region for subindex I_1 is the locus of all points where the sloping straight line intercepts the I_2-axis (that is, where I_1 = 0). Therefore, the eclipsing region is not a region at all but a straight line which coincides exactly with the I_2-axis. This result shows that the overall index I can range from 0 to 50 with I_1 being totally eclipsed.

Although eclipsing is especially serious for the weighted linear sum in decreasing scale indices, there is no ambiguous region. In a decreasing scale system, ambiguity occurs if $I = 0$ without at least one subindex being zero. In Equation 32, the overall index can be zero only if both subindices are zero. Therefore, if the index shows poor environmental quality ($I = 0$), one can unequivocally expect the two subindices also to exhibit poor environmental quality.

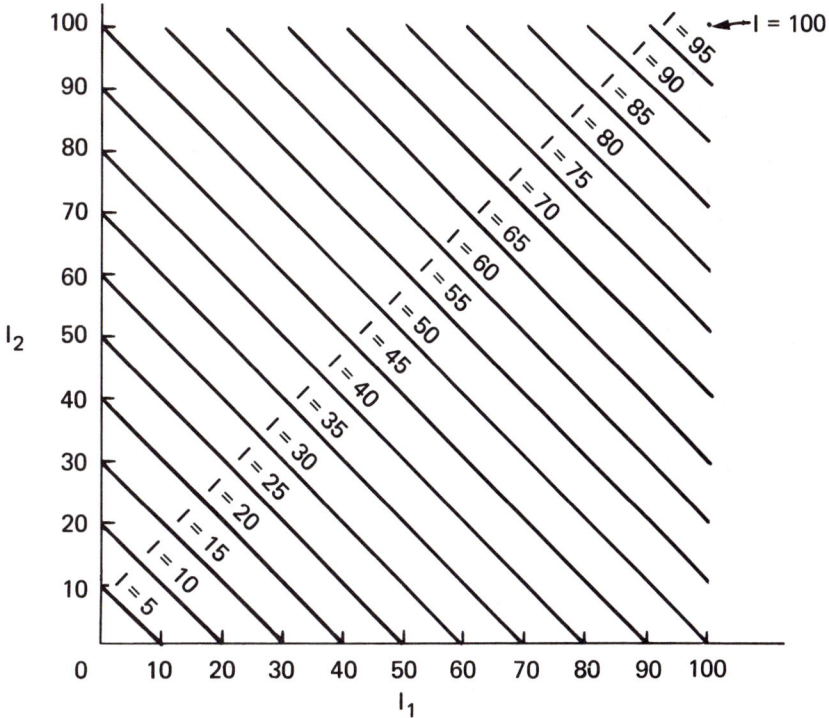

Figure 25. Plot of the weighted linear sum $I = 0.5I_1 + 0.5I_2$ in the (I_1, I_2)-plane for selected values of I.

The eclipsing region was difficult to show in Figure 25, because it coincided with the I_2-axis. Eclipsing can be illustrated more easily in the (I_1, I_2)-plane if we pick an arbitrary range of values, say $I_i \leq 10$ for subindex i, to denote poor environmental quality. Then, the region of poor environmental quality for subindex I_1 becomes the area to the left of the line $I_1 = 10$ (vertical dotted line in Figure 26). A similar region of poor environmental quality for subindex I_2 exists below the line $I_2 = 10$ (horizontal dotted line in Figure 26). The eclipsing region is the area for which either I_1 or I_2 shows poor environmental quality (the area between the dotted lines and the two axes) less the area for which I shows poor environmental quality. Thus, the area defined by $I \leq 10$ (a triangle bounded by the two axes and the line $I = 10$) must be subtracted from the larger area. The resulting eclipsing region consists of two trapezoidal areas (crosshatched areas in Figure 26). Here, I exceeds 10, but either I_1 or I_2 is 10 or less.

82 ENVIRONMENTAL INDICES

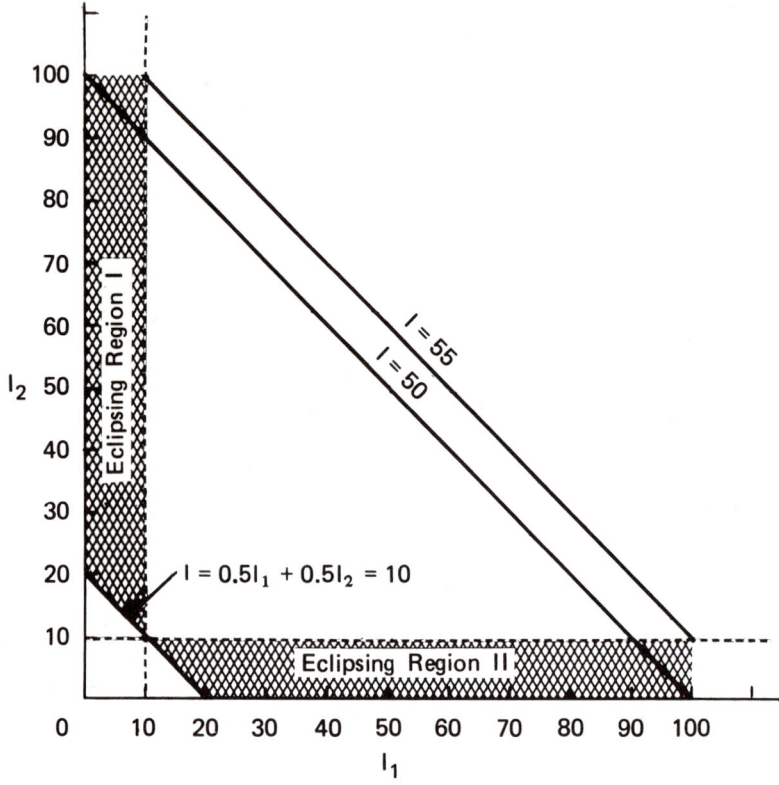

Figure 26. Plot of $I = 0.5I_1 + 0.5I_2$ in the (I_1, I_2)-plane showing decreasing scale eclipsing regions for which $I_1 \leq 10$ or $I_2 \leq 10$ while $I > 10$.

In Figure 26, it is seen that some of the lines for $I > 10$ in Equation 32 extend into the eclipsing regions. The largest value of I for which some portion of a line still extends into the crosshatched area is for the line passing through the point $I_1 = 10$ and $I_2 = 100$, giving $I = (0.5)(10) + (0.5)(100) = 55$. Consequently, in this example, eclipsing can occur over a relatively wide range of values, $10 < I \leq 55$. This result again demonstrates the seriousness of the eclipsing problem for decreasing scale indices when an additive form, such as the weighted linear sum, is used to aggregate subindices. In general, the additive forms do not appear well suited for aggregating decreasing scale subindices.

To avoid such problems, the multiplicative forms have been proposed. The most common multiplicative aggregation function is the weighted product, which has the following general form:*

---
*The symbol Π denotes the operation of multiplying together all terms immediately following it.

$$I = \prod_{i=1}^{n} I_i^{w_i} \qquad (34)$$

where
$$\sum_{i=1}^{n} w_i = 1 \qquad (34a)$$

In this aggregation function, as with all multiplicative forms, the index is zero if any one subindex is zero.

$$I = \prod_{i=2}^{n} I_i^{w_i} (0)^{w_1} = 0 \qquad (35)$$

This characteristic eliminates the eclipsing problem, because, if any one subindex exhibits poor environmental quality, the overall index will exhibit poor environmental quality. Conversely, $I = 0$ if and only if at least one subindex is zero, and this characteristic eliminates the ambiguity problem discussed earlier.

If the maximum value of each subindex in Equation 34 is 100, then the maximum value of I will be 100. This can be shown as follows. Assume that every subindex is set to its maximum value, I_{max}, a constant. That is, let $I_i = I_{max}$ for all i in Equation 34:

$$I = \prod_{i=1}^{n} [I_{max}]^{w_i} \qquad (36)$$

Taking logarithms of both sides of Equation 36,

$$\log I = \sum_{i=1}^{n} w_i \log I_{max} = \log I_{max} \sum_{i=1}^{n} w_i \qquad (37)$$

Substituting Equation 34a into Equation 37,

$$\log I = (\log I_{max})(1) = \log I_{max} \qquad (38)$$

Taking antilogarithms,

$$I = I_{max} \qquad (39)$$

Equations 35 and 39 show that, if the range of each subindex is from 0 to 100, then the range of the overall index is from $I = 0$ to $I = 100$.

For the simple case involving just two subindices, Equation 33 is written as follows:

$$I = I_1^{w_1} I_2^{w_2} \qquad (40)$$

where
$$w_1 + w_2 = 1 \qquad (40a)$$

It is instructive to examine the characteristics of this function in the (I_1, I_2)-plane for the simple case involving equal weights, $w_1 = w_2 = 0.5$:

$$I = I_1^{0.5} I_2^{0.5} \qquad (41)$$

As we did for the aggregation functions discussed earlier in this chapter, we first solve Equation 41 for I_2 as a function of I and I_1:

$$I_2 = \frac{I^2}{I_1} \qquad (42)$$

Then Equation 42 is plotted in the (I_1, I_2)-plane for selected values of I (Figure 27). This relationship gives a family of hyperbolic curves, convex to

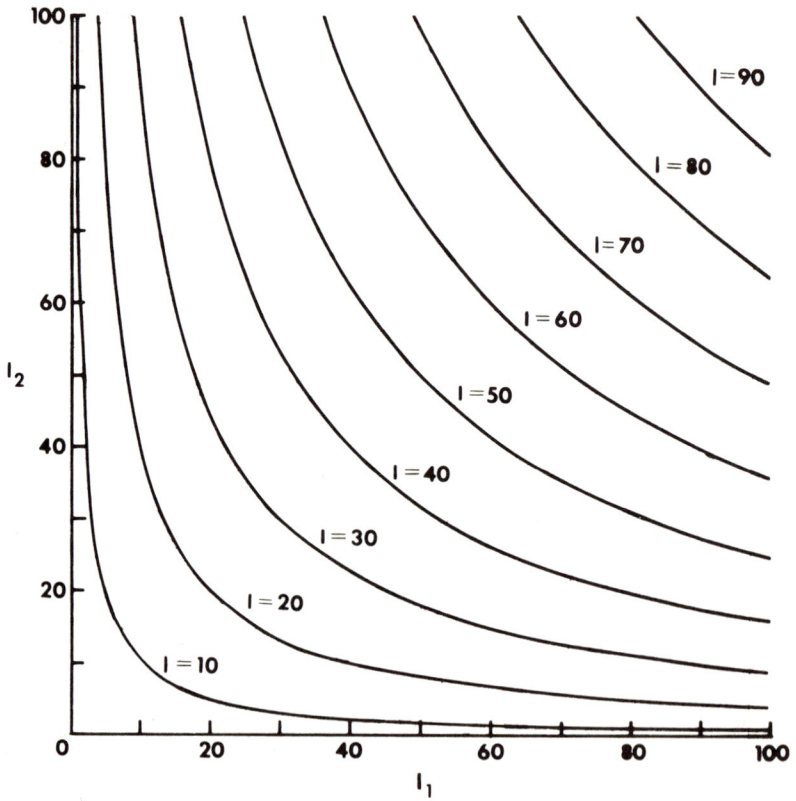

Figure 27. Plot of the multiplicative aggregation function $I = I_1^{0.5} I_2^{0.5}$ in the (I_1, I_2)-plane for selected values of I.

the origin. The curves are asymptotic to the two axes, and all curves have slopes -1 at points along a 45° line bisecting the two axes. Here, $I_1 = I_2 = I$. If the two weights were not equal, the shapes of the curves would change. The curves would still be convex to the origin, but they would not be symmetrical about the 45° line bisecting the axes.

For large values of I in Figure 27, the curvature is relatively gradual. As I becomes smaller, the curvature increases, becoming extremely abrupt near the origin. In the limit, as I approaches zero, the curve approaches a right angle coinciding with the two axes. As long as I is greater than zero, the curve never quite touches either axis. Because the curve intersects neither axis, eclipsing cannot occur, for it is impossible for either I_1 or I_2 to be zero at the same time that I exceeds zero. Note the contrast between these curves and the weighted linear sum plotted earlier in Figure 25, which intercepted both axes over the relatively large range $0 < I \leq 50$.

If, for purposes of analysis, we define "poor" environmental quality to be the range $0 \leq I_i \leq 10$ for any subindex i, a somewhat different eclipsing pattern emerges. As was shown in Figure 26, the region of poor environmental quality then is the area bounded by the dotted lines $I_1 = 10$ and $I_2 = 10$. Now, eclipsing is said to occur if $I_i \leq 10$ for any i while $I > 10$. For the weighted product given in Equation 42, two eclipsing regions emerge (Figure 28). One eclipsing region is bounded by the curve $I = 10$ and the line $I_1 = 10$ (upper crosshatched area) and the other eclipsing region is bounded by the curve $I = 10$ and the line $I_2 = 10$ (lower crosshatched area). Some of the curves, such as the curve for $I = 20$, pass into the eclipsing regions. Portions of any curve in the eclipsing region denote combinations of I_1 and I_2 for which a subindex is 10 or less but the index is above 10. The largest value of I for which eclipsing can occur is for $I = 10\sqrt{10} = 31.62$, which is a curve passing through the points (10,100) and (100,10).

This analysis shows that the weighted product, when used to aggregate decreasing scale subindices, may eclipse subindices which show moderately poor environmental quality ($I_i \leq 10$ for any i). This function does not, however, eclipse subindices which show extremely poor environmental quality ($I_i = 0$ for any i).

If the weights in Equation 34 are set equal, $w_i = w$ for all i, then Equation 34a can be written as follows:

$$\sum_{i=1}^{n} w_i = nw = 1 \qquad (43)$$

For this situation, $w = 1/n$, and Equation 34 becomes the geometric mean of subindices:

$$I = \left[\prod_{i=1}^{n} I_i \right]^{1/n} \qquad (44)$$

86 ENVIRONMENTAL INDICES

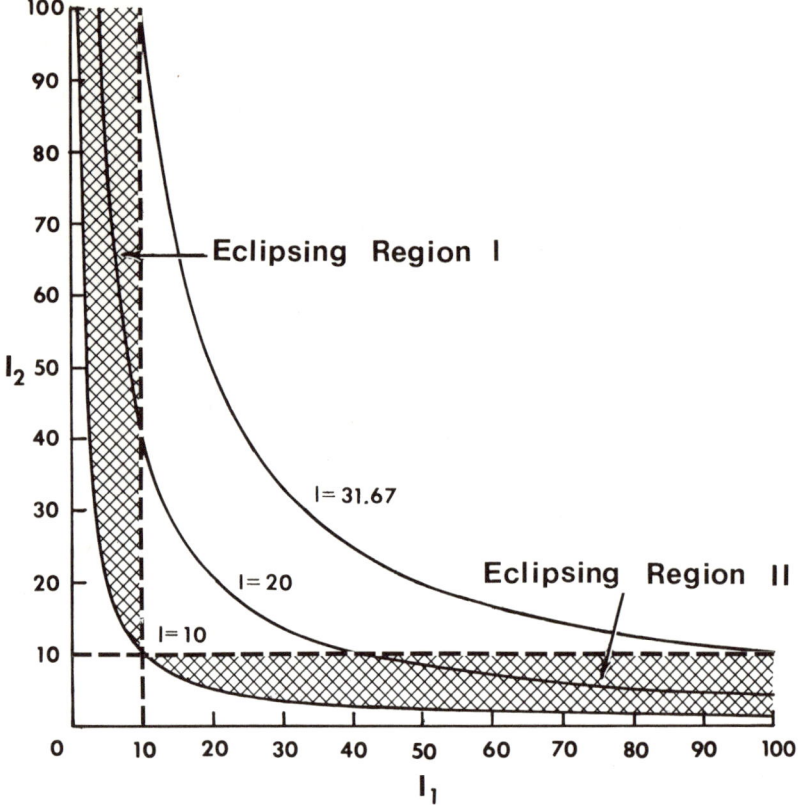

Figure 28. Plot of the weighted product $I = I_1^{0.5} I_2^{0.5}$ in the (I_1, I_2)-plane showing eclipsing regions for which $I_1 \leqslant 10$ or $I_2 \leqslant 10$ but $I > 10$.

Thus, the geometric mean is a special case of the weighted product aggregation function. Because the two weights for the example given in Equation 41 are equal, the curves plotted in the (I_1, I_2)-plane in Figure 27 are for the geometric mean.

A common version of the weighted product is the geometric aggregation function:

$$I = \left[\prod_{i=1}^{n} I_i^{g_i} \right]^{1/\gamma} \tag{45}$$

where

$$\gamma = \sum_{i=1}^{n} g_i \tag{45a}$$

However, if each subindex in Equation 45 is raised to the power $1/\gamma$, Equation 45 can be written in the same form as Equation 34, with each weight defined as follows:

$$w_i = \frac{g_i}{\gamma} = \frac{g_i}{\sum_{i=1}^{n} g_i} \qquad (46)$$

Thus, Equation 45 is identical to Equation 34, except for the notation. As in Equation 34, the sum of the weights is unity:

$$\sum_{i=1}^{n} w_i = \frac{\sum_{i=1}^{n} g_i}{\sum_{i=1}^{n} g_i} = 1 \qquad (47)$$

Often, environmental indices using the weighted product include as many as 9 or 10 subindices. Because the weights in Equation 34 sum to unity, inclusion of a large number of subindices has the effect of making each weight relatively small. Unfortunately, as the weights become smaller, the curvature changes from a gradual shape to an abrupt shape. To examine the character of this nonlinearity, we plot I versus subindex I_1 using the following equation (Figure 29):

$$I = I_1^{w} \qquad (48)$$

If $w = 1$, Equation 48 gives a straight line. As w decreases from 1 to 0.25, the line shows increasing sharpness of curvature; for $w = 0.1$, the curvature is extremely abrupt. In the extreme case, $w = 0$, a true step function will result. This happens because zero raised to the exponent $w = 0$ gives zero, while any positive number I_1, no matter how small, when raised to the exponent $w = 0$, gives 1:

$$I = (0)^0 = 0 \qquad (49)$$

$$I = (I_1)^0 = 1 \qquad (50)$$

$$\text{for } I_1 > 0$$

Thus, if w is very close to zero, the subindex is transformed such that it can take on essentially just two states, 0 and 1. It becomes, in effect, a dichotomous subindex, a step function.

Suppose, for example, that a weighted product environmental index is developed with nine subindices and equal weights. Then, $w_i = 1/9$ for $i = 1, 2, \ldots, 9$ in Equation 34. To plot I as a function of just one subindex, I_1, we set all the other subindices to reflect good environmental quality (that is, $I_i = 100$ for $i = 2, 3, \ldots, 9$). The following relationship results:

$$I = KI_1^{1/9} \qquad (51)$$

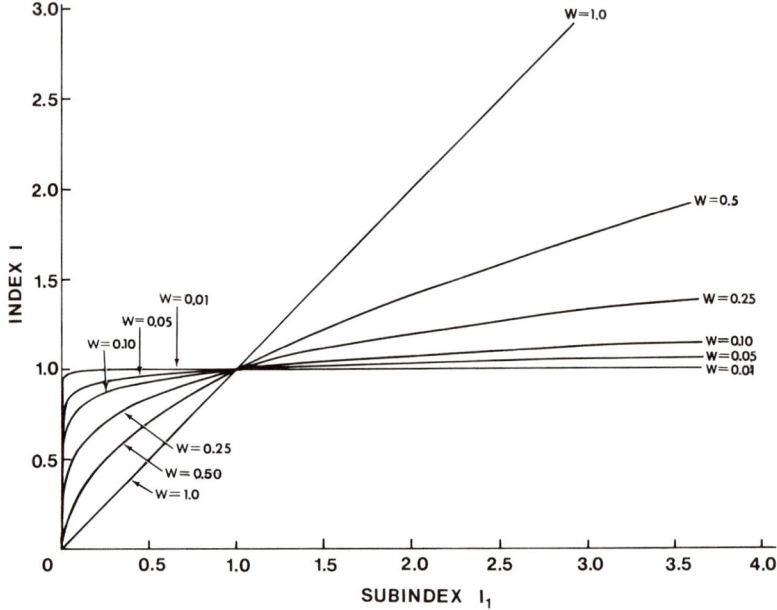

Figure 29. Plot of the aggregation function $I = I_1^w$ showing increasing sharpness of curvature as w approaches zero.

where
$$K = (100)^{8/9} = 59.95$$

This function, when plotted, shows considerable sharpness of curvature over the range from $I_1 = 0$ to $I_1 = 100$ (Figure 30). If $I_1 = 0$, for example, then $I = 0$; however, if I_1 increases from 0 to only 2, I will increase from 0 to 64.8, a striking change. By contrast, if I_1 increases over the relatively large range from 20 to 100, I will increase rather gradually from approximately 80 to 100. Therefore, an increase of just a few units when I_1 is zero causes I to increase over half its total range.

This graph shows that the weighted product tends to transform each subindex when the weights are small, introducing considerable nonlinearity in the process. This transformation can, in some instances, cause objectionable distortion, because the user may not be able to see any obvious relationship between each subindex and the index computed from it. For example, if $I_1 = 20$ for pollutant variable X_1, the index in Equation 51 reflects an entirely different value, $I = 83.6$, creating a potential for confusion. The weights used in the water quality index proposed by Brown et al.[3] range from 0.08 to 0.17 (Chapter IV), and this index, therefore, is subject to this nonlinearity problem.

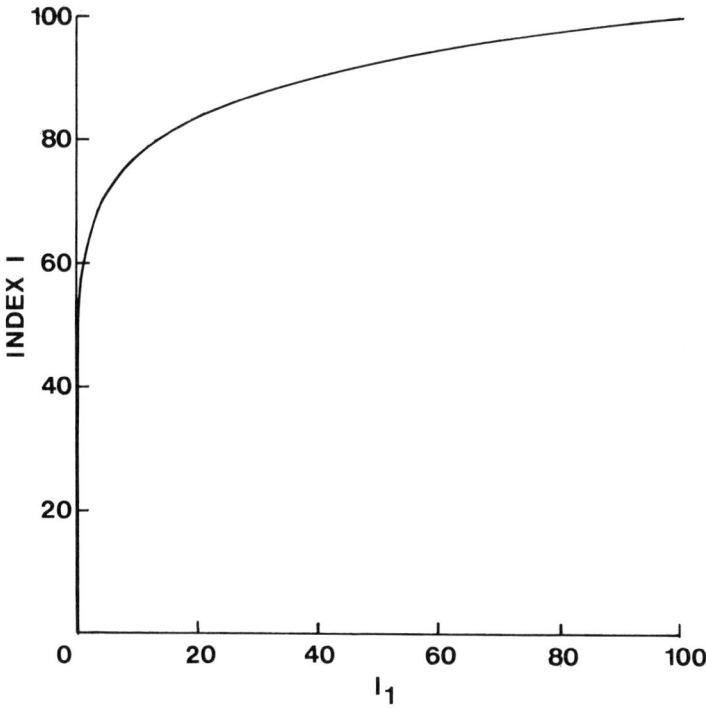

Figure 30. Plot of $I = KI_1^w$, where $K = 59.95$ and $w = 1/9$, showing nonlinearity of the weighted product aggregation function.

Minimum Operator

The minimum operator, when applied to decreasing scale subindices, performs in a fashion similar to the increasing scale maximum operator. The general form of the minimum operator is as follows:

$$I = \min\{I_1, I_2, \ldots, I_i, \ldots, I_n\} \qquad (52)$$

For the case involving two decreasing scale subindices,

$$I = \min\{I_1, I_2\} \qquad (53)$$

This function is plotted in the (I_1, I_2)-plane in a manner similar to that used to plot Equation 30 in Figure 24. The index is set to various constants, such as $I = 10$, and we plot the locus of points representing all combinations of I_1 and I_2 that satisfy this equation:

$$10 = \min\{I_1, I_2\} \qquad (54)$$

90 ENVIRONMENTAL INDICES

The result is two straight lines, joined at right angles, and parallel to the two axes (Figure 31). As I varies, Equation 53 generates a family of right-angle functions. Unlike the box-shaped functions generated by the maximum operator (Figure 24), the minimum operator gives functions which are convex to the origin. All are symmetrical to a 45° line bisecting the two axes. Here, $I_1 = I_2 = I$.

For very small values of I, the weighted product plotted in Figure 27 approaches the shape of the minimum operator in the (I_1, I_2)-plane. Like the weighted product, the minimum operator functions never touch the two axes. Therefore, eclipsing cannot occur, and no ambiguous region exists. Consequently, the minimum operator appears to be a good candidate for aggregating decreasing scale subindices. However, none of the published environmental indices employ the minimum operator, and its potential apparently remains unexplored (see Chapter IV).

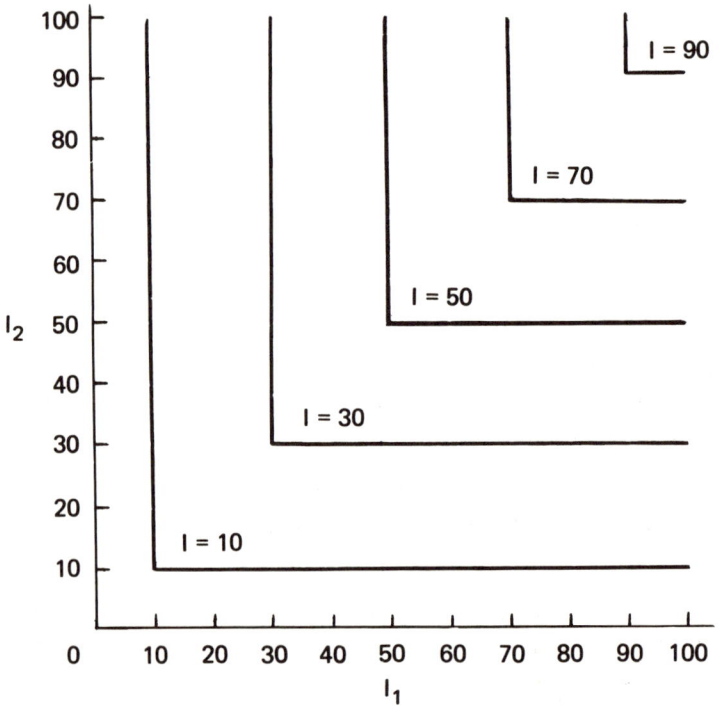

Figure 31. Plot of I = min I_1, I_2 in the (I_1, I_2)-plane for selected values of I.

SUMMARY OF INDEX STRUCTURES

The examples presented in this chapter illustrate how many different index structures can be accommodated within one general conceptual framework. Two basic forms exist: (1) those in which the index numbers increase with the degree of pollution (increasing scale indices), and (2) those in which the index numbers decrease with the degree of pollution (decreasing scale indices). This framework is better suited to representing absolute indices than relative indices.

In this general framework, calculation of an index consists of two fundamental steps: (1) calculation of subindices for the pollutant variables used in the index, and (2) aggregation of the subindices into the overall index. Subindices can be classified as one of four general types:

- Linear
- Segmented Linear
- Nonlinear
- Segmented Nonlinear

In the linear function, the subindex value is directly proportional to the pollutant variable. It is simple but provides little flexibility. The segmented linear function, which consists of two or more straight line segments joined at breakpoints, offers more flexibility. It is especially useful for incorporating administratively recommended limits, such as the NAAQS. An important segmented linear function is the step function, which exhibits just two states and therefore is called a dichotomous function. Subindices also may consist of a staircase of steps, giving a multiple-state function. The nonlinear subindices are functions which exhibit curvature when plotted on linear paper. Some of these are mathematically explicit—they are represented by explicit equations—and others are implicit forms which can be read only from a graph. Some nonlinear subindices are unimodal: they have a single maximum or minimum. Segmented nonlinear functions consist of line segments similar to the segmented linear function; however, at least one segment is nonlinear. Usually, each segment is represented by a different equation which applies over a specific range of the pollutant variable.

Four general types of aggregation functions are available (Table I). The simplest of these are the additive forms, such as the (unweighted) linear sum of subindices. In an increasing scale index, the linear sum unfortunately exhibits an ambiguous region; that is, the overall index can report "poor" environmental quality when no subindex exhibits poor environmental quality. The weighted linear sum avoids the ambiguity problem but introduces a more serious problem, eclipsing. Eclipsing occurs when at least one subindex exhibits poor environmental quality, but the overall index does not exhibit poor environmental quality. More complicated forms, such as the root-sum-power, can alleviate the eclipsing problem, but they are unwieldy to apply. The maximum operator, in which just the maximum subindex is reported,

Table I. Characteristics of Aggregation Functions

Aggregation Function	Increasing Scale Indices	Decreasing Scale Indices
Additive Forms		
Linear Sum	Ambiguity; no eclipsing	Eclipsing; no ambiguity
Weighted Sum	Eclipsing; no ambiguity	Eclipsing; no ambiguity
Root-Sum-Power	Minimizes eclipsing and ambiguity as p approaches ∞	Eclipsing; no ambiguity
Maximum Operator	No eclipsing; no ambiguity	Not applicable
Multiplicative Forms		
Weighted Product	Not applicable	No eclipsing; no ambiguity. Nonlinear if weights are small
Minimum Operator	Not applicable	No eclipsing; no ambiguity

offers a solution to this problem. It is ideally suited to determining if a NAAQS is violated and by how much. Neither the additive forms nor the minimum operator, however, are very well suited for aggregating decreasing scale subindices. For example, if zero denotes good environmental quality, any subindex that is zero will usually be totally eclipsed in the additive forms. The multiplicative forms have been proposed to circumvent these problems. The weighted product, when applied to decreasing scale subindices, does not exhibit eclipsing. If one subindex is zero, the overall index will be zero. Some eclipsing can occur, however, for values that correspond to "moderately poor" environmental quality. If the weights are small, the weighted product is very nonlinear. The geometric mean is a familiar special case of the weighted product. The minimum operator, which performs for decreasing scale indices in a fashion analogous to the increasing scale maximum operator, has not appeared in any published environmental indices. It offers, however, a possible means for aggregating decreasing scale indices without eclipsing or ambiguity.

PROBLEMS FOR STUDY

1. In your own words, describe the problem of ambiguity as it affects environmental indices, illustrating your discussion with examples of (a) an increasing scale index, and (b) a decreasing scale index.
2. In your own words, describe the problem of eclipsing as it affects environmental indices, illustrating your discussion with examples of (a) an increasing scale index, and (b) a decreasing scale index.
3. In a particular air pollution index, a segmented linear function is used to represent the subindex for carbon monoxide. The breakpoints, (X, I), are as follows, with X indicating the measured concentration in ppm: (0,0), (9,100), (15,200), (30,300), and (40,400). (a) Calculate the slope and I-intercept for each segment and write the equations. [Answers: $I = 11.11X$; $I = 16.67X - 50$; $I = 6.67X + 100$; $I = 10X$] (b) Plot the segmented linear function for this subindex.
4. Two general forms of the segmented linear function are given in this chapter, Equation 6 and Equation 11. Show that the two are equivalent by deriving one from the other.
5. Walski and Parker[5] have proposed a water quality index which uses a parabolic function for temperature. In the temperature subindex, $20°$ Celsius represents optimum water quality ($I = 1$), while $0°$ and $40°$ Celsius represent poor water quality ($I = 0$). (a) Using Equation 13, plot this subindex function. (b) Show how you would modify this equation to create an increasing scale subindex function that varies between $I = 0$ and $I = 100$. (c) Modify this function so that $5°$ and $35°$ Celsius represent poor water quality ($I = 100$), and plot the result.
6. Using the example of a segmented nonlinear function for pH given in Figure 15, calculate the values of I at each breakpoint using both equations for

the pair of segments which are joined at the breakpoint. [Answers: (A) 14, (B) 4, 4, (C) 0, 0, (D) 4, 4, (E) 14]

7. Consider the two-variable case of the weighted linear sum aggregation function, Equation 19, in which the weights are unequal. Plot this aggregation function in the (I_1, I_2)-plane for (a) $w_1 = 0.2$, $w_2 = 0.8$, (b) $w_1 = 0.55$, $w_2 = 0.45$, (c) $w_1 = 0.9$, $w_2 = 0.1$. (d) Show that the resulting straight lines always pass through the point (100,100) for I = 100.

8. Consider the multiplicative aggregation function plotted in Figure 27, Equation 41. (a) Find the values of I_1 for which the curved lines in the figure intercept the line $I_2 = 100$. [Answers: 1, 4, 9, 16, 49, 64, 81] (b) Derive an equation for these values. (c) Show that all curves in Figure 27 pass through the point $I_1 = I_2 = I$. (d) Show that the slope of each curve is -1 at this point.

9. This problem concerns the weighted product aggregation function given in Equation 40 in which the two weights, w_1 and w_2, are unequal. (a) Show that the equation for the two-variable weighted product, when plotted in the (I_1, I_2)-plane, is as follows:

$$I_2 = \left(\frac{I}{I_1^{w_1}}\right)^{1/w_2}$$

(b) Show that the slope of these curves in the (I_1, I_2)-plane is $-w_1/w_2$ at points along a 45° line bisecting the two axes; that is, for $I_1 = I_2$. (c) Show that the slope of these curves at any point is given by the following equation:

$$\text{slope} = -\frac{w_1}{w_2}\left(\frac{I}{I_1}\right)^{1/w_2}$$

10. The water quality index proposed by Brown et al.[3] (Chapter IV) is briefly mentioned in this chapter. The weights in this index, which incorporates a weighted product aggregation function, are as follows:

Pollutant Variable	Weight
Dissolved Oxygen	0.17
Fecal Coliform	0.15
pH	0.12
BOD_5	0.10
Nitrates	0.10
Phosphates	0.10
Temperature Deviation	0.10
Turbidity	0.08
Total Solids	0.08

(a) Using Equation 50, calculate K for each of the five distinct weights in this index, 0.17, 0.15, 0.12, 0.10 and 0.08. [Answers: 45.71, 50.12, 57.54, 63.10, 69.18] (b) Plot I as a function of I_1 for all five cases. (c) Assume all subindices are 50 except I_1 (that is, $I_i = 50$ for i = 2, 3, ..., 9); plot I as a function of I_1 for all five cases in this situation. (d) Discuss the possible nonlinearity problems that this index may have.

11. After examining the arguments used in plotting Figure 24 from Equation 30, explain why the minimum operator, $I = \min \{I_1, I_2\}$, should give a family of right-angle functions such as those plotted in Figure 31.

REFERENCES

1. Hunt, William F., William M. Cox, Wayne R. Ott and Gary C. Thom. "A Common Air Quality Reporting Format, Precursor to an Air Quality Index," *Proceedings of the Fifth Annual Environmental Engineering and Science Conference*, Louisville, Kentucky, March 3-4, 1975, pp. 99-121.
2. Horton, Robert K. "An Index-Number System for Rating Water Quality," *J. Water Poll. Control Fed.* 37(3):300-306 (March 1965).
3. Brown, Robert M., Nina I. McClelland, Rolf A. Deininger and Ronald G. Tozer. "A Water Quality Index–Do We Dare?," *Water Sewage Works* (October 1970), pp. 339-343.
4. McClelland, Nina I. "Water Quality Index Application in the Kansas River Basin," U.S. Environmental Protection Agency, Kansas City, Missouri, EPA-907/9-74-001 (February 1974).
5. Walski, Thomas M. and Frank L. Parker. "Consumers Water Quality Index," *J. Environ. Eng. Div.*, American Society for Civil Engineering (June 1974), pp. 593-611.
6. Prati, L., R. Pavanello and F. Pesarin. "Assessment of Surface Water Quality by a Single Index of Pollution," *Water Res.* 5:741-751 (1971).
7. Thom, Gary C. and Wayne R. Ott. *Air Pollution Indices: a Compendium and Assessment of Indices Used in the United States and Canada* (Ann Arbor, Michigan: Ann Arbor Science Publishers, Inc., 1976).
8. Landwehr, Jurate Maciunas. "Water Quality Indices–Construction and Analysis," Ph.D. Dissertation, University of Michigan, University Microfilms No. 75-10, 212 (1974).

CHAPTER III

AIR POLLUTION INDICES

A number of air pollution indices have been proposed in journals, conference proceedings, and research reports. Additional air pollution indices have been developed by state and local air pollution control agencies and have been implemented to report air quality data routinely to the public. So many different air pollution reporting schemes were in use in the United States in the mid-1970's that the federal government found it necessary to adopt a nationally uniform air pollution index, the Pollutant Standards Index (PSI).

This chapter first reviews the air pollution indices published in the literature, comparing their structural characteristics. It then discusses the many varieties of air pollution indices used by state and local air pollution control agencies. Indices in the literature are compared with those in use by means of an index classification system. Finally, the historical development and evolution of PSI are described with sufficient technical detail to enable the reader to apply PSI to air quality data. All necessary subindex graphs, equations, nomograms and tables are presented for calculating PSI. Examples are also given of PSI's performance with real air quality data. Finally, an international system for designing air pollution indices that has evolved from PSI, the UNIPEX system, is presented.

UNIFORM ADMINISTRATIVE LIMITS

Unlike the water pollution field, where many different water quality standards have been adopted by state and local governments, nationally uniform administrative limits—such as the NAAQS, the federal episode criteria, and the Significant Harm level—have brought considerable uniformity to the air pollution field. In addition, the federal government requires that air quality measurements for compliance with the NAAQS be carried out with approved federal measurement techniques, thereby encouraging uniformity of measurement approaches.

NAAQS

The NAAQS cover six different air pollutants (Table I).[1] The primary NAAQS, which are intended to protect against the adverse effects of air pollutants on health, specify somewhat higher concentration levels than the secondary NAAQS, which protect against the adverse effects of air pollution on welfare (soiling of buildings, reduced visibility, effects on animals and vegetation, etc.). Generally, effects on welfare, although considered less serious than effects on health, occur at lower pollutant concentrations. For four of these air pollutants, the primary and secondary NAAQS are identical. Although NAAQS have been promulgated for (nonmethane) hydrocarbons, the value is intended only as a guide to assist in implementing the oxidant NAAQS, because of the role of hydrocarbons in the formation of photochemical oxidants (see pages 16 and 17).

Each NAAQS is based largely on the information on health and welfare effects summarized in the Air Quality Criteria document published for the air pollutant (cited in Chapter I). Because air pollution effects are generally always a function of the duration of exposure, an averaging time is specified along with each NAAQS concentration. To determine if a given observation exceeds the 24-hour primary SO_2 NAAQS, for example, the arithmetic average value of the SO_2 concentration measured over a 24-hour period is compared with the value of 365 $\mu g/m^3$ (0.14 ppm) given in Table I. Some primary NAAQS specify concentrations for both short averaging periods (24 hours or less) and long averaging periods (1 year). The shorter averaging period is intended to protect against the occurrence of acute health effects, and the longer period is intended to protect against the occurrence of chronic health effects.

To carry out the mandates of the Clean Air Act,[2] the Environmental Protection Agency has divided the United States into over 240 Air Quality Control Regions (AQCR).[3] Each AQCR generally consists of a geographical area which shares a common air pollution problem. The Los Angeles Intrastate AQCR, for example, includes Los Angeles County, San Bernardino County, Orange County, Riverside County, and Santa Barbara County. Under the law, each state is required to prepare a comprehensive plan, called a State Implementation Plan (SIP), describing how air pollution control agencies will control the sources of air pollution in the AQCR to meet the NAAQS in a reasonable time frame.[4]

In each AQCR, the NAAQS must not be exceeded more than once a year. Therefore, to determine if a given AQCR is in compliance with the NAAQS, the second highest value measured over the year usually is compared with the NAAQS value. Although a secondary annual NAAQS has been promulgated for particulate matter, this value is not really a standard, but is used as a guide to assess whether a SIP can attain the 24-hour particulate NAAQS.

Table I. National Ambient Air Quality Standards[1]

Pollutant and Averaging Time	Primary NAAQS[a]		Secondary NAAQS[a]	
	($\mu g/m^3$)	(ppm)	($\mu g/m^3$)	(ppm)
Carbon Monoxide (CO)				
8 hr	10[b]	9	same as primary	
1 hr	40[b]	35	same as primary	
Nitrogen Dioxide (NO_2)				
1 yr	100	0.05	same as primary	
Hydrocarbons[c]				
3 hr	160	0.24	same as primary	
Photochemical Oxidants[d]				
1 hr	160	0.08	same as primary	
Particulate Matter (TSP)				
1 yr[e]	75	-	60[f]	-
24 hr	260	-	150	-
Sulfur Dioxide (SO_2)				
1 yr	80	0.03	-	-
24 hr	365	0.14	-	-
3 hr	-	-	1300	0.5

[a]Gravimetric units ($\mu g/m^3$ or mg/m^3) are measured at 25°C and 1 atmosphere. Volumetric units (ppm) do not correspond exactly to gravimetric units because of round-off.
[b]Milligrams per cubic meter (mg/m^3).
[c]Nonmethane hydrocarbons; measured from 6 to 9 AM local time. This is not a standard; it is intended as a guide to facilitate meeting the NAAQS for photochemical oxidants.
[d]Measured as ozone.
[e]Annual geometric mean of 24-hr averages. The standards for all other averaging times are arithmetic averages.
[f]This is not a standard; it is intended as a guide to assess whether state implementation plans will attain the 24-hr particulate NAAQS.

For each NAAQS pollutant, the federal government has specified a "reference measurement principle" (Table II) to be used to determine if air quality is in compliance with the NAAQS.[1] The federal government evaluates performance data submitted by manufacturers of commercial instruments to determine if instruments using this principle meet certain performance criteria. If the instrument performs satisfactorily, it is approved as a Federal Reference Method (FRM). Once approved, it is suitable for determining compliance with the NAAQS. Instruments which do not use the reference measurement principle also can be used to monitor compliance with the NAAQS, but they must first be approved by the federal government as "equivalent" methods. The reference measurement principle for carbon

Table II. Federal Reference Methods for Measuring Compliance with the NAAQS[4]

Pollutant	Reference Measurement Principle	Federal Reference Methods
Carbon Monoxide (CO)	Nondispersive Infrared Spectrometry	Bendix 8501-5CA Beckman 866 Mine Safety Appliances 202S
Nitrogen Dioxide (NO_2)	Chemiluminescence[a]	
Particulate Matter (TSP)	High-Volume Sampling	High-Volume Sampler
Photochemical Oxidants (O_3)	Chemiluminescence	Meloy 0A325-2R Meloy 0A350-2R Bendix 8002 McMillan 1100-1 McMillan 1100-2 McMillan 1100-3 Monitor LABS 8410E
Sulfur Dioxide (SO_2)	Manual Pararosaniline Colorimetry	Manual Pararosaniline Technicon I[b] Technicon II[b] Lear Siegler SM1000[c] Meloy SA185-2A[c] Thermo Electron 43[c] Philips PW9755[c] Philips PW9700[c] Monitor LABS 8450[c]

[a]Pending a decision by EPA.
[b]Automated version of reference measurement principle.
[c]Approved as "equivalent" methods.

monoxide, for example, is nondispersive infrared (NDIR) spectrometry, and an approved FRM for CO is the Model No. 866 NDIR instrument manufactured by Beckman Instruments, Inc.[4]

Episode Criteria

Each SIP must include a system for controlling air pollution during air pollution episodes. To assist air pollution control agencies in developing these systems, EPA has published recommended episode criteria.[5] The episode criteria originally were published in the *Federal Register* in 1971 in the form of "example regulations."[6] Most U.S. air pollution control agencies have adopted systems which closely resemble the original episode criteria, although the criteria were intended only as a guide.

Each of the air pollution stages specified by the episode criteria—Alert, Warning and Emergency—calls for different actions to reduce air pollution, such as limiting the use of fuels, restricting incineration, reducing traffic, and curtailing manufacturing activities. The recommended emission reduction plan for the Alert level, for example, includes the following actions[5]:

1. There shall be no open burning by any persons of tree waste, vegetation, refuse, or debris in any form.
2. The use of incinerators for the disposal of any form of solid waste shall be limited to the hours between 12 noon and 4 PM.
3. Persons operating fuel-burning equipment which require boiler lancing or soot blowing shall perform such operations only between the hours of 12 noon and 4 PM.
4. Persons operating motor vehicles should eliminate all unnecessary operations.

The plan also calls for "substantial reduction" of air pollution from coal- or oil-fired electric power generating facilities "by utilization of fuels having low ash and sulfur content" and reduction of air pollution from a variety of manufacturing activities. The recommended emission reduction plan for the Emergency level calls for more stringent control actions. These include the immediate curtailment of operations of a variety of places of employment, including manufacturing firms, government offices, wholesale and retail firms, printing establishments, laundries, theaters, construction activities, and mining and quarrying facilities. The language calls for elimination of air pollution from manufacturing firms, such as those producing chemicals, petroleum products, and primary metals ". . . by ceasing, curtailing, postponing or deferring production and allied operations to the extent possible without causing injury to persons or damage to equipment." The Emergency level also includes a ban on motor vehicles: "The use of motor vehicles is prohibited except in emergencies with the approval of local or state police."[5]

All concentrations specified in the episode criteria for triggering these control actions (Table III) have short averaging times (1 hour or 24 hours).

Table III. Recommended Federal Episode Criteria[5]

Pollutant and Averaging Time	Alert ($\mu g/m^3$)	Alert (ppm)	Warning ($\mu g/m^3$)	Warning (ppm)	Emergency ($\mu g/m^3$)	Emergency (ppm)
Carbon Monoxide (CO)						
8 hr	17[a]	15	34[a]	30	46[a]	40
Nitrogen Dioxide (NO_2)						
24 hr	282	0.15	565	0.3	750	0.4
1 hr	1130	0.6	2260	1.2	3000	1.6
Photochemical Oxidants						
1 hr	200	0.1	800	0.4	1000	0.5
Particulate Matter (TSP)						
24 hr	375	–	625	–	875	–
Coefficient of Haze (COH)						
24 hr	3.0[b]	–	5.0[b]	–	7.0[b]	–
Sulfur Dioxide (SO_2)						
24 hr	800	0.3	1600	0.6	2100	0.8
Product of Particulates and Sulfur Dioxide						
TSP x SO_2[c]						
24 hr	6.5×10^4		2.61×10^5		3.93×10^5	
COH x SO_2[d]						
24 hr	0.2		0.8		1.2	

[a] Milligrams per cubic meter (mg/m^3).
[b] COH units.
[c] Units are $(\mu g/m^3)^2$.
[d] Units are COH units x ppm.

Often, an episode stage is not called unless the meteorological forecast indicates that pollutant concentrations are likely to remain at these levels for at least 12 hours or to increase. Unlike the NAAQS, the episode criteria include limits for the Coefficient of Haze (COH). Because COH is based on measurement of the reduction in light transmission when air is passed through an exposed paper tape, the device is called the tape sampler and the values sometimes are called a measure of Smoke Shade or of Soiling Index. It is considered a relatively crude measurement, because factors besides the mass of the particles collected on the tape, such as size and color, can affect the measurement. In addition, poor correlations often have been observed between tape sampler and high-volume sampler measurements when the two systems were operated side by side. EPA has selected the high-volume sampler as the reference measurement principle and as the FRM for measuring total suspended particulates and does not consider data collected with the tape sampler to be acceptable for monitoring compliance with the NAAQS. However, tape samplers are considered acceptable for general air quality surveys, for providing rapid information during air pollution episodes, and for reporting daily air quality indices. Because the tape sampler is inexpensive and can produce data rapidly in 1-hour averaging periods, it is one of the most popular air pollution measurement methods in the United States. In this book, we shall refer to particulate data generated by the high-volume sampler as "TSP data" and data generated by the tape sampler as "COH data."

Unlike the NAAQS, the federal episode criteria specify limits for the product of particulates and sulfur dioxide (TSP x SO_2 and COH x SO_2). These limits apparently were included in the episode criteria to reflect the increasingly severe health effects that occur when both of these pollutants are present together; that is, to reflect synergism.

Significant Harm Level

In addition to the levels contained in the episode criteria, the federal government has promulgated Significant Harm levels (Table IV). These are upper limits which should *never be reached* in an AQCR. Like the NAAQS, the Significant Harm levels are based largely on scientific information on the effects of air pollution on health. There is evidence that ambient concentrations at these levels would be extremely injurious to health, causing widespread symptoms in the general population and, possibly, premature death of the ill and elderly.

The ultimate purpose of each episode system required in a SIP is to prevent ". . . air pollution from reaching levels that would cause imminent and substantial endangerment to the health of persons."[5] Thus, the purpose of

104 ENVIRONMENTAL INDICES

Table IV. Federal Significant Harm Levels[7]

Pollutant and Averaging Time	Concentrations	
	($\mu g/m^3$)	(ppm)
Carbon Monoxide (CO)		
8 hr	57.5[a]	50
1 hr	144[a]	125
Nitrogen Dioxide (NO_2)		
24 hr	938	0.5
1 hr	3750	2.0
Photochemical Oxidants		
1 hr	1200	0.6
Particulate Matter (TSP)		
24 hr	1000	-
Coefficient of Haze (COH)		
24 hr	8.0[b]	-
Sulfur Dioxide (SO_2)		
24 hr	2620	1.0
Product of Particulates and Sulfur Dioxide		
TSP x SO_2[c]		
24 hr	4.9 x 10^5	
COH x SO_2[d]		
24 hr		1.5

[a] Milligrams per cubic meter (mg/m^3).
[b] COH units.
[c] Units are $(\mu g/m^3)^2$.
[d] Units are COH units x ppm.

the episode system is to avoid reaching the Significant Harm levels. Presumably, the increasingly stringent control actions triggered by the episode criteria will prevent concentrations from ever rising to the Significant Harm level for a given air pollutant.

Like the episode criteria, the Significant Harm levels include COH and the product of sulfur dioxide and particulates (TSP x SO_2 and COH x SO_2). They do not include hydrocarbons.

HISTORICAL BACKGROUND

At the time that PSI was developed, a number of air pollution indices had already been developed and published in the air pollution literature. Many more were being used routinely in U.S. and Canadian cities to report air quality data to the public on a daily basis. The search for a single uniform index was inspired largely by the great variety of indices which had come into common use. As we shall see, PSI resulted from an evolutionary process which attempted to combine the best features of previous indices—particularly those which had been tested by routine applications in cities—into a single index structure.

A comprehensive review of air pollution indices was sponsored in 1976 by the Council on Environmental Quality and the Environmental Protection Agency. The result, *Air Pollution Indices: A Compendium and Assessment of Air Pollution Indices in the United States and Canada* by Thom and Ott,[8] discusses and compares the structural characteristics of existing indices—those in the literature and those used by air pollution control agencies. Before discussing the evolution of PSI, it is useful to consider the characteristics of earlier published indices as well as those used by air pollution control agencies. Readers wishing to explore this topic in greater detail should consult the original literature references and the 1976 *Compendium*, which has been published in book form.[8]

Indices in the Literature

Although the number of indices published in the literature is not great, the indices are notable for their diversity. Each index embodies unique features or reflects a slightly different concept in index design.

Green's Index

One of the earliest air pollution indices to appear in the literature was proposed by Green[9] in 1966. Green's index included just two pollutant variables, sulfur dioxide and Coefficient of Haze. In each subindex equation, the pollutant variable was raised to a power less than unity:

$$SO_2 \text{ (ppm)}: \quad I_1 = 84X^{0.431} \qquad (1)$$
$$COH: \qquad I_2 = 26.6X^{0.576} \qquad (2)$$

The index was computed as the arithmetic mean of the two subindices:

$$I = (1/2)(I_1 + I_2) \qquad (3)$$

One of the three descriptors was reported along with the index to provide a qualitative report of air quality levels: "desired level" ($I = 25$), "alert level" ($I = 50$), and "extreme level" ($I = 100$).

Green's index was intended more as a system for triggering control actions during air pollution episodes than a means for reporting air quality data to the public. Thus, the index also included two additional alert levels: a second alert ($I = 60$) and a third alert ($I = 68$). For example, the third alert was to be declared if $I = 68$ and if adverse weather conditions were predicted to continue. The third alert included the following actions:

At the discretion of government leaders, restrictions may be enforced to limit certain industrial operations and to reduce the use of automobiles and trucks to only essential uses.

If the index still continues to rise and there is no break predicted in the weather, increasingly severe restrictions may be imposed and it may be advisable to declare a general closing of industrial and commercial operations. All buildings not occupied are to be without heat, thus further reducing additional pollution from fuel consumption.[9]

Because the index did not include any pollutants besides SO_2 and COH, it was very limited in its possible applications, and the author cautioned that "it is pertinent only to the colder seasons," when most of the fossil fuels are burned. Nevertheless, it is an interesting early index structure. Because the arithmetic mean is a weighted linear sum, Green's index exhibited the eclipsing problem discussed in the context of Equation 21 in Chapter II (see pages 70 and 71).

CPI

The Combusion Products Index (CPI) was suggested by Rich[10] in 1967 as an aid for determining the potential severity of air pollution within a fixed geographical area. This source-oriented index was based on the tons of fuel burned—a measure of the total pollutants emitted within a city—and the volume of air in which the combustion products are mixed:

$$CPI = \frac{F}{V} \qquad (4)$$

where   F = quantity of fuel burned in the area (obtained by inventorying fuel deliveries)
V = ventilation volume

If we consider a hypothetical square city beneath a box-shaped volume of air, the ventilation volume is V = uwh, where u = average wind speed, w = width of the city, and h = mixing height. Suppose, for example, that the city is 10 miles square, with a mixing height on a particular date of h = 0.3 miles, and an average wind speed of u = 5 miles per hour. For this situation, the mixing volume is V = (5 miles/hour) (10 miles) (0.3 miles) = 15 cubic miles/hour, or (15)(24) = 360 cubic miles/day. If the total fuel consumed in this city were 24,000 tons/day, then the index calculated using Equation 4 gives CPI = 24,000/360 = 66.7 tons/cubic mile.

The CPI is, of course, a very simple concept which applies primarily to locations downwind of a city and to average meteorological conditions. It does not take into account emission control measures which may be implemented, such as sulfur limitations in fuel, changes in combustor design, and filtering systems. However, in a very limited way, it provides a general indication of a city's air pollution potential under specified conditions.

MURC

A simple ambient air quality index, the Measure of Undesirable Respirable Contaminants (MURC), was published in 1968.[11] MURC was implemented routinely in the city of Detroit to report air quality data to the public and was broadcast between 8:30 and 9:00 AM each day on local radio stations.

MURC is based on just one pollutant variable, COH. Observed COH values are raised to an exponent less than unity:

$$MURC = 70X^{0.7} \tag{5}$$

where X = COH units. COH values ranging from 0.3 to 2.15 give MURC values ranging from approximately 30 to 120. Five different descriptors are reported for various ranges of the MURC index (Table V).

Table V. Descriptors Reported with the MURC Index[11]

Range of Index	Descriptors
0-30	Extremely Light Contamination
31-60	Light Contamination
61-90	Medium Contamination
91-120	Heavy Contamination
121 and over	Extremely Heavy Contamination

The nonlinear function for MURC given in Equation 5 apparently was developed from an examination of correlations between COH data and TSP data. The author of the article states that the function was chosen to "... reflect a good average approximation of the actual weight of suspended particulate in the atmosphere . . .," as measured by the high-volume sampler.[11] However, the correlation between COH and TSP data varies from location to location and depends on meteorological conditions, the size and composition of the particles, and other factors. Therefore, if MURC were to be used outside Detroit, Equation 5 probably would have to be modified.

AQI

In 1969, Fensterstock et al.[12] developed the Air Quality Index (AQI) to assess "the relative severity of air pollution among various major metropolitan areas." This was the first air pollution index to estimate air pollutant concentrations from data on source emissions and meteorological conditions in each city. This approach was selected to compensate for the limited amount of air quality data then available for some cities.

AQI employed a modified version of an atmospheric diffusion model developed previously by Miller and Holzworth.[13] The index was applied to 29 metropolitan areas (which included 44% of the total U.S. population) using emission data on carbon monoxide, particulates and sulfur dioxide. The overall index was computed as the sum of three subindices:

$$AQI = \sum_{i=1}^{3} w_i I_i \quad (6)$$

where I_1 = estimated CO subindex.
 I_2 = estimated TSP subindex
 I_3 = estimated SO_2 subindex

Using AQI, the investigators ranked the 29 cities from most severe air pollution to least severe air pollution. For this ranking, the investigators chose $w_1 = w_2 = w_3 = 1$.

Indices based on diffusion model calculations require many simplifying assumptions and are therefore subject to error. The same accuracy problem occurs for the crude box model used in the CPI. For AQI, three major assumptions were used:

- The urban area is square, and the wind is always directed parallel to one of its sides.
- The pollutant sources emit continuously and are distributed uniformly over the urban area.
- Throughout the year, meteorological conditions remain constant at neutral (Class D) stability.

Although this approach was approximate, the investigators compared the TSP subindex estimates with measurements of TSP from the cities and found good agreement. The other two subindices could not be validated in a similar fashion, because air quality data were unavailable for the cities.

This approach was intended to illustrate a possible analytical technique rather than a practical methodology. It offers an intersting example of a possible way to incorporate source data in an air pollution index when ambient data may be limited or unavailable. AQI is intended not for daily air quality reports but for estimating the overall air pollution potential of a metropolitan area.

Ontario API

In 1970, Shenfeld[14] reported that the government of Ontario, Canada, had developed an Air Pollution Index (API) which was implemented routinely in Toronto in March of that year. The index was intended both to provide the public with daily information about air quality levels and to trigger control actions during air pollution episodes. During episodes, the API was used to communicate to owners of air pollution sources throughout the city the need to decrease pollutant emissions. Thus, the API was designed to "give warning of, and to prevent, the adverse effects of a build-up of air pollution which may occur during prolonged periods of stagnant weather."[14]

The Ontario API included two pollutant variables, COH and SO_2, and used linear subindex functions:

$$COH \qquad I_1 = 30.5X_1 \qquad (7)$$
$$SO_2 \text{ (ppm)} \qquad I_2 = 126X_2 \qquad (8)$$

The pollutant variables X_1 and X_2 were both 24-hour running average concentrations. That is, 24 1-hour values were averaged together, one beginning at each hour of the day. The aggregation function, a nonlinear additive form, was the sum of the two subindices raised to the power 1.35 and multiplied by 0.2:

$$API = 0.2(I_1 + I_2)^{1.35} \qquad (9)$$

An API value less than 32 was considered "acceptable" air quality. When the API reached 50 and adverse weather conditions were forecast to continue for at least 6 hours, a "first alert" was declared calling for curtailment of some air pollution sources. If this curtailment did not succeed in reducing air pollution levels and the API continued to rise to 75, a "second alert" was called. If the API reached 100, the Ontario "air pollution episode threshold level," officials could curtail all activities not essential to public health or safety.

PINDEX

In 1970, Babcock[15] published PINDEX, the first air pollution index in the literature which included the six major air pollutants and was based on air quality standards. It also included solar radiation, a variable designed to reflect the generation of photochemical air pollution.

Babcock recognized that development of a meaningful air pollution index is a serious technical challenge:

> Evaluating overall air pollution can be a complex undertaking. Urban air pollution consists of an often ill-defined mixture of several pollutants emitted from different energy and industrial processes. Additional secondary pollutants are created in the atmosphere. Synergisms can occur between certain pollutants. Despite these complexities, efforts should be made to total the effects of the individual pollutants.[15]

He felt that indices should both communicate air quality information to the public and assist officials in evaluating alternative air pollution control policies:

> Overall air pollution measures serve at least two purposes. First, they can be used to give the layman a more meaningful assessment of air pollution

110 ENVIRONMENTAL INDICES

> severity. The layman wants to know "how bad it is." He may even object to the control agency hiding behind individual part-per-million numbers which the layman does not understand....
>
> Second and perhaps more important, a combined air pollution measure or index enables evaluation of the trade-offs involved in alternative air pollution control policies or in evaluation of control equipment which, for instance, reduces levels of certain pollutants while increasing levels of others....[15]

At the time that PINDEX was originally developed, the NAAQS had not yet been promulgated, so PINDEX was based on air quality standards proposed by the State of California. Each subindex used a simple linear function, although more involved calculations were made to depict the generation of photochemical air pollution. For the major air pollutants, each subindex used the linear equation given in Equation 3 of Chapter II (page 52), with slope $\alpha = 1/X_s$, where X_s was the air quality standard. The overall index was computed using a linear sum:

$$\text{PINDEX} = \sum_{i=1}^{7} I_i \qquad (10)$$

Babcock[15] applied PINDEX to air quality data from a number of U.S. cities. When PINDEX rankings of the severity of air pollution in these cities were compared with similar rankings by the federal government, significant differences were observed. These differences were attributed to different assumptions about the relative importance of each pollutant variable in the two rankings.

ORAQI

Oak Ridge National Laboratory published the Oak Ridge Air Quality Index (ORAQI) in 1971.[16] It was based on the 24-hour average concentrations of five air pollutants: carbon monoxide, nitrogen dioxide, photochemical oxidants, particulate matter, and sulfur dioxide. ORAQI embodies many of the concepts in PINDEX and appears to have evolved from it. Like PINDEX, each subindex was calculated as the ratio of the observed pollutant concentration to its respective standard (Equation 3 of Chapter II):

$$I_i = \left(\frac{X}{X_s}\right)_i \qquad (11)$$

ORAQI assumed that all air quality standards have 24-hour averaging times. For pollutants which did not have 24-hour NAAQS, the developers extrapolated the federal standards to equivalent 24-hour values using an approach suggested by Larsen[17] and McGuire and Noll.[18] The resulting values for X_s (Table VI), therefore, were not directly comparable with existing primary or secondary NAAQS (Table I).

Table VI. Extrapolated 24-Hour Standards Used in ORAQI[16]

Pollutant Variable	X_S ($\mu g/m^3$)	(ppm)
Carbon Monoxide	7800	7.0
Nitrogen Dioxide	400	0.2
Photochemical Oxidants	59	0.03
Particulate Matter (TSP)	150	-
Sulfur Dioxide	266	0.1

As reported by Babcock and Nagda,[19] the ORAQI aggregation function was a nonlinear form in which the sum of the subindices times a constant was raised to the power 1.37:

$$ORAQI = \left[5.7 \sum_{i=1}^{5} I_i \right]^{1.37} \quad (12)$$

The constants in the equation were selected so that ORAQI = 10 when all concentrations are at their "naturally occurring or background level" and ORAQI = 100, approximately, when all concentrations are at the standards (that is, I_i = 1 for all i).

To facilitate computation of ORAQI, a nomogram was developed (Figure 1). To use the nomogram, one reads the appropriate scale for each pollutant variable at the left side of the figure ("summing columns"). The resulting values are added and the total is entered on the "measured total" scale. For the example given in Table VII, the column labeled "nomogram units" shows the values obtained from the summing columns in Figure 1. The measured total is 20, and a line then is drawn from the measured total scale to the appropriate point on the scale marked "unmeasured pollutants." ORAQI is read from the point on the line which intersects the air quality index scale. Here, ORAQI = 28.

The index also can be calculated directly from Equation 12. These calculations are illustrated in the first three numerical columns of Table VII. The sum of the five indices is 1.99, giving the following:

$$ORAQI = [(5.7)(1.99)]^{1.37} = 11.35^{1.37} = 28$$

Like the nomogram, the ORAQI equation must be modified if data for any pollutant are missing.

Like the (unweighted) additive aggregation functions discussed in Chapter II, ORAQI is subject to eclipsing. In Table VII, one pollutant variable, TSP, is above its secondary standard. Yet ORAQI is only 28, well below the value of 100. For this index value, the descriptor reported with the index is "good" (Table VIII). Thus, violation of the particulate standard is totally eclipsed, and "good" air quality is reported, giving an erroneous picture of the air pollution problem.

112 ENVIRONMENTAL INDICES

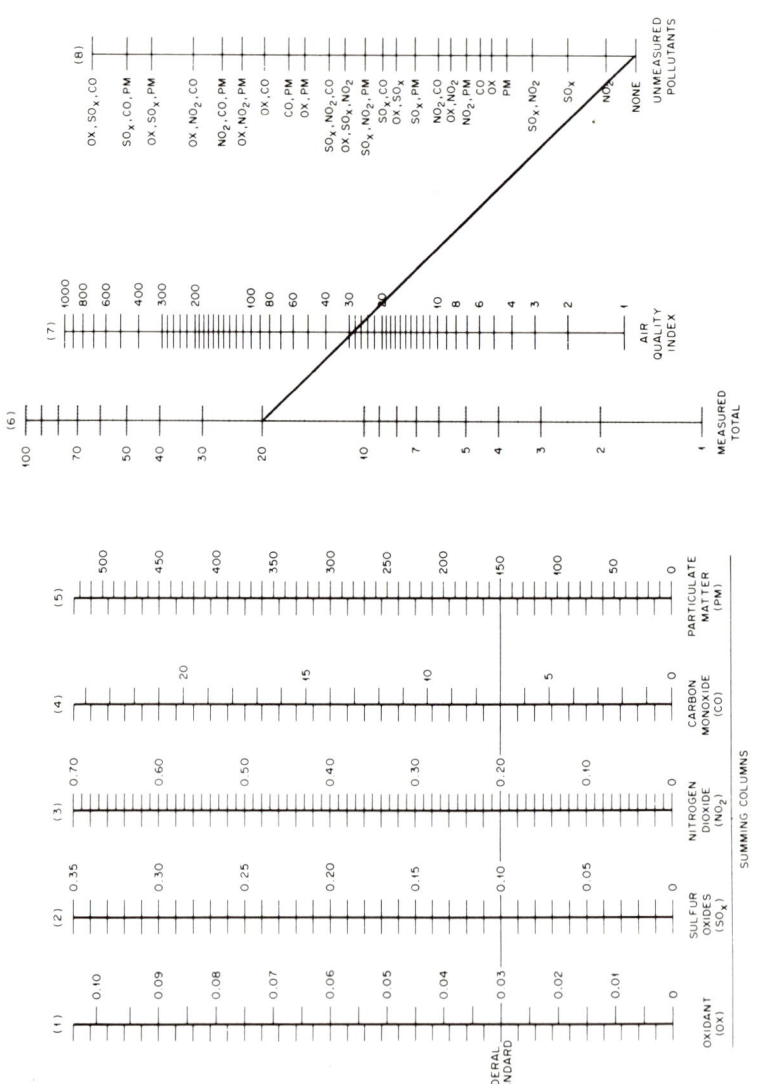

Figure 1. Nomogram for determining the Oak Ridge Air Quality Index (ORAQI).[16]

Table VII. Example of ORAQI Computation

Pollutant Variable	Observed Concentration	Ratio	Calculated Subindex	Nomogram Units
CO	2.0 ppm	2/7	0.29	3
NO_2	0.02 ppm	0.02/0.2	0.10	1
O_3	0.01 ppm	0.01/0.03	0.33	3
TSP	175 $\mu g/m^3$	175/150	1.17	12
SO_2	0.01 ppm	0.01/0.1	0.10	1
		Total	1.99	20

Table VIII. Descriptor Words for ORAQI[16]

Range of Index	Descriptor
under 20	Excellent
20-39	Good
40-59	Fair
60-79	Poor
80-99	Bad
100 and over	Dangerous

After the index was developed, about 2500 copies of the ORAQI report were distributed to persons who had requested information about the index. In 1973, Thomas[20] mailed a questionnaire to 1120 of these persons to find out how they had used the index and whether it satisfied their needs. The list of 1120 consisted of 249 members of educational institutions, 193 employees of governmental agencies concerned with environmental pollution control, 187 members of environmental action groups and conservation organizations, 177 scientists and engineers, 165 members of the general public, 93 members of the press (including radio and television broadcasters), and 56 employees of industrial and consulting firms. He received 370 replies, a response of 33%. Of the 303 respondents who indicated they had read the ORAQI report, 62% thought the index needed some additional explanation, 29% felt the index was satisfactory as is, and 9% felt the index was ineffective. Most respondents felt that the index was not too difficult to calculate (if the data were available), but some respondents, particularly those from the communications professions, felt that ORAQI was too difficult to explain to the public. The possible applications cited by respondents included preparation of environmental reports for municipal and state agencies, preparation of campaign material for candidates for public office, communication of air quality data to the public through the mass media, and correlation of air pollution data with health data. As we shall see, few U.S. air pollution control agencies have ever implemented ORAQI on a daily basis, preferring simpler air pollution indices for this purpose.

MAQI

In 1972, CEQ sponsored an effort by the MITRE Corporation to develop environmental indices. The final report by Bisselle, Lubore and Pikul[21] describes two air pollution indices. One was called the National Air Quality Index (NAQI) but was later named the MITRE Air Quality Index (MAQI). The other was called the Extreme Value Index (EVI). Along with ORAQI, the two MITRE indices were applied to data on air quality for selected U.S. cities in CEQ's third annual report.[22]

MAQI and EVI were unique in that they included standards for several different averaging times in each subindex. If, for example, a pollutant had NAAQS with two different averaging times, the results were combined in a root-sum-square operation. The MAQI subindex for carbon monoxide, I_1, was calculated using the following subindex equation:

$$I_1 = \sqrt{\left(\frac{X_8}{S_8}\right)^2 + \delta \left(\frac{X_1}{S_1}\right)^2} \qquad (13)$$

where X_8 = CO concentration (maximum 8-hr value)
S_8 = 8-hr NAAQS (S_8 = 9 ppm)
X_1 = CO concentration (maximum 1-hr value)
S_1 = 1-hr NAAQS (S_1 = 35 ppm)
δ = 1 if $X_1 \geqslant S_1$; 0 otherwise

Because MAQI was designed to compare air quality measured at different air monitoring stations, the subindex equations use the maximum value for the year and not the daily concentration. In Equation 13, X_1 is the single highest hourly value for the year (maximum of the year's 8760 hourly values). All subindices were based on the secondary NAAQS, which, for many air pollutants, are the same as the primary NAAQS (Table I).

The ratio of the hourly CO concentration to the NAAQS is multiplied by a dichotomous factor, δ. This factor apparently gives added emphasis to the violation of a NAAQS; that is, it reduces eclipsing. For example, the maximum observed values of carbon monoxide at the Washington, DC, Continuous Air Monitoring Project (CAMP) air monitoring station in 1965 were X_8 = 15 ppm and X_1 = 31 ppm. The CO subindex is calculated as:

$$I_1 = \sqrt{\left(\frac{15}{9}\right)^2 + 0\left(\frac{31}{35}\right)^2} = 1.67$$

The maximum observed CO concentration at the Chicago CAMP station in 1965 were X_8 = 44 ppm and X_1 = 54 ppm, giving the following subindex value:

$$I_1 = \sqrt{\left(\frac{44}{9}\right)^2 + 1\left(\frac{59}{35}\right)^2} = 5.17$$

Notice that the 1-hour CO concentration is 90% higher in Chicago than in Washington, DC, but the subindex is 210% higher due to the effect of the dichotomous factor.

The nitrogen dioxide subindex, I_2, was based on the annual NAAQS for NO_2:

$$I_2 = \frac{X_a}{S_a} \quad (14)$$

where X_a = NO_2 concentration (annual arithmetic mean)
S_a = annual mean NAAQS (S_a = 0.05 ppm)

The subindex for photochemical oxidants, I_3, also was based on the concentration for a single averaging time:

$$I_3 = \frac{X_1}{S_1} \quad (15)$$

where X_1 = oxidant concentration (maximum 1-hr value)
S_1 = 1-hr NAAQS (S_1 = 0.08 ppm)

The subindex for TSP, I_4, included two averaging times:

$$I_4 = \sqrt{\left(\frac{X_a}{S_a}\right)^2 + \left(\frac{X_{24}}{S_{24}}\right)^2} \quad (16)$$

where X_a = TSP concentration (annual geometric mean)
S_a = secondary annual NAAQS (S_a = 60 $\mu g/m^3$)
X_{24} = TSP concentration (maximum 24-hr value)
S_{24} = 24-hr secondary NAAQS (S_{24} = 150 $\mu g/m^3$)

The primary and secondary NAAQS for SO_2 specify three averaging times, so the MAQI SO_2 subindex includes three terms:

$$I_5 = \sqrt{\left(\frac{X_a}{S_a}\right)^2 + \delta_1\left(\frac{X_{24}}{S_{24}}\right)^2 + \delta_2\left(\frac{X_3}{S_3}\right)^2} \quad (17)$$

where X_a = SO_2 concentration (annual arithmetic mean)
S_a = annual NAAQS (S_a = 0.03 ppm)
X_{24} = SO_2 concentration (maximum 24-hr value)
S_{24} = 24-hr NAAQS (S_{24} = 0.14 ppm)
X_3 = SO_2 concentration (maximum 3-hr value)
S_3 = 3-hr NAAQS (S_3 = 0.5 ppm)
δ_1 = 1 if $X_{24} \geqslant S_{24}$; 0 otherwise
δ_2 = 1 if $X_3 \geqslant S_3$; 0 otherwise

The annual and 24-hour secondary SO_2 NAAQS originally used in MAQI no longer are in effect, so, instead, the primary NAAQS for SO_2 are listed above.

MAQI combines the five subindices given by Equations 13 through 17 using a root-sum-square aggregation function:

116 ENVIRONMENTAL INDICES

$$\text{MAQI} = \left[\sum_{i=1}^{5} I_i^2 \right]^{1/2} \qquad (18)$$

When plotted in the (I_1, I_2)-plane, the root-sum-square exhibits an ambiguous region but no eclipsing region (see Figure 21 of Chapter II, page 74). If, for example, any pollutant concentration is above its NAAQS for a particular averaging time, then the ratio of the concentration to the standard will exceed one (that is, X/S ⩾ 1). If any ratio exceeds one, the subindex for that pollutant will exceed one (that is, I_i ⩾ 1 for pollutant i), and MAQI also will exceed one. Thus, eclipsing cannot occur because violation of any NAAQS causes MAQI to exceed its "critical value" of one. However, it is possible for MAQI to exceed one even though no pollutant variable exceeds a NAAQS. For example, if the ratios in Equations 13 through 17 are all 0.9, then MAQI = $\sqrt{4.5}$ = 2.1, giving a false impression that a NAAQS has been violated. Note that there are nine ratios in MAQI, one each in Equations 14 and 15, two each in Equations 13 and 16, and three in Equation 17. If all pollutant variables are exactly at the NAAQS values (for all averaging times), then MAQI = $\sqrt{9}$ = 3. This gives an ambiguous region between MAQI = 1 and MAQI = 3. When MAQI lies in this range, a NAAQS may be violated, but the user cannot be certain. Only when MAQI is 3 or more can he be absolutely certain that a NAAQS has been violated.

EVI

One additional problem with MAQI was its dependence on the highest concentration over the year, as noted by its developers:

> It is quite possible that the observed maxima are related to an isolated meteorological situation and not to the general pollutant levels in the ambient air. In addition, the maxima may not change considerably from year to year even though the general air quality is improving.[21]

For this reason, they developed a second index, the Extreme Value Index (EVI). EVI was based not on the maximum concentration for the year but on the "accumulated extreme" values obtained by summing the number of ppm-hours for which the observed concentration exceeded its NAAQS. For example, EVI's CO subindex was calculated as follows:

$$I_1 = \sqrt{\left(\frac{A_8}{S_8}\right)^2 + \left(\frac{A_1}{S_1}\right)^2} \qquad (19)$$

where A_8 = accumulation of observed 8-hr CO concentrations which exceed the NAAQS
S_8 = 8-hr NAAQS (S_8 = 9 ppm)
A_1 = accumulation of observed 1-hr CO concentrations which exceed the NAAQS
S_1 = 1-hr NAAQS (S_1 = 35 ppm)

The other subindices were calculated in a similar fashion. The equations resembled those used in MAQI, except that accumulated extreme values were substituted for the maximum concentration. EVI included four of the five pollutants used in MAQI, and subindices were aggregated using a root-sum-square aggregation function:

$$\text{EVI} = \sqrt{I_1^2 + I_2^2 + I_3^2 + I_4^2} \qquad (20)$$

where I_1 = CO subindex
 I_2 = photochemical oxidants subindex
 I_3 = TSP subindex
 I_4 = SO_2 subindex

MAQI and EVI were intended to be used side-by-side, one reporting the highest levels for the year and the other reporting the frequency of adverse levels. Using these two indices together is conceptually similar to the way in which the EPA Region X Environmental Quality Profile reports air quality. The profile reports the frequency of adverse conditions and the severity of adverse conditions on separate charts (see Figure 6 of Chapter I, page 19). The frequency/severity measures are two important, but very different, dimensions of environmental quality (see page 20).

STARAQS

Usually, it is important to report an air pollution index with as current information as possible, particularly during air pollution episodes when conditions may be changing rapidly. However, the NAAQS for some air pollutants specify averaging periods for 8 hours, 24 hours, or longer. In an index based on the NAAQS, one way to report hourly index values is to extrapolate the 24-hour NAAQS to a 1-hour period, giving an "equivalent" 1-hour concentration.

In 1973, Miller[23] published an air quality index model for making this extrapolation. To apply this methodology, called the Short Time Averaging Relationships to Air Quality Standards (STARAQS), one first analyzes historical data for each air pollutant at each monitoring station. For a given air pollutant, one examines data over the previous year, for example, calculating the ratio of the highest hourly concentration to the 24-hour average value. To compute the equivalent 1-hour SO_2 NAAQS for a particular station, the 24-hour SO_2 NAAQS is multiplied by the median of the ratios. If the median of the ratios for the year were 2.1, the 1-hour equivalent SO_2 NAAQS would be calculated as 2.1 x 0.14 ppm = 0.29 ppm. This value then would be used in the index calculation for data from this monitoring station. Because CO already has a 1-hour NAAQS of 35 ppm, the extrapolated 1-hour NAAQS for CO is used only if it is less than 35 ppm. For NO_2, which has only an annual average NAAQS, the equivalent 1-hour NAAQS is based on the highest 1% of hourly average NO_2 concentrations measured during the previous year.

An air pollution index embodying the STARAQS model was implemented in Knoxville, Tennessee, by the Knox County Department of Air Pollution Control. The index included six pollutants: CO, NO_2, oxidants, TSP, COH and SO_2. Each subindex was computed as the simple linear ratio of observed 1-hour concentrations to the equivalent 1-hour NAAQS value, with 100 corresponding to the equivalent 1-hour NAAQS. Subindices were aggregated using the maximum operator; that is, only the maximum subindex was reported.

Miller[23] also used the STARAQS approach to forecast whether or not a 24-hour NAAQS would be violated from the 1-hour readings. Although the median (50 percentile) of peak-to-mean ratios was used for most pollutants in the STARAQS model, any other fractile (for example, the upper 10 percentile) probably also could be used. It would seem useful to conduct additional research extending this work, determining the statistical implications with greater rigor. With such a statistical basis, it may be possible to calculate confidence limits, with specified probabilities, for each prediction.

EQI

In 1974, Inhaber[24] suggested a set of environmental indices for Canada called the Environmental Quality Index (EQI). The EQI included an air quality index, a water quality index, and a land quality index. The air quality index[25] consisted of three subindices, each intended to reflect a particular component of air quality: (1) the Index of Specific Pollutants, (2) the Index of Interurban Air Quality, and (3) the Index of Industrial Emissions.

The Index of Specific Pollutants, I_a, was based on ambient measurements of CO, NO_2, oxidants, TSP, COH and SO_2 in urban areas. Subindex calculations were based on the Canadian air quality "objectives" (standards), which are analogous to the U.S. NAAQS. For example, the TSP subindex was calculated as the ratio of the annual geometric mean concentration to the Canadian objective of 70 $\mu g/m^3$. Each of the remaining subindices was calculated by combining the annual and daily concentrations in a root-sum-square formula and expressing the result as a ratio to the air quality objective.

The Index of Interurban Air Quality, I_b, was intended as a rough measure of air quality outside central metropolitan areas. It was based on visibility readings at airports, corrected for humidity. Visibility was used as an approximate measure of the concentration of particles in the air. Because visibility standards do not exist, this index was computed as the ratio of the visibility at airports in the northern areas of Canada (background levels) to visibilities at other airports.

Finally, the Index of Industrial Emissions, I_c, was intended to reflect a third component of air quality, industrial emissions, usually outside downtown areas. This index was computed using Canada's nationwide inventory of sources and air pollutant emissions. It combined subindices of industrial emissions of particulate matter and sulfur dioxide using a root-mean-square calculation method. The calculation then was weighted by the population residing in the counties where the emissions occurred relative to the total population of Canada.

The three indices were aggregated in the EQI's air quality index using a (weighted) root-mean-square aggregation function:

$$I_{air} = \sqrt{0.5 I_a^2 + 0.3 I_b^2 + 0.2 I_c^2} \qquad (21)$$

The root-mean-square function was used to emphasize large subindex values; that is, to reduce eclipsing:

> ... to combine indices or subindices, the root-mean-square method was used. This method combines the advantages of simplicity with a greater sensitivity to extreme values of indices of environmental conditions than ordinary linear averaging. In other words, if indices were averaged linearly, a large value (indicating an undesirable condition) would tend to be "lost" when combined with other indices of low values. The root-mean-square method tends to emphasize these large values mathematically, and so produces somewhat larger values of a combined index than averaging does.[24]

In the previous chapter, it was possible to show graphically how the root-mean-square aggregation function tends to reduce (but not eliminate) the problem of eclipsing. When plotted in the (I_1, I_2)-plane, the root-mean-square (Figure 23 of Chapter II, page 76) exhibited a smaller eclipsing region than the arithmetic mean (Figure 18 of Chapter II, page 70). Although the area of the eclipsing region was smaller, the area still was sufficiently large to cause serious eclipsing problems.

Although the EQI has not been adopted routinely by the Canadian Government, it was applied by its author on a pilot basis to examine Canadian air quality trends. It offers another example of a possible approach.

Discussion

By the mid-1970's, a total of 11 different air pollution indices had appeared in the literature (Table IX). Some ranged from 0 to less than 10, others ranged from 0 to 100, and still others ranged from 0 to 1000. All had increasing scales. They differed strikingly in terms of the number and type of variables included, mathematical structures, and tendency to show ambiguity or eclipsing (Table X). Most employed linear subindices, usually expressing the ratio of pollutant concentration to a standard or other administrative limit. However, a number of the indices (5 out of 11) employed nonlinear

120 ENVIRONMENTAL INDICES

Table IX. Summary of Air Pollution Indices Published in the Literature

Index	Reference	Publication Date	Range
Green's Index	Green[9]	1966	0-100
Combustion Products Index (CPI)	Rich[10]	1967	a
Measure of Undesirable Respirable Contaminants (MURC)	*Air Engineering*[11]	1968	0-120+
Air Quality Index (AQI)	Fensterstock et al.[12]	1969	0-150+
Ontario Air Pollution Index (API)	Shenfeld[14]	1970	0-580+
PINDEX	Babcock[15]	1970	0-2.2+
Oak Ridge Air Quality Index (ORAQI)	Babcock et al.[16]	1971	0-1000
MITRE Air Quality Index (MAQI)	Bisselle et al.[21]	1972	0-3+
Extreme Value Index (EVI)	Bisselle et al.[21]	1972	a
STARAQS	Miller[23]	1973	a
Environmental Quality Index (EQI)-Air	Inhaber[25]	1974	0-1+

[a]Range is from 0 to a large positive number.

aggregation functions, such as the sum raised to a power, the root-sum-square, or the root-mean-square. In the nonlinear forms, a scientific rationale was not usually provided for the equations, and exponents or powers were not explicitly based on research on health effects or on damage functions. More than half of the published indices exhibited either ambiguous or eclipsing regions.

Many of these indices seem to have emerged spontaneously and do not appear to be the end product of any detailed scientific inquiry. However, there is some evidence of an evolution of index structures. The later indices tend to be more complex than the earlier indices, usually including more pollutant variables. Later indices also usually are based on ambient air quality standards. Although some evolution was taking place, it is not evident from the literature that any single, preferred index structure was emerging.

The majority of the published indices apparently were intended for comparing air quality in different cities or for examining long-term trends over time. Because such indices usually require air quality data for an entire year (for example, annual average or yearly maximum values), we shall refer to them as "long-term" indices. "Short-term" indices require data for much shorter time periods, such as an 8-hour or 24-hour period. MURC, the Ontario API, and STARAQS are short-term indices; they were designed to report

Table X. Mathematical Characteristics of Air Pollution Indices Published in the Literature

Index	Pollutant Variable	Subindices	Aggregation Function	Comments
Green's Index[9]	COH, SO_2	Power	Arithmetic Mean (Weighted Sum)	Eclipsing Region
CPI[10]	Fuel Burned, Ventilation	Estimated	Ratio	Based on Box Model
MURC[11]	COH	Power	None	Daily Index
AQI[12]	CO, TSP, SO_2	Estimated	Linear Sum	Based on Diffusion Model
Ontario API[14]	COH, SO_2	Linear	Nonlinear Sum	Daily Index
PINDEX[15]	CO, NO_2, HC,[a] TSP, SO_2, Solar Radiation	Linear	Linear Sum	Ambiguous Region
ORAQI[16]	CO, NO_2, OX,[b] TSP, SO_2	Linear	Nonlinear Sum	Eclipsing Region
MAQI[21]	CO, NO_2, OX, TSP, SO_2	Linear	Root-Sum-Square	Ambiguous Region
EVI[21]	CO, OX, TSP, SO_2	Linear	Root-Sum-Square	Ambiguous Region
STARAQS[23]	CO, NO_2, OX, TSP, COH, SO_2	Linear	Maximum Operator	Employs Equivalent 1-hr NAAQS
EQI (air)[25]	CO, NO_2, OX, TSP, COH, SO_2, Visibility, Population, Emissions	Linear	Root-Mean-Square	Eclipsing Region

[a] HC = Hydrocarbons.
[b] OX = Photochemical oxidants.

air quality data to the public on a daily basis or to assist in triggering air pollution episode control actions.

A few additional articles have been published about short-term indices, such as the New Jersey daily air quality index,[26] but the literature about short-term indices has been relatively limited. Cullen, Flaherty and Barnett[27] have discussed the problems that citizens have in understanding short-term indices. They noted that the previously developed indices often proved ". . . to be either too technical for the public to understand, or too abstract to give physical significance to the true status of the ambient air condition."

122 ENVIRONMENTAL INDICES

To make air quality reports more meaningful to the public, they suggest that air quality be reported in very familiar terms, such as "equivalent number of cigarettes smoked." They believe that simplicity is the key to success for a daily air quality index: "While it may be argued that indices tend to oversimplify and inaccurately describe a very complex subject, simplicity, it turns out, is the required criterion for a public air quality index to succeed."[27]

Reidy and Dziewulski[28] surveyed environmental agencies in 16 cities and found that "... no two agencies use identical methods in reporting air quality levels to the public." They concluded that there was need for a nationally uniform approach for reporting air pollution levels to the news media and, in turn, to the public. Although their survey covered cities which represent about 10% of the U.S. population, they did not obtain detailed information about the technical characteristics of the short-term indices currently in use.

Despite these efforts, many questions still remained about short-term indices. How many short-term indices were there and how widespread was their use? What were their structural characteristics? Were they similar to the long-term indices that have been published? Was there much uniformity across the nation?

Indices in Use

The 1976 *Compendium* of air pollution indices by Thom and Ott[8] included a detailed survey of air pollution indices. The survey was designed to answer a variety of questions about the short-term indices in use:

> Although many technical papers proposing specific indices appear in the literature, no detailed study has been available to describe the characteristics of the many indices that are actually being used for public reporting. How many air pollution indices are there in the United States? What are the experiences of metropolitan agencies with these indices? Have the indices proposed in the literature been adopted by State and local air pollution control agencies? What pollutants do the indices include? How are the indices calculated and how are the individual pollutants weighted? What reporting formats are used to convey this information to the public?[8]

As Ott and Thom[29] indicated in a paper containing highlights from the survey, one goal was to use the characteristics of existing indices to determine the structure that a common air pollution index, if one could be developed, should possess:

> When we initially reviewed the air pollution indices proposed in the literature, we found a confusing array of different index types. Further, although many indices had been developed, few of those proposed in the literature were being used in practice—either by governmental decision makers or by State and local agencies seeking to report air quality levels to

the public. Instead, the agencies tended to develop their own indices. Two important questions were evident: (1) What are the technical characteristics of all the air pollution indices proposed or in use? (2) Does any "common index" emerge from these characteristics and, if so, what does it look like?[29]

The survey population, consisting of 55 city and county air pollution control agencies with 10 or more staff members (Table XI), was selected from the annual directory of air pollution control agencies published by the Air Pollution Control Association.[30] State agencies known to operate statewide air pollution index systems also were surveyed. In addition, the survey population included Canadian provinces with air pollution control agency staffs of 10 or more persons and one Canadian city known to use an index.

Telephone calls were made to these agencies from August to December of 1974 to solicit information on the nature and characteristics of each agency's air pollution index, if there was one. Of the 55 U.S. metropolitan agencies, 28 supplied written material, and additional information was obtained from the telephone interview. Typically, the material covered the nature of the index, its manner of calculation, and the way in which it was reported.

The results revealed that 35 (64%) of the 55 U.S. air pollution control agencies used an air pollution index. In 14 of these cities, the index was operated as part of a general statewide or regional index system. Six states were operating this type of system: Connecticut, the District of Columbia, Minnesota, New Jersey, New York and Ohio. These statewide indices served 53 additional cities, but because the air pollution agencies in these cities had staffs of less than 10 persons, they are not listed in Table XI.

The agencies in Baltimore, Maryland, Boston, Massachusetts, and Portland, Oregon, are operated by the state but are not part of statewide index systems. In Portland, the state agency not only reports the air pollution index but it operates the entire city air pollution control agency as well. In Baltimore, on the other hand, the state reports the index, but the local air pollution agency is organizationally separate from the state agency. In Canada, only Montreal operates a city air pollution control agency, so all provinces with staffs of 10 or greater were included in the survey population. Alberta and Ontario operate province-wide air pollution indices. Eight cities within these provinces use daily air pollution indices. Most indices have been implemented since 1970.

Index Classification System

To facilitate comparison of the structural characteristics of the many indices used by these agencies, Thom and Ott[8] devised an index classification system based on four criteria:

124 ENVIRONMENTAL INDICES

Table XI. Utilization of Air Pollution Indices in 55 City/County Air Pollution Control Agencies with Staffs of 10 or More[8]

City/County	Agency Size	Material Received	Index In Use	Comments
Birmingham, AL	17	●		
Phoenix, AZ	25	●		Discontinued Index
Anaheim, CA	24		●	Replaced Index
Los Angeles, CA	380	●	●	Replaced Index
Riverside, CA	26			
San Bernardino, CA	53			
San Diego, CA	53			
San Francisco, CA	220	●	●	Replaced Index
Denver, CO[a]	54	●	●	
New Haven, CT[b]	11	●	●	Replaced Index
Washington, DC[b]	14	●	●	
Bradenton, FL	11			
Jacksonville, FL	15	●	●	
Miami, FL	50	●	●	
Sarasota, FL	21			
Tampa, FL	16	●	●	
Atlanta, GA	14	●	●	
Chicago, IL	175	●	●	
Gary, IN	18			
Indianapolis, IN	15			
Louisville, KY	39	●	●	
Baltimore, MD[a]	90	●	●	Replaced Index
Montgomery Co., MD[b]	10		●	
Boston, MA[a]	87	●		
Springfield, MA	12			
Detroit, MI	77	●	●	
St. Paul, MN[b]	13	●	●	
Kansas City, MO	15			
St. Louis, MO	35			
Albuquerque, NM	15			
Albany, NY[b]	237		●	
Buffalo, NY[b]	44	●	●	
Mineola, NY[b]	37		●	
New York City, NY	382		●	
Rochester, NY[b]	12		●	
Charlotte, NC	14			
Akron, OH[b]	13		●	
Cincinnati, OH[b]	65	●	●	Replaced Index
Cleveland, OH[b]	80	●	●	
Dayton, OH[b]	45		●	Replaced Index
Toledo, OH[b]	25		●	
Oklahoma City, OK	15	●		Discontinued Index

Table XI., continued

City/County	Agency Size	Material Received	Index In Use	Comments
Portland, OR[a]	20	●	●	
Philadelphia, PA	94	●	●	Replaced Index
Pittsburgh, PA	82	●	●	
Chattanooga, TN	22		●	
Memphis, TN	14		●	
Nashville, TN	17	●	●	
Dallas, TX	21	●	●	
El Paso, TX	10			
Houston, TX	76	●		
Pasadena, TX	45			
Fairfax Co., VA[b]	12		●	
Seattle, WA	39	●	●	
Milwaukee, WI	25			

[a]City index is operated by state but is not part of statewide index system.
[b]City index is part of statewide or regional index system.

- number of variables included in the index
- calculation method used to compute the index
- calculation mode (combined or uncombined)
- descriptor categories reported with the index

The calculation "method" denotes the manner in which the overall index—subindices plus aggregation function—is calculated. Four calculation methods were identified:

  A. **Nonlinear:** At least one subindex or the aggregation function is nonlinear.
  B. **Segmented Linear:** At least one subindex is a segmented linear function.
  C. **Linear:** Subindices and aggregation function are linear.
  D. **Actual Concentrations:** Concentrations are reported in scientific units.

Methods A, B and C use mathematical functions that were described in Chapter II. Method D was introduced because some agencies reported observed pollutant concentrations along with descriptor words such as "good," "fair," etc., in a manner similar to the dimensionless indices.

Another component of the index classification system is the "mode," which is a slightly different way of describing the aggregation functions presented in Chapter II. Three modes were identified:

1. **Individual:** Subindices are reported separately for each pollutant variable.
2. **Maximum:** Subindices are aggregated using the maximum operator.
3. **Combined:** Subindices are aggregated using an additive form.

The individual mode was reserved for those indices which report the subindices separately and do not attempt to aggregate them. The combined mode refers to both linear and nonlinear additive forms. Because no air pollution indices were found which use multiplicative forms, there was no multiplicative mode. The mode was indicated in the system by appending a subscript to the calculation method classification.

To be included in the Thom-Ott index classification system, an index must have "descriptor categories," or words reported for different ranges of the index. For example, 0-100, "good"; 101-300, "poor"; over 300, "extremely poor." Three types of descriptor categories were identified:

A. **Standards:** Based on federal, state or local air quality standards.
B. **Standards and Episode Criteria:** Based both on air quality standards and episode criteria.
C. **Arbitrary:** Based neither on air quality standards nor episode criteria.

If, for example, the ranges of the descriptor categories correspond to the NAAQS and do not involve any other administrative limits, the index has Type A descriptor categories.

The result of this classification system is a four-character code which can be used to describe any air pollution index. The code for ORAQI, for example, is "$5A_3A$" (Figure 2). The number "5" indicates that the index includes five pollutant variables; "A_3" denotes the calculation method and mode (*i.e.*, it is nonlinear and the variables are combined using an additive form); "A" refers to the basis for the descriptor categories (*i.e.*, the categories reported with this index are based on the NAAQS).

Results of Classification. The 11 indices published in the literature can be classified readily using this system (Table XII). Eight of the 11 use a combined calculation mode (Type 3), and seven of these use a nonlinear (Type A) calculation method. Nine of the 11 include SO_2, and nine include some measure of particulate matter—either TSP or COH. Because the indices in Table XII are listed in chronological order, we can clearly see the tendency for later indices to include more variables than earlier indices, as indicated by the number of dots in each row.

The 35 indices used by city or county air pollution control agencies were classified by Ott and Thom[29] in a similar fashion (Table XIII). If TSP and COH are lumped together as measures of particulate matter, then the latter was the most common air pollutant included by these agencies in their indices. Of the 35 agencies, 33 (94%) included either TSP or COH; COH was

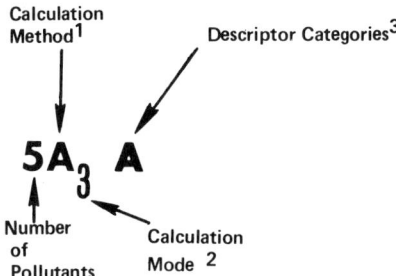

| 1/ | A: Nonlinear
B: Segmented linear
C: Linear
D: Actual Concentrations | 2/ | 1: Individual
2: Maximum
3: Combined | 3/ | A: Standards
B: Standards and
    Episode Criteria
C: Arbitrary |

Figure 2. Example of Thom-Ott index classification system applied to ORAQI air pollution index.[8]

Table XII. Classification of Indices Reported in the Literature[29]

		Variables [a]						
Index	Classification	CO	NO_2	OX	TSP	COH	SO_2	Other
Green's Index[9]	$2A_3C$					●	●	
CPI[10]	$2C_1C$							b
MURC[11]	$1A_1C$					●		
AQI[12]	$3C_3C$	●			●		●	
Ontario API[14]	$2A_3B$					●	●	
PINDEX[15]	$7C_3C$	●	●	●	●		●	c
ORAQI[16]	nA_3A (n = 1 to 5)	●	●	●	●		●	
MAQI[21]	nA_3A (n = 1 to 5)	●	●	●	●		●	
EVI[21]	nA_3A (n = 1 to 4)	●		●	●		●	
STARAQS[23]	$6B_1A$	●	●	●	●	●	●	
EQI (air)[25]	$8A_3A$	●	●	●	●	●	●	d
	Total	7	5	6	7	5	9	

[a]OX, photochemical oxidants; COH, coefficient of haze; TSP, total suspended particulates.
[b]Fuel burned and ventilating volume.
[c]Hydrocarbons and solar energy.
[d]Visibility and industrial emissions.

Table XIII. Characteristics of the 35 U.S. City/County Air Pollution Indices[29]

City/County	Classification	CO	NO$_2$	OX	TSP	COH	SO$_2$	Vis	PS	No. of Descriptor Categories
Anaheim, CA	3D$_1$B	•		•			•			3
Los Angeles, CA	3D$_1$B	•		•			•			3
San Francisco, CA	4D$_2$C	•	•	•		•				6
Denver, CO	2B$_2$C	•				•				5
New Haven, CT	3B$_2$B			•		•	•			5
Washington, DC	5B$_2$B	•	•	•		•	•			7
Jacksonville, FL	6C$_3$C	•	•	•	•	•	•			None
Miami, FL	5C$_3$C	•	•	•		•		•		5
Tampa, FL	5A$_3$A	•	•	•		•	•			6
Atlanta, GA	3C$_3$C	•			•		•			None
Chicago, IL	4D$_1$B	•		•		•	•			4
Louisville, KY	5C$_3$B	•	•	•		•	•			4
Baltimore, MD	5B$_2$B	•	•	•		•	•			8
Montgomery Co., MD	5B$_2$B	•	•	•		•	•			7
Detroit, MI	1A$_1$C					•				5
St. Paul, MN	3A$_3$B	•			•		•			4
Albany, NY	3D$_1$A	•				•	•			3
Buffalo, NY	3D$_1$A	•				•	•			3
Mineola, NY	3D$_1$A	•				•	•			3
New York City, NY	5D$_2$A	•	•	•		•	•			4
Rochester, NY	3D$_1$A	•				•	•			3
Akron, OH	5B$_2$B	•	•	•	•		•			12
Cincinnati, OH	5B$_2$B	•	•	•	•		•			12
Cleveland, OH	5B$_2$B	•	•	•	•		•			12
Dayton, OH	5B$_2$B	•	•	•	•		•			12
Toledo, OH	5B$_2$B	•	•	•	•		•			12
Portland, OR	1D$_1$C								•	5
Philadelphia, PA	4B$_2$B	•		•	•		•			5
Pittsburgh, PA	2C$_3$B					•	•			6
Chattanooga, TN	1D$_1$C				•					4
Memphis, TN	1A$_1$C					•				4
Nashville, TN	1C$_1$C					•				4
Dallas, TX	2C$_3$B		•		•					4
Fairfax Co., VA	5B$_2$B	•	•	•		•	•			7
Seattle, WA	2B$_2$B				•	•				3
Total		26	16	20	11	22	26	1	1	

[a]OX, photochemical oxidants; COH, coefficient of haze; TSP, total suspended particulates; Vis, visibility; PS, particle scattering (integrating nephelometer).

used by 22 agencies and TSP by 11 (Jacksonville, Florida, used both). The popularity of the COH measurement probably is due to the convenience of the approach and the short averaging times that are possible. CO and SO$_2$ were the next most common pollutants to be included in these indices—26 agencies (74%) for each. The next most popular variables were oxidants (20 agencies, 57%) and NO_2 (16 agencies, 45%). Visibility was included in one agency's index, and particle scattering was the only variable making up another agency's index.

When the air pollution indices used by the states (or regions) of the U.S. and by the Canadian provinces were examined (Table XIV), a similar pattern emerged. The most common pollutant variables were CO, SO_2 and particulates (TSP or COH). The least common pollutant was NO_2, with only two agencies—Ohio and the District of Columbia—reporting it in their indices. The two Canadian province indices included different numbers of pollutants; the smaller agency, Alberta, reported five air pollutants, while the larger, Ontario, reported only two.

Table XIV. Characteristics of the Six Statewide and Two Province-Wide Air Pollution Indices[29]

State/Province	Classification	Variables						No. of Descriptor Categories
		CO	NO_2	OX	TSP	COH	SO_2	
State								
Connecticut	$3B_2B$			●		●	●	5
District of Columbia	$5B_2B$	●	●	●		●	●	7
Minnesota	$3A_3B$	●			●		●	4
New Jersey	$4D_1A$	●		●		●	●	4
New York	$3D_1A$	●				●	●	3
Ohio	$5B_2B$	●	●	●	●		●	12
Total		5	2	4	2	4	6	
Province								
Alberta	$5A_3C$	●	●	●		●	●	5
Ontario	$2A_3B$					●	●	5

Are the indices used by the air pollution control agencies similar to those in the literature? Ott and Thom[31] used the classification system to compare the structural characteristics of the published indices with those in use and found striking differences (Tables XV, XVI, and XVII). The majority of the published indices (7 of 10) used a nonlinear (Type A) calculation method (Table XV); none reported actual concentration data, and only one used a segmented linear function (Type B). By contrast, a majority of the U.S. city and county agencies favored the linear (Type B or C) calculation methods (20 of 35 agencies). Only 4 of the 35 agencies used a nonlinear calculation method, and approximately one-third (11 of 35 agencies) reported actual concentration values. The state and province indices were approximately

Table XV. Calculation Methods of Air Pollution Indices[31]

Method	Indices in the Literature	City/County Indices	State/Province Indices
A. Nonlinear	7	4	3
B. Segmented Linear	1	13	3
C. Linear	3	7	0
D. Actual Concentrations	0	11	2
Total	11	35	8

Table XVI. Calculation Modes of Air Pollution Indices[31]

Mode	Indices in the Literature	City/County Indices	State/Province Indices
1. Individual	3	12	3
2. Maximum	0	15	3
3. Combined	8	8	2
Total	11	35	8

Table XVII. Descriptor Categories of Air Pollution Indices[31]

Category	Indices in the Literature	City/County Indices	State/Province Indices
A. Standards	5	6	2
B. Standards and Episode Criteria	1	19	5
C. Arbitrary	5	10	1
Total	11	35	8

equally divided among the nonlinear, segmented linear, and actual concentration calculation methods, with none choosing the purely linear method.

There were similar differences in calculation modes. A majority of the published indices (8 of 11) used a combined (Type 3) mode, and none used a maximum (Type 2) mode. By contrast, the maximum mode was the most frequent choice among city and county agencies, with only a minority (8 of 35) choosing a combined mode. The state and province indices appeared equally divided among the three modes, but the sample was too small to allow meaningful comparisons.

Finally, differences were evident in descriptor categories (Table XVII). Most air pollution indices reported in the literature had descriptor categories that were based either on ambient air quality standards (Type A) or were arbitrary (Type C), but one was based on both standards and episode criteria (Type B). By contrast, the greatest share of air pollution indices used by city and county agencies had descriptor categories that were based on standards and episode criteria. State and province indices revealed a pattern that was similar to the city and county indices.

If the index classifications in Tables XIII and XIV are analyzed in detail, the classification system reveals that 14 "basic types" of indices were being used by the states, provinces, cities and counties (Table XVIII). A basic type refers to the mathematical structure of the index—calculation method, mode, and descriptor categories—but not to the number of variables included. In Table XVIII, the basic types have been grouped according to their calculation method in order to simplify the presentation. Thus, 14 fundamentally different index structures were being used in the United States and Canada in the mid-1970's.

Although many differences were evident, there also were important similarities among the indices in use. What were the common features of indices that had been adopted by U.S. city and county air pollution agencies? Table XIX summarizes the characteristics of these indices, and the most common feature in each grouping is denoted by an arrow to the right of the table. Of the 35 city and county indices, 13 (37%) included five variables in their index calculations. This was due mainly to the fact that five agencies in Ohio and four in the Baltimore, Maryland-Washington, DC, area used indices that included five variables. In fact, each of these nine agencies used the "$5B_2B$" type of index.

Despite the large influence of these nine agencies on the data, the segmented linear function (Type B) and actual concentrations (Type D) were the most popular calculation methods (37 and 32%, respectively). Only 11% of the cities used a nonlinear form, and one of these, St. Paul, used ORAQI (Type A) but in addition reported actual concentration values (Type D). If St. Paul is reclassified as Type D (shown in parentheses, Table XIX), then only three cities (9%) used a nonlinear calculation method.

132 ENVIRONMENTAL INDICES

Table XVIII. The 14 Basic Types of Indices and Their Users[29]

Type	Users
A_1C	Detroit, Oklahoma City,[a] Memphis
A_3A	Tampa
A_3B	Minnesota, Ontario
A_3C	Alberta
B_2B	Baltimore,[b] Philadelphia,[b] Seattle, Connecticut,[b] Ohio,[c] Washington,DC
B_2C	Denver
C_1C	Nashville
C_3B	Louisville, Pittsburgh, Dallas
C_3C	Jacksonville, Miami, Atlanta, Phoenix[a]
D_1A	New Jersey, New York State
D_1B	Anaheim,[b] Los Angeles,[b] Chicago
D_1C	Portland, Chattanooga
D_2A	New York City
D_2B	San Francisco[b]

[a]Discontinued index.
[b]Replaced index.
[c]Replaced index in Cincinnati and Dayton

Table XIX. Classification Breakdown for the 35 U.S. City/County Air Pollution Indices[29]

Number of Variables	No.[a]	Percent[a]
1	5	14
2	4	11
3	9	26
4	3	9
5	13	37 ←
6	1	3
Calculation Method		
A. Nonlinear	4 (3)	11 (9)
B. Segmented Linear	13	37 ←
C. Linear	7	20
D. Actual Concentrations	11 (12)	32 (34)
Calculation Mode		
1. Individual	12 (14)	34 (40)
2. Maximum	15	43 ←
3. Combined	8 (6)	23 (17)
Descriptor Categories		
A. Standards	6	17
B. Standards and Episode Criteria	19	54 ←
C. Arbitrary	10	29

[a]Numbers in parentheses show the results when indices can be classified two ways (see text). Arrow denotes most common characteristic in each grouping.

AIR POLLUTION INDICES 133

The maximum mode was the most commonly used manner of aggregating subindices, accounting for 43% of the agencies, while 34% selected the individual mode. The remaining 23% of the city and county agencies, a clear minority, used the combined mode in their indices. However, Jacksonville and St. Paul used both the individual and combined modes. If they are reclassified as using the individual mode (as shown in parentheses, Table XIX), then only six (17%) of the agencies used the combined mode, and 83% used the two uncombined modes.

A total of 25 agencies (71%) used index descriptor categories based either on standards (Type A, 17%) or on standards and episode criteria (Type B, 54%). The number of descriptor categories used in the indices is shown on a histogram (Figure 3). There appeared to be a definite preference for three or four descriptor categories. However, there was little tendency for different agencies to use the same words for their descriptor categories. A total of 44 different words were identified in the descriptor categories of the index systems used in the U.S. (Table XX).

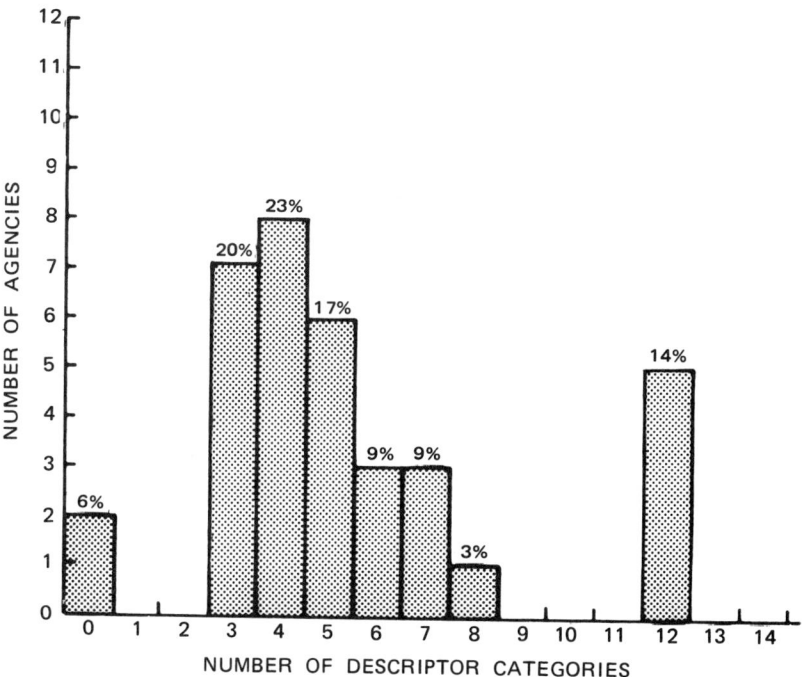

Figure 3. Histogram of the number of descriptor categories used in the 35 U.S. city and county air pollution indices.[29]

Table XX. Frequency Distribution of the 44 Words Used for Descriptor Categories[29]

Heavy	11	Normal	2
Good	10	Stage 1	2
Light	9	Stage 2	2
Moderate	7	Stage 3	2
Unhealthy(ful)	7	Very Heavy	2
Unsatisfactory	7	Warning	2
Emergency	6	Above Average	1
Poor	5	Acute	1
Alert	4	Average	1
Extremely Heavy	4	Below Average	1
Fair	4	Endangerment	1
Medium	4	Extremely Poor	1
Satisfactory	4	Harmful	1
Severe	4	High	1
Clean	3	Red Alert	1
Extremely Light	3	Significant	1
Very Poor	3	Slight	1
Acceptable	2	Very Dangerous	1
Dangerous	2	Very Good	1
Excellent	2	Very Light	1
Hazardous	2	Watch	1
Low	2	Yellow Alert	1

For the six states and two Canadian provinces, detailed analysis was more difficult due to the small sample size. However, examination of Table XIV shows that all of the states reported at least three pollutants in their indices; two states reported all five NAAQS pollutants. Because Minnesota reported individual pollutant concentrations in addition to its combined index (ORAQI), all state indices could be classified as using the segmented linear or actual concentration calculation methods, while the calculation mode was either individual or maximum. Thus, both the states and cities made only limited use of the more complex, nonlinear combined indices (Type A_3). On the other hand, in Canada there appeared to be a preference for this type of index, with Alberta and Ontario using nonlinear combined indices.

By combining the results from the index classification (Table XIX) with the analysis of descriptor words (Table XX), it was possible to specify the most frequently encountered, or "preferred," characteristics of the indices used by air pollution control agencies (Table XXI). Ott and Thom[29] believed that these characteristics, for the first time, provided a structural basis for development of a nationally uniform air pollution index:

Table XXI. Summary of Preferred Structural Characteristics
of Air Pollution Indices Used by 35 Agencies[29]

Structural Characteristic	Preferred Structure	Number of Agencies	Percent
Number of Variables	5	13	37
Calculation Method	B. Segmented Linear	13	37
Calculation Mode	2. Maximum	15	43
Descriptor Categories	B. Standards and Episode Criteria	19	54
Number	4	8	23
	3	7	20
Words	Good	10	36[a]
	Moderate	7	26[a]
	Unhealthy(ful)	7	22[a]
	Unsatisfactory	7	21[a]

[a]Percent use in 28 indices.

This summary is particularly useful when considering the possibility of developing a uniform national air pollution index. Such an index, if it is to gain wide acceptance, must have a structure which closely resembles that of the greatest number of indices currently in use. Table XXI shows that such an index would be classified as a $5B_2B$ index. This index would (1) be based on the five NAAQS pollutants (excluding hydrocarbons), (2) be a segmented linear function, (3) be calculated using the maximum mode, (4) have four descriptor categories based on the NAAQS and Episode Criteria, using the more frequently occurring words such as Good, Moderate, Unhealthy (ful) and Unsatisfactory.[29]

As we shall see, the Pollutant Standards Index subsequently recommended by the federal government was based on this philosophy and on these structural characteristics.

Criteria for a Uniform Index

Besides examining the structural characteristics of existing air pollution indices, Thom and Ott[8] documented the informal comments about indices that they received from their respondents. From these comments, they hoped to gain insight into additional qualities that a nationally uniform index should possess. They also carried out a case study of a region where neighboring communities reported air quality data to the public using three different approaches. Air pollution control agencies in the three-state area of Steubenville, Ohio; Pittsburgh, Pennsylvania; and Wheeling, West Virginia, were attempting to adopt a common air quality reporting format to reduce public confusion. Readers wishing to examine these findings in greater detail should consult the 1976 *Compendium*.[8]

From the viewpoints expressed by the many respondents of their survey and from findings of the case study, Ott and Thom[29] proposed 10 criteria that a nationally uniform air pollution index should possess:

1. *Easily Understood by the Public*

The index should transform the scientific concentration units of each pollutant into a nondimensional number which is easily understood by the public. To facilitate comprehension of the index, the basis for the transformation should be identical for each pollutant and should relate to a nationally uniform set of air quality standards.

2. *Include Major Pollutants and Be Capable of Including Future Pollutants*

The index should include all those pollutants which have been identified as major pollutants by the federal government and for which nationally uniform standards and episode criteria have been established. Furthermore, the structure of the index should be flexible so that when new pollutant standards are set they can be included in the index without modifying its basic form.

3. *Relate to Ambient Air Quality Standards and Episode Criteria*

The index should relate to the NAAQS which have been established for CO, SO_2, TSP and oxidants. The index values also should be an indicator of the relationship between air pollution levels and these national air quality goals.

4. *Relate to Federal Episode Criteria*

The recommended federal Alert, Warning and Emergency criteria and the Significant Harm levels, which have been established for CO, NO_2, oxidants, TSP, SO_2 and the TSP x SO_2 product, should be an additional basis for the index. This will provide a uniform system of both public information and agency administrative procedures during air pollution episodes.

5. *Calculated in a Simple Manner Using Reasonable Assumptions*

Pollutant concentrations should be easily convertible into their corresponding index values using simple equations or appropriate plots of the index values versus pollutant concentration. The index equation and/or curves should be based on the relationships described in Criteria 3 and 4.

6. *Based on a Reasonable Scientific Premise*

Because the index should relate to the NAAQS and federal episode criteria (Criteria 3 and 4), its basis should be as solid as the scientific basis on which the standards and episode criteria were established.

7. *Not Be Inconsistent with Perceived Air Pollution Levels*

Index values become inconsistent with perceived air pollution levels when, in combined indices, high pollutant concentrations are eclipsed by lower concentrations. To circumvent this problem, the index should calculate separate values for each pollutant, thus enabling each index value to be reported separately, if desired.

8. *Spatially Meaningful*

The spatial meaningfulness of an index depends on how the data are selected or manipulated prior to the index computation. For example, the pollutant concentration may be averaged over several monitoring stations, or the highest pollutant concentration from any station in a city may be selected for the index calculation. The latter method should be used because it prevents the masking of high pollutant concentrations, which occurs in the averaging process, and allows index values to be associated with locations of specific monitoring sites.

9. *Exhibit Day-to-Day Variation*

The structure of the index should allow for noticeable variation from day to day (and hourly, if desired). When the index is calculated for each pollutant separately (Criterion 7) and the maximum value is reported, variation also is possible in the pollutant reported. Combined indices do not possess these important characteristics. Reporting the maximum actual pollutant concentrations or the ratio of the pollutant concentration to the NAAQS gives pollutant variation. However, because both of these systems report the index value as a fraction, neither exhibits the large numerical variation necessary for the values to be clearly understood by the public.

10. *Can Be Forecast a Day in Advance (Optional)*

When the index is based on the NAAQS and episode criteria (Criteria 3 and 4), the pollutant concentrations are determined using the Federal

Reference Methods. However, the FRM for SO_2 and TSP require 24-hour averaging times, preventing the index from being reported more frequently than once every 24 hours. More current information, or even index forecasts, are of much greater interest and use to the public. More current index values can be obtained using other equivalent measurement methods having shorter averaging times. A 24-hour index forecast will be difficult to make without extensive meteorological data, but qualitative index forecasting is possible using the National Weather Service's Operation Manual for Air Pollution Weather Forecasts.[32] This system can provide an 18-hour forecast of index values using the following word descriptors: "remain the same," "decrease," or "increase."

Compendium Findings

A major conclusion of the 1976 *Compendium*[8] was that the air pollution indices which had been developed and adopted by air pollution control agencies throughout the nation showed striking diversity:

> Public awareness of air pollution problems has increased the need for timely information about changes in air pollution levels. Every day, air quality conditions in our nation's cities are presently being reported to millions of Americans by local agencies and news media. In more than half of our large cities, the public receives this information—on television, on the radio, and in print—through the use of various air pollution indices....
>
> This study has revealed a great diversity and lack of consistency in the way air quality conditions are reported to the public by means of air pollution indices. States, provinces and U.S. cities use daily informational indices which differ from each other and which greatly differ from the more complex, long-term trend indices that appear in the scientific literature. State and local air pollution control agencies clearly prefer the simpler types of indices. Nevertheless, the variation in these simpler indices is striking.[8]

When pollutant variables and descriptor categories were taken into account, the study found that no two indices were exactly the same:

> Of 55 U.S. metropolitan air pollution control agencies surveyed, 35 (64%) routinely use some form of air pollution index. The metropolitan indices are simpler in form than those appearing in the literature, but there are dramatic differences from city to city. The index classification system reveals *14 basically different index types* among the 35 U.S. cities and six states currently using daily air pollution indices. With two minor exceptions, when descriptor words are taken into account, *no two indices are exactly the same.* Each has a different mathematical formulation and a different meaning to the public. Thus, an index value of 100 reported in Washington, DC, means something entirely different from a value of 100 reported in Cleveland, OH.[29]

This diversity was viewed as a source of numerous problems:

> ...lack of uniformity among different indices creates serious problems. Not only does the diversity raise questions about the meaningfulness of today's indices, but an interested member of the public who travels from city to city will readily become confused about air pollution levels in each city. Informed members of the public, when they wish to compare air pollution levels in different cities, cannot do so using today's air pollution indices. Regulatory officials, with so many disparate indices in existence, are unable readily to use these numbers to draw a national picture of air pollution levels and trends. Further, the diversity itself suggests that no consistent scientific rationale has been employed in developing these indices. Thus, the present lack of uniformity among air pollution indices has at least three undesirable consequences: (1) it creates potential confusion, (2) it raises questions of technical validity, and (3) it prevents the indices from being used to gain insight into national air pollution problems and changes in these problems over time.[29]

One problem created by the diversity of indices is apparent if we consider the descriptor words that would be reported with an index value of 25 (or 25 ppm for carbon monoxide) in a variety of cities (Table XXII):

Table XXII. Descriptors Reported with an Index Value of 25 in 13 Cities[8]

City	Calculation Method[a]	Air Pollution Descriptor
Tampa, FL	A	Moderate
Denver, CO	B	Fair
Washington, DC	B	Fair
Baltimore, MD	B	Fair
Cincinnati, OH	B	Excellent
Miami, FL	C	Normal
Louisville, KY	C	Good
Los Angeles, CA	D	Stage 1
San Francisco, CA	D	Severe
St. Paul, MN	D	Unhealthy
Trenton, NJ	D	Unsatisfactory
Albany, NY	D	High
New York City, NY	D	Unhealthy

[a]For methods A, B and C, the index value of I = 25 is calculated from one or more pollutant concentrations; for method D, individual pollutant concentrations are reported and an index value of 25 corresponds to 25 ppm carbon monoxide.

Because of this variability, the individual who travels to different cities may easily become confused about air pollution levels in each city.... In 13 cities, a reported index value of 25 (or 25 ppm for carbon monoxide) would be accompanied by any of 10 descriptor words. If a citizen does not differentiate between index types, he would encounter descriptor words in different cities ranging from "unhealthy" to "fair" to "excellent," all describing the same index value of 25.[8]

An important recommendation of the *Compendium* was that a federal interagency task force be established to consider its findings and to examine the feasibility of adopting a nationally uniform air pollution index:

> To evaluate the feasibility of establishing a standardized index and standardized index monitoring criteria, it is recommended that a Federal Interagency Task Force on Air Quality Indicators be established. This report should serve as the starting point for deliberations of this task force. The task force also should consider the development of an *Index Monitoring Guidelines* document to assist local agencies that wish to adopt such a standardized index system or reporting format.[8]

In response to this recommendation, CEQ established the Federal Interagency Task Force on Air Quality Indicators.

A NATIONALLY UNIFORM AIR POLLUTION INDEX

A major reason for preparing the 1976 *Compendium*[8] of air pollution indices, with its detailed survey of air pollution control agency practices, was to find the common characteristics that a nationally uniform air pollution index, if one could be developed, should possess. Work on the *Compendium* was begun in May 1974; the final report was completed in December 1975 and distributed in early 1976. In March 1975, before any specific uniform index structure had emerged from the *Compendium* findings, a paper by Hunt et al.[33] proposed that a Common Air Quality Reporting Format be used as an interim solution until a more meaningful index could be developed. The Reporting Format expressed the linear ratio of each pollutant to its NAAQS and included brief language discussing the health effects of the various pollutant levels. No formal aggregation function was used, and the results were presented in tabular form.

In July 1975, when the major findings of the *Compendium* had become apparent within the federal government, CEQ established the Federal Interagency Task Force on Air Quality Indicators.[34] The Task Force, chaired by Dr. James J. Reisa, included representatives from the Environmental Protection Agency, the National Oceanic and Atmospheric Administration, and the National Bureau of Standards. One function of the Task Force was to consider candidate index structures and to select, if possible, a recommended uniform air pollution index.

As the *Compendium* findings were being analyzed in the summer of 1975, it became apparent that they pointed the direction toward the structural characteristics that a uniform index should possess. Using the 10 criteria that had been identified, Thom and Ott[8] developed such a candidate index, the Standardized Urban Air Quality Index (SUAQI), which they included as Appendix E of the *Compendium*. They also provided the Task Force with a second, simpler index to consider, and this was included as Appendix F. The second index was based on the Common Air Quality Reporting Format proposed by Hunt et al.[33]

The overall process by which SUAQI was developed is illustrated in a flow diagram (Figure 4). In the top half of the diagram, the index classification system was applied to the indices in the literature and those in common use. Using this system, the most commonly occurring characteristics of the indices used by air pollution control agencies, or the "preferred" index characteristics," were readily identified (Table XXI). In the bottom half of the diagram, the comments from the index users and nonusers, along with information gained from the three-state case study, were evaluated to arrive at the 10 criteria for a uniform index.

An index which matches the preferred characteristics of the indices in use and meets the 10 criteria would have the following structural characteristics:

- includes six variables (CO, NO_2, oxidants, TSP, SO_2 and TSP x SO_2)
- incorporates segmented linear functions in subindices
- calculated using a maximum mode
- uses the NAAQS, episode criteria, and Significant Harm levels as breakpoints
- includes four descriptor categories

SUAQI meets these requirements. The SUAQI breakpoints (Table XXIII) may be viewed as a consolidation of the administrative limits given in Tables I, III and IV. The resulting segmented linear functions can be plotted all on one graph (Figure 5). Four descriptor categories were used in this index: 0-50, "good"; 51-100, "satisfactory"; 101-199, "unhealthful"; 200-500, "hazardous."

To formally present this concept in the scientific literature, Thom and Ott[35] renamed SUAQI as the Pollutant Standards Index and submitted it in November 1975 for publication in *Atmospheric Environment*:

> Using these criteria, we have developed and now propose the Pollutant Standards Index (PSI), or ψ, as a leading candidate for a nationally uniform air pollution index. We have designed PSI to overcome the serious objections that some air pollution control agencies have raised to existing indices. PSI is a segmented linear function based on the National Ambient Air Quality Standards and the Federal Episode and Emergency Criteria.

142 ENVIRONMENTAL INDICES

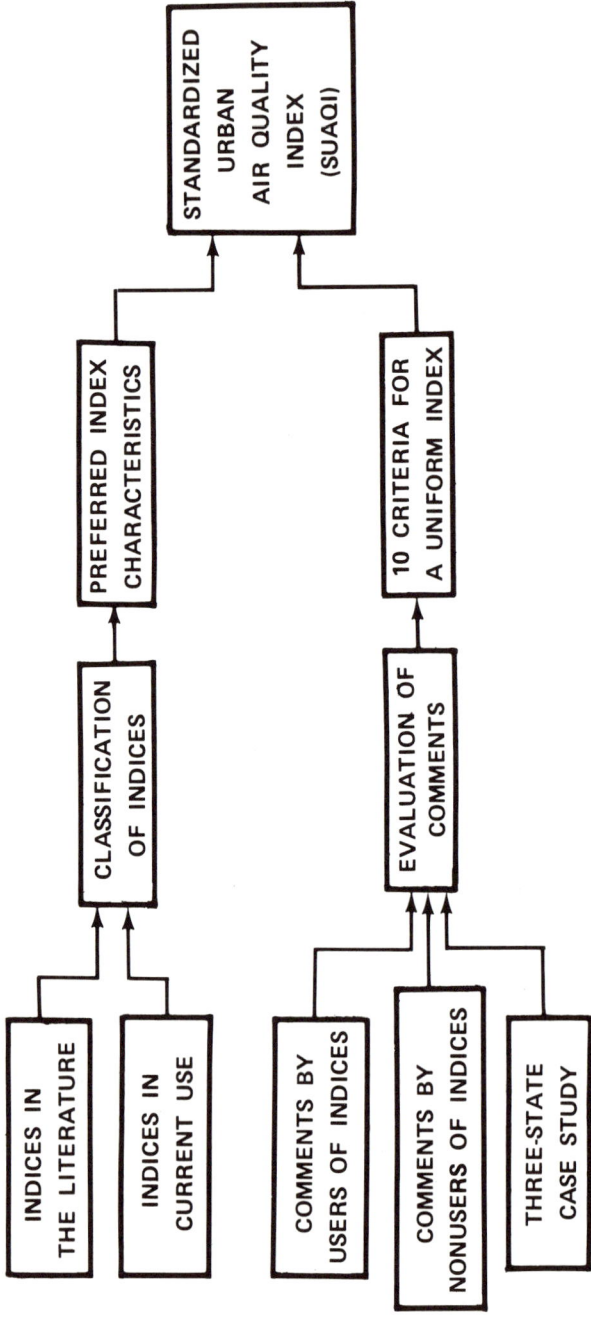

Figure 4. Development process of the Standardized Urban Air Quality Index (SUAQI).

It is a "maximum" type index, reporting only the highest numerical value of all subindex values for each pollutant. There are four descriptor words reported with the index values.[35]

Table XXIII. SUAQI Breakpoints Initially Proposed by Thom and Ott[8,29]

Federal Level	SUAQI	CO (ppm)	NO$_2$ (ppm)	OX (ppm)	SO$_2$ (ppm)	TSP (μg/m^3)	SO$_2$ x TSP (μg/m^3)^2
		\multicolumn{6}{c}{Averaging Time (hr)}					
		8	1	1	24	24	24
50% of Primary NAAQS	50	4.5	a	0.04	0.07	150[b]	a
Primary NAAQS	100	9	a	0.08	0.14	260	a
Alert	200	15	0.6	0.10	0.3	375	65,000
Warning	300	30	1.2	0.40	0.6	625	261,000
Emergency	400	40	1.6	0.50	0.8	875	393,000
Significant Harm	500	50	2.0	0.60	1.0	1,000	490,000

[a] No NAAQS exists—SUAQI is not reported below the Alert level.
[b] Secondary NAAQS.

Ott and Thom[29] concluded that the SUAQI/PSI index, if adopted on a nationwide basis, would have a variety of benefits:

> We believe that the SUAQI/PSI formulation offers the greatest chance for acceptance by state and local air pollution control agencies. Adoption of a nationally uniform air pollution index should (1) facilitate comparisons of air quality levels in different cities, (2) reduce confusion among residents of cities about daily air quality levels, and (3) provide policy makers with a uniform measure for evaluating the impact of regulatory actions.

The Pollutant Standards Index (PSI) ultimately proposed by EPA to the Interagency Task Force was based on the SUAQI/PSI concept.

Development of PSI

The Federal Interagency Task Force on Air Quality Indicators requested that EPA prepare a technical guidelines document containing its suggested uniform air pollution index. The document subsequently would be used to assist state and local air pollution agencies in implementing the index. To prepare the guidelines document, EPA formed a second group, the Working Group to Develop an Air Quality Index. This internal group included members from EPA's Offices of Air and Waste Management, Research and

144 ENVIRONMENTAL INDICES

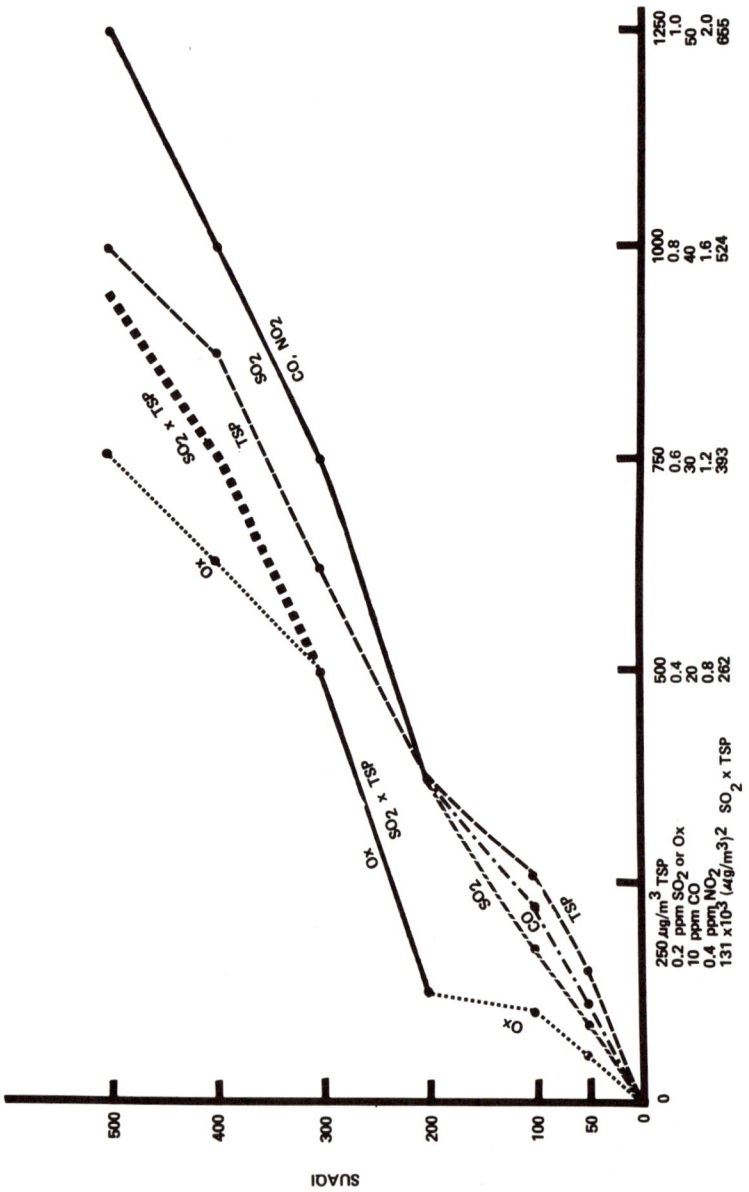

Figure 5. SUAQI subindex functions initially proposed by Thom and Ott.[8,29]

Development, and Planning and Management. Its chairman was Raymond Smith, and William F. Hunt, Jr., coordinated its work activities.

The Working Group began with the SUAQI/PSI formulation. Successive drafts of the guidelines were used as vehicles for proposing various changes in the index and obtaining comments from reviewers on these alterations. An important early contribution of the Working Group was the addition of language describing the health effects associated with various index levels. This language was designed to be reported with the index to help the public understand the meaning of the numbers. At the Working Group's direction, this language was prepared specifically for the index by medical specialists at EPA's Health Effects Research Laboratory in Research Triangle Park, North Carolina.

As the index continued to evolve, various changes were considered in its descriptor categories. In a version of PSI presented by Thom et al.[36] in April 1976, for example, six descriptor categories were used: "good," "satisfactory," "potentially unhealthful," "unhealthful," "very unhealthful," and "hazardous." In all versions, an index value of PSI = 100 corresponded to the NAAQS, and PSI = 500 corresponded to the Significant Harm level, with 200, 300 and 400 corresponding to the Alert, Warning and Emergency levels in the federal episode criteria. These values were identical to the values used in SUAQI. In the case of the Alert level for oxidants, however, a concentration of 400 $\mu g/m^3$ [0.2 ppm] was selected instead of the published value of 200 $\mu g/m^3$ [0.1 ppm]. This was done because available scientific evidence on the health effects of oxidants suggested that 400 $\mu g/m^3$ represented the Alert level better than 200 $\mu g/m^3$. Like SUAQI, the "good"-"satisfactory" breakpoint (PSI = 50) corresponded to 50 percent of the NAAQS.

The April 1976 version of PSI[36] was distributed widely for review and comment. The reviewers included selected state and local air pollution control agencies, the Association of Local Air Pollution Control Officials (ALAPCO), the State and Territorial Air Pollution Program Administrators (STAPPA), the EPA Regional Offices, EPA's research laboratories, the National Oceanic and Atmospheric Administration, and the American Lung Association. The reviewers recommended a variety of changes, many of which were incorporated into the final version of the index.

For two air pollutants, SO_2 and TSP, primary NAAQS have been promulgated for both 24-hour and 1-year averaging periods (see Table I). In addition, TSP has a secondary NAAQS 24-hour averaging period. In the April 1976 version of the index,[36] the SO_2 breakpoint between "good" and "satisfactory" (PSI = 50) was set at approximately 50% of the 24-hour primary NAAQS (180 $\mu g/m^3$). Some reviewers felt that this value was too high and would create problems for the index users. Because the annual

SO_2 NAAQS is relatively low (80 $\mu g/m^3$), they felt that a large proportion of the days in a given year might be reported by the index in the "good" or "satisfactory" ranges, only to discover at the end of the year that the annual average primary NAAQS had been violated. To eliminate this possible inconsistency, the EPA Working Group substituted the annual average SO_2 concentration of 80 $\mu g/m^3$ for the breakpoint at PSI = 50, interpreting this value on a 24-hour basis instead of an annual basis. For TSP, the April 1976 version of PSI, like SUAQI, originally used the secondary NAAQS (150 $\mu g/m^3$) for PSI = 50. Following logic similar to that used to alter the SO_2 breakpoint at PSI = 50, the TSP annual average (geometric mean) NAAQS of 75 $\mu g/m^3$ was substituted for the 24-hour TSP breakpoint at PSI = 50. These changes helped insure that, if the annual NAAQS for TSP is violated, more than half the days of the year would be above PSI = 50.*

To determine the impact of these changes on the performance of the index, Ott[37] developed a special computer program, INDEX.PLOT, and Hunt and Ott[38] carried out a computer analysis of the index. For any important suggested changes in the index breakpoints, the equations in the computer program were changed, and PSI's performance was examined using actual air quality data from different U.S. cities. The computer program generated daily index readings for a year of data in each city, enabling both the daily patterns and the overall statistical properties of the index to be evaluated.

The many reviewers also suggested a variety of changes in the proposed PSI descriptor categories. In the April 1976 version of PSI,[36] the words "potentially unhealthful" were intended to mean that concentrations immediately above the NAAQS (but below the Alert level) could cause adverse health effects to selected members of the population, such as the ill and elderly, but not necessarily to the entire population. Some reviewers felt that this language was cumbersome and subject to possible misinterpretation, so the simpler term used in SUAQI, "unhealthful," was adopted. To clarify the meaning of this term, the language describing the health effects associated with this descriptor category stated that aggravation of symptoms could be expected in "susceptible persons." Because many reviewers felt that the range between 50 and 100 should convey a slight or moderate air pollution problem, the word "moderate" was substituted for "satisfactory."

---

*For most TSP air quality data, the annual geometric mean concentration lies very close to the *median* of 24-hour values. By definition, one-half of the 24-hour values lie below the median, and one-half lie above the median. Therefore, if the observed geometric mean concentration at a particular monitoring station is equal to the NAAQS value, approximately one-half of the 24-hour average values should be above 75 $\mu g/m^3$, and one-half below 75 $\mu g/m^3$. Thus, if the geometric mean exceeds its NAAQS value, more than half of the 24-hour values should be above 75 $\mu g/m^3$, with corresponding PSI values above 50.

The overall process by which PSI evolved from the SUAQI/PSI formulation is summarized in a flow diagram (Figure 6). The first step was to add language on the health effects associated with various descriptor categories. The comments from technical reviewers of the index were used as a basis for making certain changes in breakpoints. For each change, the computer evaluation enabled the performance of the index to be examined. These successive iterations can be viewed as a total system in which the computer evaluation serves as a "feedback loop." Many variations in index breakpoints were examined. At one point, for example, three descriptor categories were used below PSI = 100. Although the computer evaluation revealed considerable daily variation in index reports using this scheme, the additional complexity did not appear justified.

On successive iterations, fewer and fewer changes appeared necessary. Finally, a version of PSI was agreed upon which satisfactorily met the major objections of all reviewers (Table XXIV). This version was submitted to the Federal Interagency Task Force on Air Quality Indicators in August of 1976. It was published in the *Federal Register*[39] in September 1976 and appeared in the final report of the Task Force.[40] The index also appeared in an EPA guidelines document written by Hunt *et al.*[41]

In summary, the Pollutant Standards Index, as it finally evolved, was based on (1) the preferred characteristics of existing indices used by city and county air pollution control agencies (Table XXI), (2) the 10 criteria for a uniform index proposed by Thom and Ott[8] in the *Compendium* of indices, (3) a computer evaluation performed by Hunt and Ott[38] using actual air quality data from selected cities, and (4) comments from numerous technical reviewers within and outside EPA and across the nation.

Structure of PSI

Like SUAQI, on which it is based, PSI is structurally a "$5B_2B$" type of index. It includes five pollutants (CO, NO_2, oxidants, TSP and SO_2); its subindices consist of segmented linear functions; it reports the maximum subindex ("maximum mode"); and its descriptor categories are based on the NAAQS, federal episode criteria and Significant Harm levels. Five descriptor categories are used: "good" (0-50); "moderate" (51-100); "unhealthful" (101-199); "very unhealthful" (200-299); and "hazardous" (300 and above). Photochemical oxidants are reported as ozone (O_3).

The EPA guidelines document by Hunt *et al.*[41] presents the index—breakpoints, descriptor categories, and language describing the health effects of various ranges—in a single table (Table XXIV). The segmented linear functions are shown graphically by plotting the breakpoints for the five pollutants on linear graph paper (Figures 7 through 12). Because NO_2 has

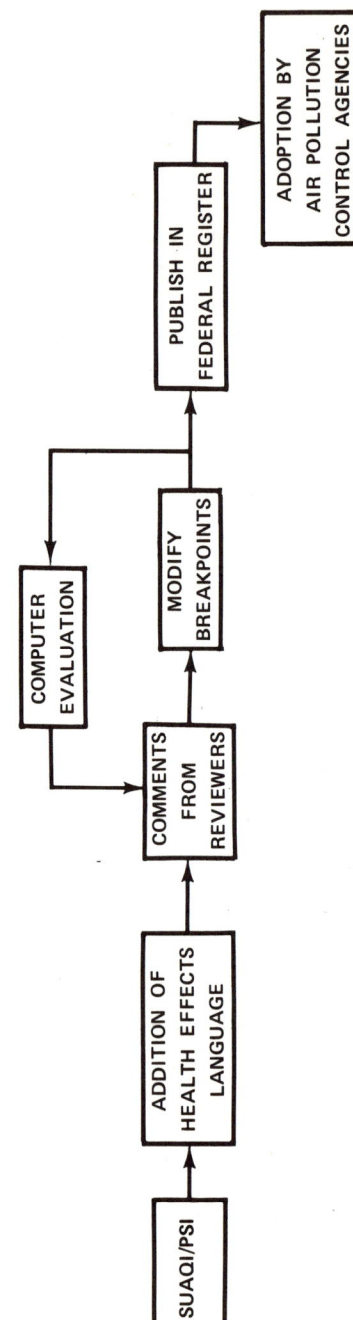

Figure 6. Development process of the Pollutant Standards Index (PSI).

Table XXIV. Comparison of PSI Values with Pollutant Concentrations, Descriptor Words, General Health Effects and Cautionary Statements[41]

Index Value	Air Quality Level	Pollutant Levels					Health Effect Descriptor	General Health Effects	Cautionary Statements
		TSP (24-hr) ($\mu g/m^3$)	SO_2 (24-hr) ($\mu g/m^3$)	CO (8-hr) (mg/m^3)	O_3 (1-hr) ($\mu g/m^3$)	NO_2 (1-hr) ($\mu g/m^3$)			
500	Significant Harm	1000	2620	57.5	1200	3750	Hazardous	Premature death of ill and elderly. Healthy people will experience adverse symptoms that affect their normal activity.	All persons should remain indoors, keeping windows and doors closed. All persons should minimize physical exertion and avoid traffic.
400	Emergency	875	2100	46.0	1000	3000	Hazardous	Premature onset of certain diseases in addition to significant aggravation of symptoms and decreased exercise tolerance in healthy persons.	Elderly and persons with existing diseases should stay indoors and avoid physical exertion. General population should avoid outdoor activity.
300	Warning	625	1600	34.0	800	2260	Very Unhealthful	Significant aggravation of symptoms and decreased exercise tolerance in persons with heart or lung disease, with widespread symptoms in the healthy population.	Elderly and persons with existing heart or lung disease should stay indoors and reduce physical activity.

150 ENVIRONMENTAL INDICES

Table XXIV, continued.

Index Value	Air Quality Level	Pollutant Levels					Health Effect Descriptor	General Health Effects	Cautionary Statements
		TSP (24-hr) ($\mu g/m^3$)	SO_2 (24-hr) ($\mu g/m^3$)	CO (8-hr) (mg/m^3)	O_3 (1-hr) ($\mu g/m^3$)	NO_2 (1-hr) ($\mu g/m^3$)			
200	Alert	375	800	17.0	400[a]	1130	Unhealthful	Mild aggravation of symptoms in susceptible persons, with irritation symptoms in the healthy population.	Persons with existing heart or respiratory ailments should reduce physical exertion and outdoor activity.
100	NAAQS	260	365	10.0	160	[b]	Moderate		
50	50% of NAAQS	75[c]	80[c]	5.0	80	[b]			
0		0	0	0	0	[b]	Good		

[a] 400 $\mu g/m^3$ was used instead of the O_3 Alert level of 200 $\mu g/m^3$ (see text).
[b] No index values reported at concentration levels below those specified by "alert level" criteria.
[c] Annual primary NAAQS.

AIR POLLUTION INDICES 151

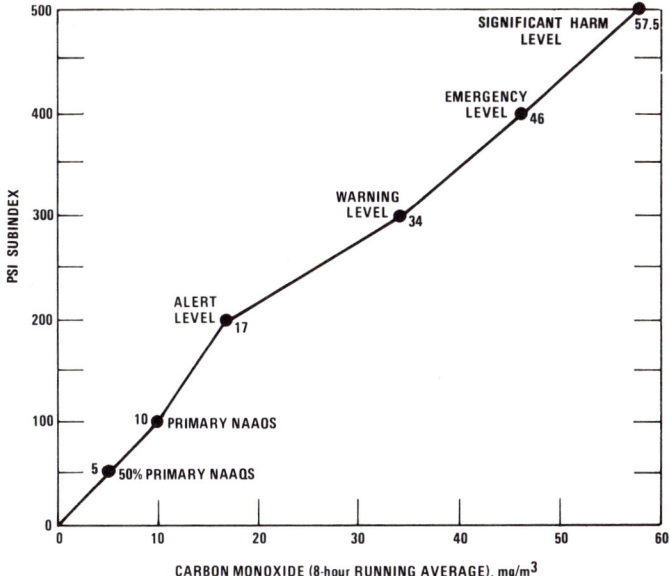

Figure 7. PSI subindex function for carbon monoxide.[41]

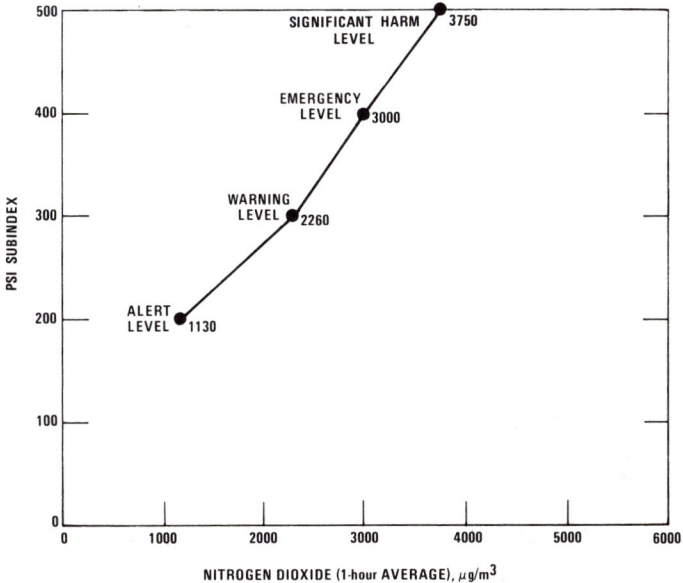

Figure 8. PSI subindex function for nitrogen dioxide.[41]

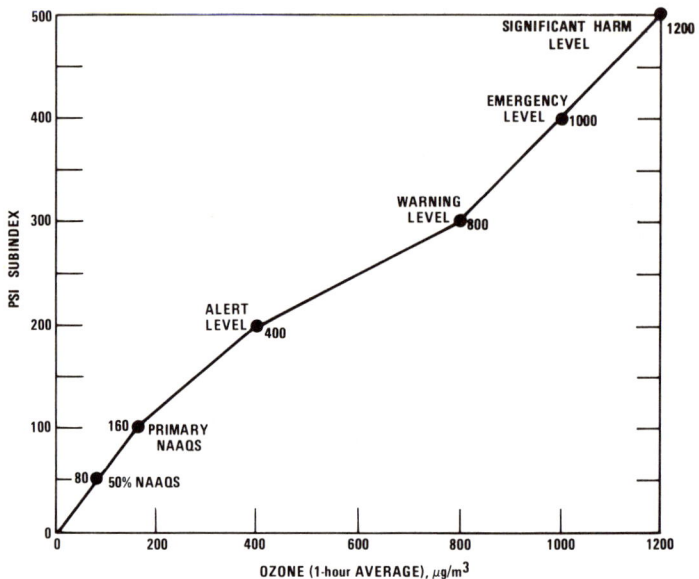

Figure 9. PSI subindex function for ozone.[41]

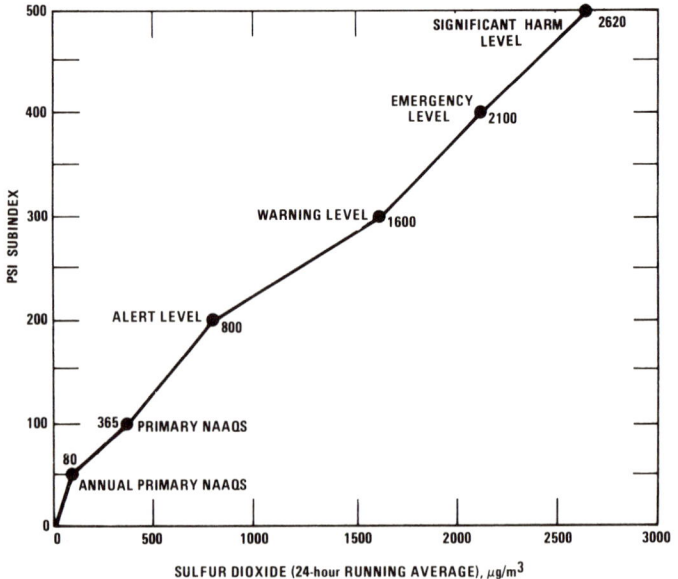

Figure 10. PSI subindex function for sulfur dioxide.[41]

AIR POLLUTION INDICES 153

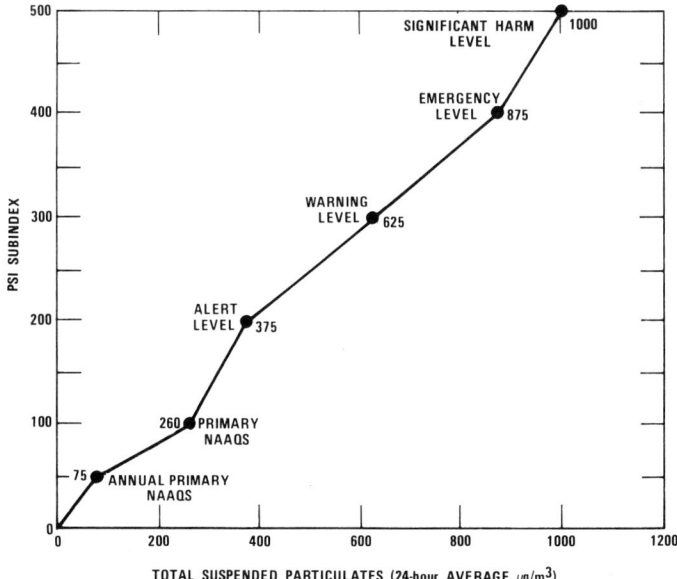

Figure 11. PSI subindex function for total suspended particulates.[41]

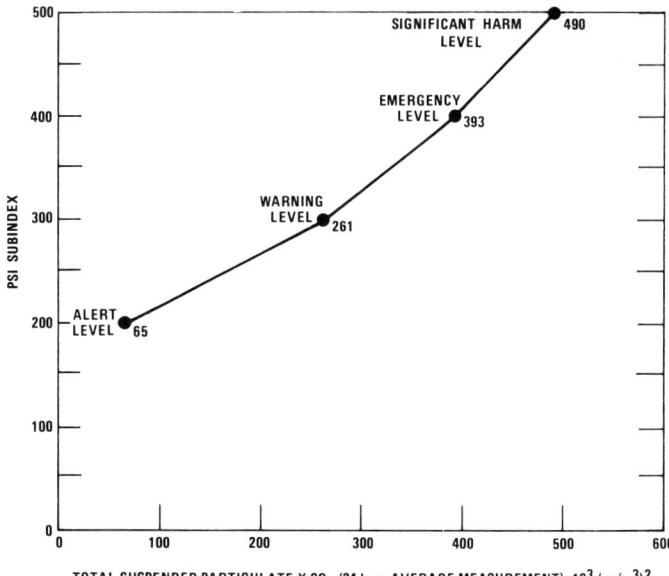

Figure 12. PSI subindex function for the product of total suspended particulates and sulfur dioxide.[41]

only an annual NAAQS, the NO_2 subindex function begins at the Alert level (PSI = 200), and the line segments do not extend to the origin. A sixth variable—the product of TSP and SO_2—is computed and included in the index aggregation function, which is the maximum operator. Because PSI is intended to be flexible, other pollutant variables can be included as future NAAQS or administrative limits are adopted.

To calculate PSI, one enters the observed concentration on the horizontal axis of the appropriate graph and reads the corresponding subindex value from the vertical axis. Then, PSI, or ψ, is reported as the maximum of the six subindices:

$$\psi = \max\{I_1, I_2, \ldots, I_6\} \tag{22}$$

The pollutant responsible for the maximum subindex is called the "critical pollutant." For completeness, the EPA guidelines document[41] recommends that the critical pollutant be reported along with the index. For example, "Today PSI is 120, and the pollutant responsible is carbon monoxide." To gain the advantages of the "individual" mode, EPA recommends that all subindices which exceed 100 (thereby indicating violation of a NAAQS) also be reported.

Data Used in PSI

For routine index reporting, it is important for the most recent concentration values to be used in the index calculations. Two pollutant variables, NO_2 and photochemical oxidants (measured as ozone), have 1-hour averaging times, permitting index reports to be made on an hourly basis if desired. For CO and SO_2, which have 8-hour and 24-hour PSI averaging periods, respectively, running averages are used. The Federal Reference Method used to measure CO gives continuous readings, permitting 8-hour running averages to be calculated quite easily. Because a new 8-hour average value is computed every hour, relatively up-to-date index reports can be provided. The FRM for SO_2 is based on manual pararosaniline colorimetry, which gives a 24-hour average value and requires laboratory analysis. Serious logistical problems can arise if an index number is to be calculated two or three times per day from measurements at different locations using the FRM for SO_2. Fortunately, equivalent measurement methods are available for SO_2, such as automated pararosaniline colorimetry, which give continuous readings from which 24-hour running averages can be computed easily.

TSP presents a more serious measurement problem, because the high-volume sampler has a minimum averaging period of 24 hours, and EPA specifies that the sampling period should begin and end at midnight. Further, the high-volume sampler requires an additional 24-hour period during which the sample is dried. Thus, the TSP data used in an index may be as much as 2 days old by the time the index is calculated. EPA has recommended that the drying period be eliminated for index reporting and more convenient sampling times be established (such as 8 AM to 8 AM, or noon to noon), but the high-volume sampler still does not provide sufficiently current information. Other measurement methods have been suggested for index reporting, such as the tape sampler, which can give COH readings on an hourly basis.

As originally published,[41] PSI did not include a subindex for COH. However, federal episode criteria (Table III) and a Significant Harm level (Table IV) have been promulgated for COH. Using the same philosophy employed in developing the other subindex functions, a COH segmented linear function is easily plotted (Figure 13). Because EPA has not approved the tape sampler as an FRM or an equivalent method, it should not be used to report violations of a NAAQS. However, EPA considers the method acceptable for index reporting, and, therefore, COH values can be computed in PSI at or above the Alert level (PSI $\geq$ 200).

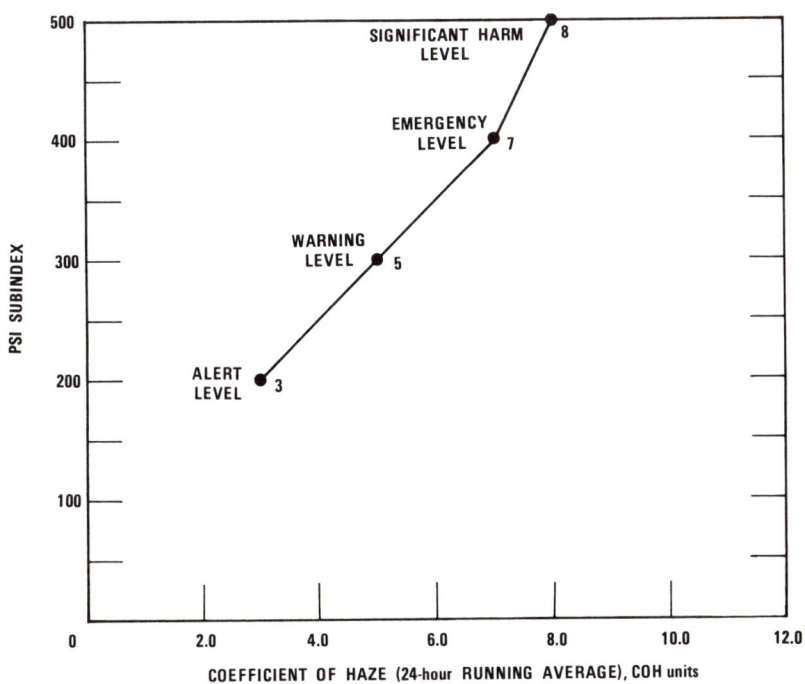

Figure 13. PSI subindex function developed by the author for Coefficient of Haze.

156 ENVIRONMENTAL INDICES

Unfortunately, TSP and COH, because they reflect different attributes of particulate matter (mass of particles versus opacity of particles), cannot be related to each other in a single equation that would apply to all locations. Rather, a separate relationship must be established by examining the historical record of data at each air monitoring site or by conducting a special field study. Usually, a high-volume sampler and a tape sampler are operated side-by-side for a sufficient period of time (ideally, a year or more), and a statistical study is conducted of correlations between the two sets of observations. Once a site-specific relationship between TSP and COH is established, the new COH breakpoints should be substituted into PSI.

Some air monitoring specialists have raised questions about the accuracy and meaningfulness of tape sampler measurements. A newer instrument, the integrating nephelometer, offers promise and appears to correlate well with visibility and TSP measurements. It measures the scattering of light from small particles and provides continuous data which can be averaged over 1-hour or 24-hour periods. EPA is carrying out further evaluations of the integrating nephelometer and other candidate measurement methods for TSP.

Calculation of PSI

Because it may be difficult to read PSI values with sufficient precision using the segmented linear graphs plotted in Figures 7 through 13, other techniques for calculating PSI have been suggested. These include computer approaches, a PSI nomogram, and a tabular method.

Calculation by Computer

To facilitate computation of PSI by computer or by a programmable hand calculator, Ott and Hunt[42] have developed the necessary equations and coefficients for the PSI subindex functions plotted in Figures 7 through 12. Using the notation given in Equation 6 of Chapter II, in which the breakpoints of the segments are represented by $(a_1, b_1), (a_2, b_2), \ldots, (a_j, b_j)$, the equation for a segmented linear function with m segments is written as follows:

$$I = \alpha_j (X - a_j) + b_j \quad (23)$$

where

$$\alpha_j = \frac{b_{j+1} - b_j}{a_{j+1} - a_j}$$

for $a_j < X \leq a_{j+1}$

and $j = 1, 2, 3, \ldots, m$

Table XXV. Coefficients for Calculating PSI by Computer, Expressed in Gravimetric Units[a,42]

j	b_j	CO a_j	CO α_j	Oxidant a_j	Oxidant α_j	NO_2 a_j	NO_2 α_j	SO_2 a_j	SO_2 α_j	TSP a_j	TSP α_j	TSP x SO_2 a_j	TSP x SO_2 α_j
1	0	0.0	10.000000	0	0.62500000	b	b	0	0.62500000	0	0.66666667	b	b
2	50	5.0	10.000000	80	0.62500000	b	b	80	0.17543860	75	0.27027027	b	b
3	100	10.0	14.285714	160	0.41666667	b	b	365	0.22988506	260	0.86956522	b	b
4	200	17.0	5.8823529	400	0.25000000	1130	0.088495575	800	0.12500000	375	0.40000000	65000	0.000510204
5	300	34.0	8.3333333	800	0.50000000	2260	0.135135135	1600	0.20000000	625	0.40000000	261000	0.000757576
6	400	46.0	8.6956522	1000	0.50000000	3000	0.133333333	2100	0.19230769	875	0.80000000	393000	0.001030928
7	500	57.5	8.6956522	1200	0.50000000	3750	0.133333333	2620	0.19230769	1000	0.80000000	490000	0.001030928

[a] For CO, a_j is expressed in mg/m^3; for oxidants, NO_2, SO_2 and TSP, a_j is expressed in $\mu g/m^3$; for TSP x SO_2, a_j is expressed in $(\mu g/m^3)^2$.
[b] Index is not calculated; no standards or episode criteria exist at these levels.

158 ENVIRONMENTAL INDICES

Table XXVI. Coefficients for Calculating PSI by Computer, Expressed in Volumetric Units (ppm)[42]

j	b_j	CO a_j	CO α_j	Oxidant a_j	Oxidant α_j	NO_2 a_j	NO_2 α_j	SO_2 a_j	SO_2 α_j	TSP x SO_2^a a_j	TSP x SO_2^a α_j
1	0	0.0	11.111111	0.0	1250.0000	b	b	0.0	1666.6667	b	b
2	50	4.5	11.111111	0.04	1250.0000	b	b	0.03	454.54546	b	b
3	100	9.0	16.666667	0.08	833.3333	b	b	0.14	625.00000	b	b
4	200	15.0	6.666667	0.20	500.0000	0.6	166.66667	0.30	333.33333	24.82	1.3361839
5	300	30.0	10.000000	0.40	1000.0000	1.2	250.00000	0.60	500.00000	99.66	1.9825535
6	400	40.0	10.000000	0.50	1000.0000	1.6	250.00000	0.80	500.00000	150.1	2.7027027
7	500	50.0	10.000000	0.60	1000.0000	2.0	250.00000	1.00	500.00000	187.1	2.7027027

[a] Total suspended particulates (TSP) are not listed because they are not reported in ppm. The units of the product TSP x SO_2 are ppm-$\mu g/m^3$. Values in the original reference differ from those presented here, because the original numbers were based on an SO_2 conversion factor at 0°C instead of 25°C.
[b] Index is not calculated; no standards or episode criteria exist at these levels.

For the first segment, j = 1; for the second segment, j = 2, and so on, until j = m. In this system, α_j is the slope of the jth segment; a_j is the value of X at the beginning of the segment; and b_j is the corresponding value of the subindex I at the beginning of the segment (see Figure 7 of Chapter II, page 57).

As discussed in Chapter II (page 58), a segmented linear function gives the same value for I at the end of one segment as it does for I at the beginning of the next segment, because the two segments are joined at the breakpoint. If the inequalities were specified in the same manner as in Equation 6 of Chapter II, a computer program based on this equation would calculate the subindex value at each breakpoint twice. To reduce unnecessary computations and simplify programming, Equation 23 specifies a strict inequality ("<") for the lower bound of X and a "less than or equal" sign ("≤") for the upper bound. With the inequalities specified in this manner, the program calculates the subindex value at the breakpoint just once.

The article by Ott and Hunt[42] gives the slope α_j, along with a_j and b_j, for all segments of the PSI subindex functions. For convenience, these values are listed in Table XXV in gravimetric units ($\mu g/m^3$ or mg/m^3) and in Table XXVI in volumetric units (ppm). To illustrate the manner in which the values in Table XXV were computed, we note that the NAAQS for CO is 10 mg/m^3 and the Alert level is 17 mg/m^3. Then, the slope for the third segment is computed using $a_3 = 10$ and $a_4 = 17$, with corresponding index values of $b_3 = 100$ and $b_4 = 200$:

$$\alpha_3 = \frac{b_4 - b_3}{a_4 - a_3} = \frac{200 - 100}{17 - 10} = 14.285714$$

Notice that the slope α_3 is computed to more significant figures than is justified by the precision of the measuring technique. For example, the FRM for carbon monoxide is capable of measuring with a precision of ± 0.5 mg/m^3 (1-hour average). However, for computer applications, the slope α has been calculated to eight significant figures to assure that successive line segments will meet each other at the breakpoints with a high degree of consistency and precision. This approach also assures computation uniformity: different persons writing their own computer programs will obtain exactly the same answers. Using today's computers, no additional cost or difficulty is incurred by carrying out computations with this many significant figures. Of course, the final PSI values usually will be rounded off to the nearest whole unit.

The values for a_j given in Tables XXV and XXVI are based on official numbers published in the *Federal Register*[1,5,7] and listed in Tables I, III and IV. In converting from one set of units to another, some rounding already has occurred in the official numbers. Therefore, an exact conversion of gravimetric units to volumetric units may give slightly different values than the

Federal Register numbers. For example, suppose that one wishes to convert the SO_2 annual NAAQS of 80 $\mu g/m^3$ to volumetric units, retaining two significant figures. From the *ideal gas law*, PV = RT, one gram molecule (mole) of any gas (6.02×10^{23} molecules) occupies the following volume at 0° Celsius (273.15° Kelvin) and 1 atmosphere pressure:

$$V = RT/P = (0.08205 \text{ atm-}\ell/\text{mole-}°K)(273.15°K) / (1 \text{ atm}) = 22.41 \text{ }\ell/\text{mole}$$

At 25° Celsius (298.15° Kelvin), the temperature usually used in such calculations instead of 0° Celsius, a mole occupies a volume of (298.15/273.15) x 22.41 = 24.46 liters (ℓ). Using 64.06 as the molecular weight for SO_2, 80 $\mu g/m^3$ is converted from gravimetric units to volumetric units as follows:

$$\frac{80 \text{ }\mu g}{m^3} \times \frac{1 \text{ m}^3}{1000 \text{ }\ell} \times \frac{24.46 \text{ }\ell}{1 \text{ mole}} \times \frac{1 \text{ mole}}{64.06 \text{ g}} \times \frac{1 \text{ g}}{10^6 \text{ }\mu g} = 0.0305 \times 10^{-6} = 0.031 \text{ ppm}$$

By comparison, the official *Federal Register* value for the SO_2 NAAQS is 0.03 ppm (Table I). If the official value were used in a FORTRAN computer program, it would appear in the computer as 0.030000000 ppm, giving slightly different results than if the exact conversion were used.

If the raw data were available only in volumetric units, they could first be converted exactly* and then compared with the gravimetric breakpoints (Table I). Another approach is to use the volumetric breakpoints, but there are two of these to choose from: the official values and the more exact, converted values. Although the converted values are more exact and therefore would be preferable, many state and local air pollution control agencies have grown accustomed to using the official ppm values. To give maximum flexibility, the FORTRAN computer program developed by Ott[37] for calculating PSI from air quality data, INDEX.PLOT, contains both the gravimetric and volumetric breakpoints. Figure 14, which shows 13 lines from the overall 700-line program, lists the algorithm for calculating the CO subindex. Before reaching this point in the program, the station codes which accompany the data are read to determine if the CO value is expressed in mg/m^3 or ppm. If CO is expressed in mg/m^3, the upper portion of the algorithm is executed, beginning at the statement labeled "11." In this algorithm, "T" denotes the CO concentration, and "SUBCO" denotes the subindex value. Each "IF" statement represents one segment of the CO segmented linear function, and the statement is executed if "T" lies within the stated range. If the CO concentration is expressed in ppm, the lower portion of the algorithm is executed, beginning at the statement labeled

---

*Using the constants given above, 1 ppm SO_2 = 2619 $\mu g/m^3$ at 25° Celsius and 1 atmosphere.

AIR POLLUTION INDICES 161

"12." Thus, PSI breakpoints with the proper units are always compared with the raw concentration values, and conversion factors are not used on the raw data.

```
C......CO IN MG/M3:
   11 IF(T.GE.0.0.AND.T.LT.10.0)      SUBCO=10.0*T
      IF(T.GE.10.0.AND.T.LT.17.0)     SUBCO=14.285714*(T-10.0) + 100.0
      IF(T.GE.17.0.AND.T.LT.34.0)     SUBCO=5.8823529*(T-17.0) + 200.0
      IF(T.GE.34.0.AND.T.LT.46.0)     SUBCO=8.3333333*(T-34.0) + 300.0
      IF(T.GE.46.0)                   SUBCO=8.695652*(T-46.0) + 400.0
      GO TO 13
C......CO IN PPM:
   12 IF(T.GE.0.0.AND.T.LT.9.0)       SUBCO=11.111111*T
      IF(T.GE.9.0.AND.T.LT.15.0)      SUBCO=16.666667*(T-9.0) + 100.0
      IF(T.GE.15.0.AND.T.LT.30.0)     SUBCO=6.6666667*(T-15.0) + 200.0
      IF(T.GE.30.0)                   SUBCO=10.0*(T-30.0) + 300.0
      GO TO 13
```

Figure 14. Portion of the FORTRAN computer program INDEX.PLOT, showing algorithm for calculating the PSI subindex for CO.[37]

The EPA guidelines document[41] does not give specific details for determining which descriptor category PSI falls into when it is exactly at a breakpoint. That is, if PSI = 100, which descriptor category is appropriate: "moderate" or "unhealthful"? We shall assume that a NAAQS is violated only if the concentration exceeds the NAAQS value, but not if it equals the NAAQS value. This is consistent with the way that EPA interprets the NAAQS values in the context of State Implementation Plans. Therefore, for PSI's breakpoints at the NAAQS and 50% of the NAAQS, a strict inequality should be used (Table XXVII). Thus, if PSI = 50, the air quality is

Table XXVII. Numerical Ranges for PSI Descriptor Categories

PSI Range	Descriptor Category
$0 \leqslant \psi \leqslant 50$	Good
$50 < \psi \leqslant 100$	Moderate
$100 < \psi < 200$	Unhealthful
$200 \leqslant \psi < 300$	Very Unhealthful
$300 \leqslant \psi < 500$	Hazardous

"good"; if PSI = 100, the air quality is "moderate." The air becomes "unhealthful" only if PSI *exceeds* 100.

For the federal episode criteria, the language published in the *Federal Register*[6] takes a somewhat different approach for the Alert, Warning and Emergency levels. Actions to avoid episodes are recommended when any of the administrative limits is *reached*. For the Alert level, for example, "An Alert will be declared when any one of the following levels is reached at any monitoring site."[6] Thus, it appears advisable to use a strict inequality for PSI's upper bounds in the "very unhealthful" and "hazardous" descriptor categories, which generally coincide with the concentrations for the Alert and Warning levels. Table XXVII shows the inequalities specified in this fashion.

INDEX.PLOT, which is available through EPA,[37] is designed for computing daily index values over an extended time period, such as a month or a year. It is especially useful for "retrospective" data analysis; that is, for comparing air quality levels at different locations or for examining air quality trends over time. INDEX.PLOT generates a time series graph of index values very inexpensively using the computer line printer. It also computes basic statistical information on PSI over the time period under study (for example, the mean, standard deviation, coefficients of variation, skewness and kurtosis, etc.). It provides summary information, by subindex, on the frequency of occurrence of various descriptor categories and critical pollutants. Finally, it uses the line printer to generate a histogram showing the number and percentage of PSI values falling into successive intervals of 5 PSI units in width, along with the cumulative frequency ("CUM.FREQ.") of index values (Figure 15).

PSI Nomogram

The computational procedures described above, which can be implemented on a digital computer or a programmable hand calculator, give precise answers quite rapidly. However, procedures which do not require manual computation, such as graphs or tables, may be more convenient for some situations. In routine field activities, where computational facilities may be unavailable, a simple graphical approach for calculating PSI is desirable. To meet this need, Wallace and Ott[43] developed a PSI nomogram. The nomogram consists of two parts, one for pollutant variables which span the full PSI range from 0 to 500 (Figure 16), and another for pollutant variables which range from above 200 to 500 (Figure 17). The second part includes scales for COH and the product COH x SO_2. One nomogram contains pollutant concentrations expressed in gravimetric units (Figures 16 and 17), and a second nomogram, also in two parts, contains pollutant concentrations expressed in volumetric units (Figures 18 and 19).

AIR POLLUTION INDICES 163

```
         *  *  *   DISTRIBUTION OF DATA (INTERVAL WIDTH = 5.0000)   *  *  *
NO. OF VALUES LESS THAN   0.0  =    0    NO. OF ZERO VALUES =   0
 INTERVAL       NUMBER    PERCENT  +-----------------------HISTOGRAM----------------------+   CUM. FREQ.
  0.0 -  5.00      0       0.0  %  I                                                      I    0.0   %
  5.00- 10.00      0       0.0  %  I                                                      I    0.0   %
 10.00- 15.00      2       0.567% I*                                                      I    0.567%
 15.00- 20.00      5       1.416% I+                                                      I    1.983%
 20.00- 25.00      4       1.133% I*+                                                     I    3.116%
 25.00- 30.00     13       3.683% I****+                                                  I    6.799%
 30.00- 35.00     15       4.249% I***** +                                                I   11.048%
 35.00- 40.00     24       6.799% I******* +                                              I   17.847%
 40.00- 45.00     28       7.932% I********   +                                           I   25.779%
 45.00- 50.00     25       7.082% I*******       +                                        I   32.861%
 50.00- 55.00     35       9.915% I**********      +                                      I   42.776%
 55.00- 60.00     36      10.198% I***********        +                                   I   52.974%
 60.00- 65.00     32       9.065% I*********             +                                I   62.040%
 65.00- 70.00     22       6.232% I******                    +                            I   68.272%
 70.00- 75.00     14       3.966% I****                         +                         I   72.238%
 75.00- 80.00     15       4.249% I****                            +                      I   76.487%
 80.00- 85.00      7       1.983% I**                                +                    I   78.470%
 85.00- 90.00      9       2.550% I***                                 +                  I   81.020%
 90.00- 95.00      6       1.700% I**                                    +                I   82.719%
 95.00-100.00      5       1.416% I*                                      +               I   84.136%
100.00-105.00      8       2.266% I**                                        +            I   86.402%
105.00-110.00      5       1.416% I*                                          +           I   87.819%
110.00-115.00      4       1.133% I*                                            +         I   88.952%
115.00-120.00      6       1.700% I**                                             +       I   90.651%
120.00-125.00      1       0.283% I                                                +      I   90.935%
125.00-130.00      5       1.416% I*                                                +     I   92.351%
130.00-135.00      2       0.567% I*                                                 +    I   92.918%
135.00-140.00      2       0.567% I*                                                  +   I   93.484%
140.00-145.00      4       1.133% I*                                                   +  I   94.617%
145.00-150.00      3       0.850% I*                                                    + I   95.467%
150.00-155.00      2       0.567% I*                                                    +I   96.034%
155.00-160.00      2       0.567% I*                                                     +I  96.600%
160.00-165.00      3       0.850% I*                                                     +I  97.450%
165.00-170.00      1       0.283% I                                                      +I  97.733%
170.00-175.00      1       0.283% I                                                      +I  98.017%
175.00-180.00      0       0.0  %  I                                                     +I  98.017%
180.00-185.00      2       0.567% I*                                                     +I  98.583%
185.00-190.00      1       0.283% I                                                      +I  98.867%
190.00-195.00      0       0.0  %  I                                                     +I  98.867%
195.00-200.00      0       0.0  %  I                                                     +I  98.867%
200.00-205.00      2       0.567% I*                                                     +I  99.433%
205.00-210.00      1       0.283% I                                                      +I  99.716%
210.00-215.00      1       0.283% I                                                     +I  100.000%
215.00-220.00      0       0.0  %  I                                                    +I  100.000%
220.00-225.00      0       0.0  %  I                                                    +I  100.000%
225.00-230.00      0       0.0  %  I                                                    +I  100.000%
230.00-235.00      0       0.0  %  I                                                    +I  100.000%
235.00-240.00      0       0.0  %  I                                                    +I  100.000%
240.00-245.00      0       0.0  %  I                                                    +I  100.000%
245.00-250.00      0       0.0  %  I                                                    +I  100.000%
                                   +-----------------------------------------------------+
```

Figure 15. INDEX.PLOT computer output showing histogram of daily PSI values for the year; based on data from Newark, NJ, 1974.[37]

To use the nomogram, one locates the appropriate point on the right side of the scale for each pollutant variable and reads the PSI subindex value from the left side of the scale. All PSI scales on the nomogram are linear; however, the scale is compressed between PSI values of 300 and 500. PSI will be reported as the highest subindex value, and any other subindex values over 100 will usually accompany the report.

When used routinely in field operations, the authors recommend that copies of the nomogram be carried on a clipboard, with a different copy marked

164 ENVIRONMENTAL INDICES

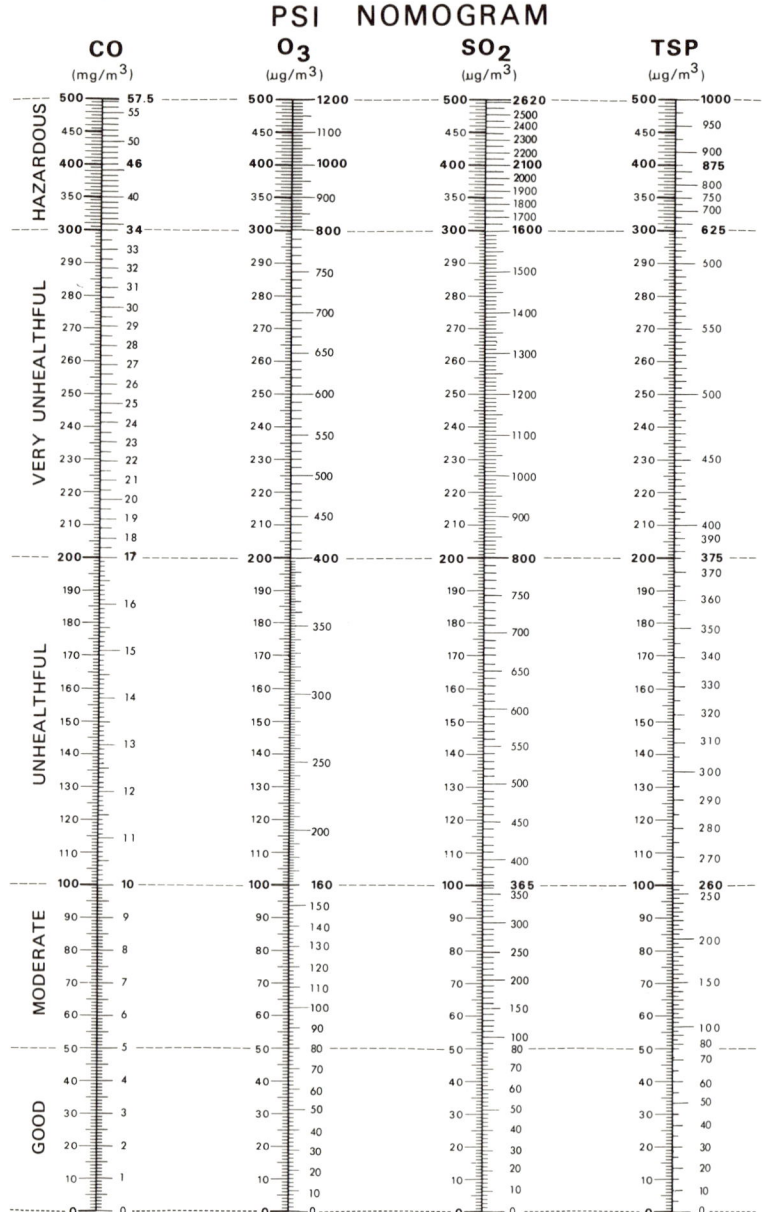

Figure 16. PSI nomogram for full-scale pollutants (CO, O_3, SO_2 and TSP), expressed in gravimetric units.[43]

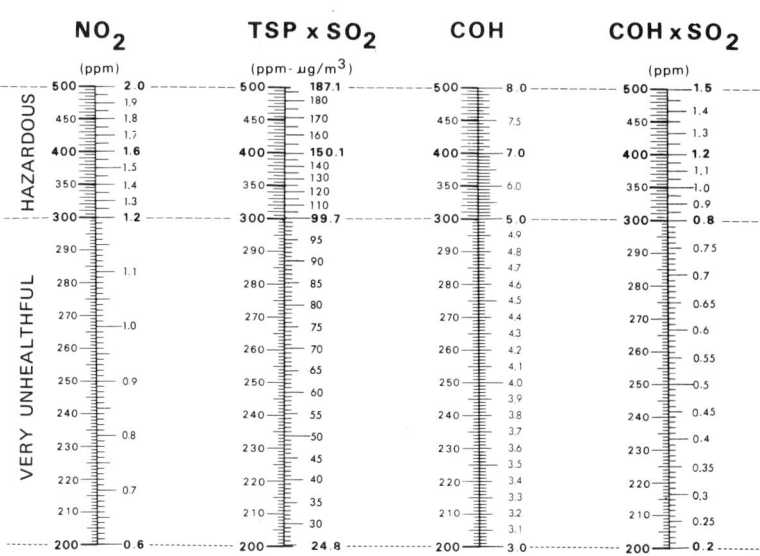

Figure 17. PSI nomogram for partial-scale pollutants (NO_2, TSP x SO_2, COH, and COH x SO_2), expressed in gravimetric units.[43]

in ink for each PSI reading. These individually marked copies can be stored and dated, giving a permanent record of each PSI report.[43]

Calculation by Tables

Another possible approach for calculating PSI is to look up the PSI values in specially developed tables. Wallace and Ott[43] have computed and published tables designed to facilitate PSI calculations. The appendix contains tables for calculating PSI in increments of 5 units (Tables A-1 and A-2). Linear interpolation is used to compute PSI in between the values listed in the table.

Wallace and Ott[43] also have proposed a tabular scheme for computing PSI values to the nearest whole unit without the need for linear interpolation. In

166 ENVIRONMENTAL INDICES

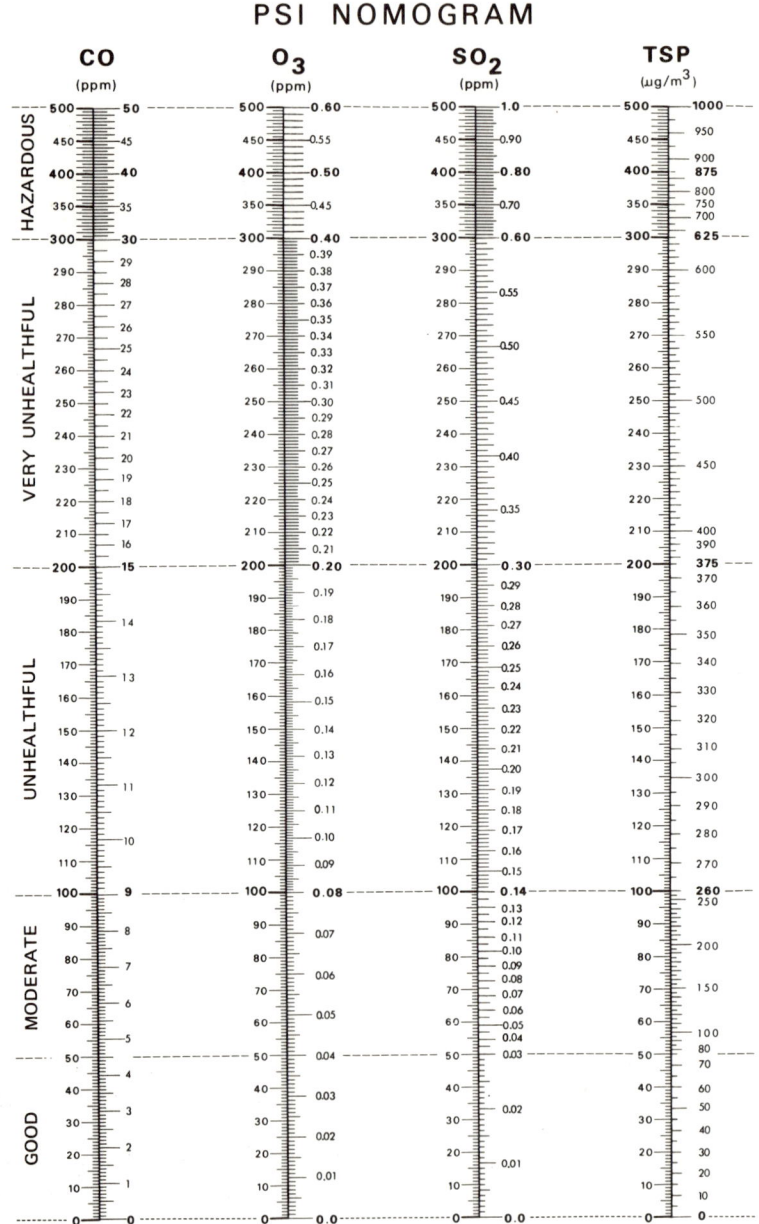

Figure 18. PSI nomogram for full-scale pollutants (CO, O_3, SO_2 and TSP), expressed in volumetric units.[43]

AIR POLLUTION INDICES 167

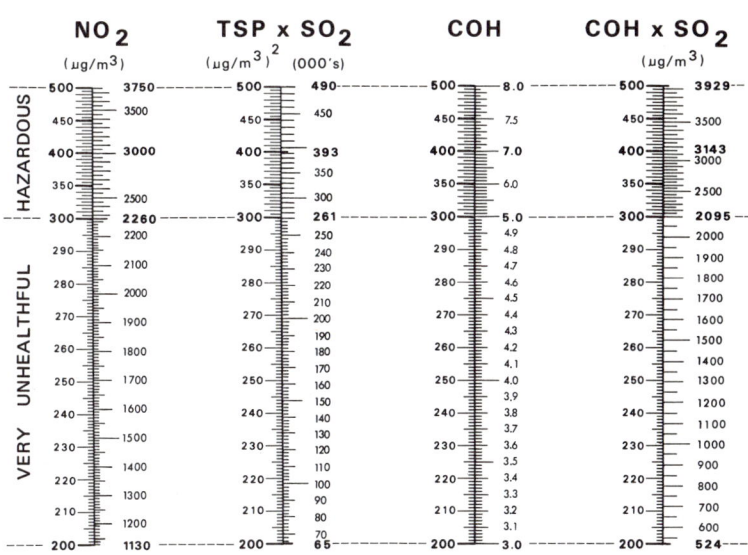

Figure 19. PSI nomogram for partial-scale pollutants (NO_2, TSP x SO_2, COH and COH x SO_2), expressed in volumetric units.[43]

this approach, computer-generated tables have been developed for PSI increments of 1 unit, beginning at 0.5. Table XXVIII shows a portion of the CO columns from one of these tables, and the first column lists PSI values that have been rounded off to the nearest whole unit. To use the table, one runs down the pollutant column until a pair of values are found immediately above and immediately below the observed concentration; then the correct value of PSI is read from the first column. For example, if the observed CO concentration is 3.1 ppm (8-hour average), it will lie between 3.05 mg/m^3 and 3.15 mg/m^3, with corresponding PSI values of 30.5 and 31.5. Then the proper value of PSI is read from the first column as 31. Thus, if an observed CO concentration, X, lies in the range 3.05 ⩽ X < 3.15, the correct PSI CO subindex will be 31. These tables, although exact, are lengthy and therefore are not included in this book. They are recommended only when very precise calculations of PSI are required in tabular form. They are included in the report by Wallace and Ott[43] that is available from the federal government. For most routine applications, the tables included in the appendix of this book should be satisfactory.

Table XXVIII. Example from Tables Designed to
Calculate PSI Without Linear Interpolation[43]

PSI (Nearest Unit)	PSI (Exact)	CO (mg/m^3)
31	30.5	3.05
32	31.5	3.15
33	32.5	3.25
34	33.5	3.35
35	34.5	3.45

Reporting PSI

Considerable flexibility is available in the way that PSI is to be reported to the public. The number of times per day that the index is reported, along with the time of the index report, is up to the individual air pollution control agency. One possibility is to report the index three times per day; for example, at 8:00 AM, noon, and 5:00 PM. During air pollution episode conditions, the agency may wish to report PSI on an hourly basis to give the public maximum information on prevailing conditions.

Flexibility also is afforded in the level of detail and content of the descriptive information that accompanies the index reports. The EPA guidelines document by Hunt et al.[41] illustrates several ways that an ozone concentration of 280 $\mu g/m^3$ (0.14 ppm) could be reported by radio, television and the news media:

1. Today, the air pollution index is 150. The air is "unhealthful." The pollutant responsible is oxidants.

2. An air pollution alert has (or has not) been called based on the forecast for the remainder of the day (and/or) tomorrow.

3. Repeat the above and add the following cautionary statement: "Persons with existing heart or respiratory ailments should reduce physical exertion and outdoor activity."

4. The report could include everything said in (1), (2), and (3) and then add that "unhealthful" air can cause mild aggravation of symptoms in susceptible persons, with irritation of symptoms in the healthy population.

5. Finally, the report could conclude with the forecast of tomorrow's air pollution level, such as "no change in the air pollution level is expected."

Additional language may be included to describe the health effects of ozone in greater detail.

AIR POLLUTION INDICES 169

Announcers on radio or television newscasts generally do not have sufficient time available to give detailed reports about the implications of various PSI descriptor categories. However, brief reports can be presented very successfully. Local TV newscasts usually include a weather report, accompanied by a forecast of meteorological conditions. Such broadcasts offer an excellent opportunity for reporting the index. A typical report might consist of a visual scale or graph, accompanied by one of the "cautionary statements" listed in Table XXIV. The EPA guidelines document[41] gives an example of a possible TV report (Figure 20). In one original version of Figure 20, the shaded segments were represented by successive colors of the spectrum: "good" (blue); "moderate" (green); "unhealthful" (yellow); "very unhealthful" (orange); "hazardous" (red). The colors appear well suited to TV images and are designed to give a subjective impression of a gradual worsening of the air pollution problem with each descriptor category. Such a color-coded system, if adopted by air pollution control agencies across the nation, would afford a simple, uniform way to graphically represent air quality data.

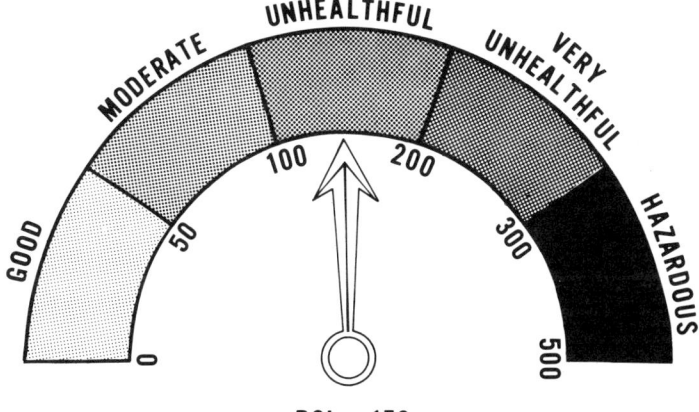

PSI = 150
POLLUTANT: Oxidants
TODAY'S HEALTH IMPLICATIONS: Respiratory ailment and heart disease patients should reduce exertion and outdoor activity.
FORECAST: No change.

Figure 20. Example of a possible PSI report for television.[41]

Recorded telephone messages can be designed to provide extremely current information to the public but, like TV, they do not offer an opportunity for presenting extensive detail. On radio, greater detail can be presented, and the announcer may wish to discuss the nature of the effects associated with each air pollutant (Chapter I, pages 16-18). In newspapers, where even more lengthy descriptive information can be presented, detailed descriptions of the characteristics of each pollutant can be included. A daily newspaper report might include a table containing all the PSI subindex values, along with subindex values for several previous days, and the next day's PSI forecast. Background information on each air pollutant could be listed below the table. Occasional articles can be included to discuss index trends, seasonal variations, and the local sources of each air pollutant, along with scientific information about its effect on health and welfare. As discussed in Chapter I, such information is available from the air quality criteria reports published by the federal government for each air pollutant. EPA also has prepared a brochure[44] that discusses the health effects associated with the pollutants in PSI. The brochure is written for the layman and is available on request from the Agency's Office of Public Awareness.

Although scientific approaches for forecasting air pollution indices have not yet been developed in a rigorous fashion, some guidance is available. McAdie and Gillies[45] report success in forecasting an air pollution index on an operational basis in Sarnia, Ontario. The *Compendium* of air pollution indices by Thom and Ott[8] reveals that a great many air pollution agencies currently are forecasting their indices. The PSI guidelines document by Hunt et al.[41] contains an appendix describing possible approaches for forecasting PSI. It notes that "the types of meteorological information that could be used for forecasting PSI have been rather well defined through past experience with forecasting methods developed in support of air pollution control activities." This meteorological information includes the following:[41]

- character and movement of air masses and fronts
- areas of air mass subsidence
- incidence, intensity and height of inversions
- mixing layer height
- prevailing wind direction
- mean wind speed (surface and mixing layer)
- ventilation (mixing layer mean wind speed x mixing height)
- precipitation
- total sky cover

Much of this information is available through services provided by the National Weather Service, and supplementary meteorological data usually are collected by local air pollution control agencies. The guidelines notes, "It is advisable for agencies planning to use the index along with a forecast procedure to have personnel on their staffs familiar with meteorological data and how these data may be applied in development of index prediction methodology."[41] With such information, qualitative forecasts of PSI can be made for periods up to a day in advance in terms of "No significant change," "Increase," or "Decrease."

Performance of PSI

During the developmental phases of PSI, the computer program INDEX.PLOT[37] was run on air quality data from eight U.S. cities.[38] This analysis examined the time series of index values, the statistical properties of the index, and the frequency distributions of descriptor categories and critical pollutants. As reported by Ott and Hunt,[42] the index was found to be a suitable tool for compactly representing a year of air quality data at a given air monitoring station. It also showed sufficient day-to-day variation to provide useful information to the public on daily air quality levels. This analysis illustrates how PSI can be used to report air quality data in a variety of ways.

Using data from the air monitoring station in downtown Los Angeles, California, the index was calculated once per day for the entire year of 1974 (Figure 21). The calculation used the appropriate averaging time for each pollutant and was based on the maximum running average concentration for each 24-hour period. Each vertical line represents the index value for a day of the year, and the letter at the top of each line denotes the critical pollutant for that day. This time series plot shows that violations of the NAAQS (values of PSI > 100) occur in Los Angeles throughout the year, with the highest PSI values occurring in December. The predominant pollutants are carbon monoxide and oxidants, with carbon monoxide dominating the index during the period from mid-October to March and oxidants dominating during the remaining months.[38]

Statistical Properties

INDEX.PLOT was run on 365 days of 1974 air quality data for each of eight cities covered in the performance evaluation. Although these data sets were reasonably complete, there were many instances in three of the cities—Chicago, Houston and Seattle—in which one or more pollutants were missing on a given day. Therefore, the findings were more meaningful for the five cities which had reasonably complete data—downtown Los Angeles, CA; Lennox, CA; Anaheim, CA; Camden, NJ; and Newark, NJ.[42]

Because the index is a function of pollutants which behave as random variables, PSI is itself a random variable. The difference between the highest and lowest PSI values (the range) did not vary greatly from city to city;

172 ENVIRONMENTAL INDICES

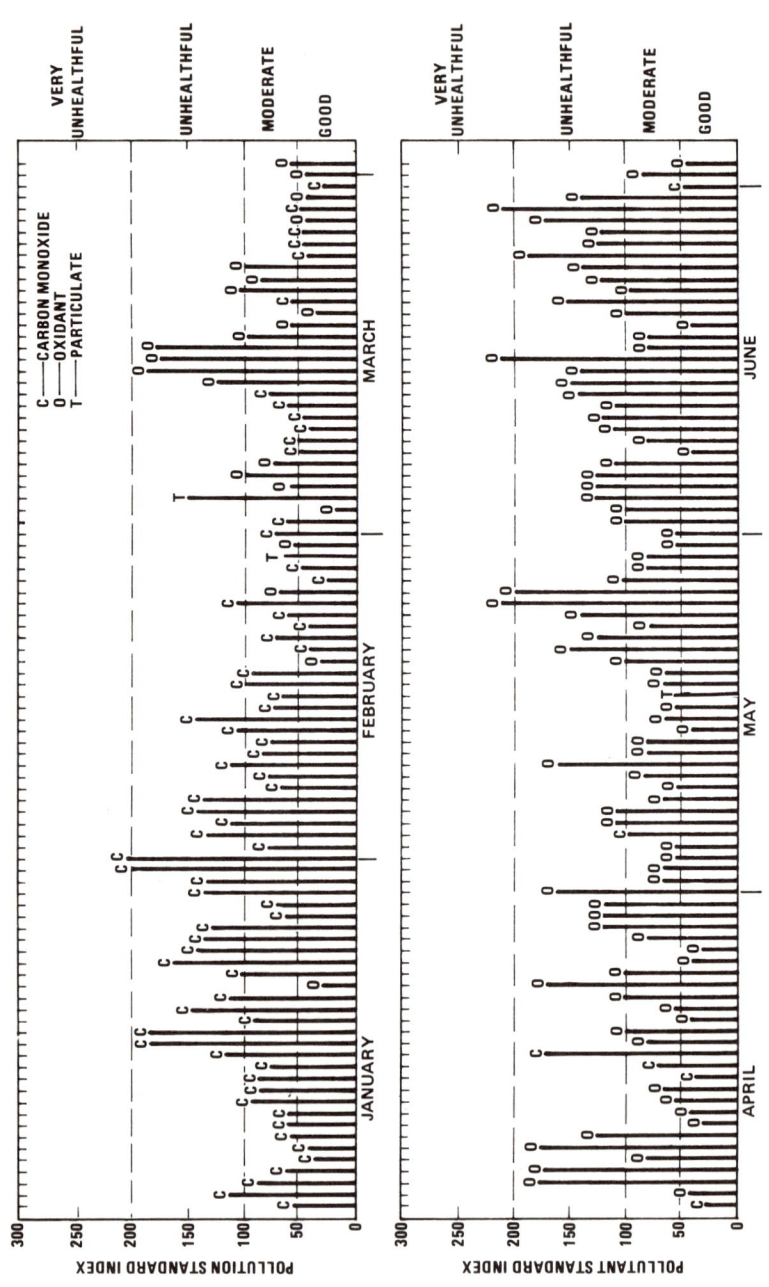

AIR POLLUTION INDICES 173

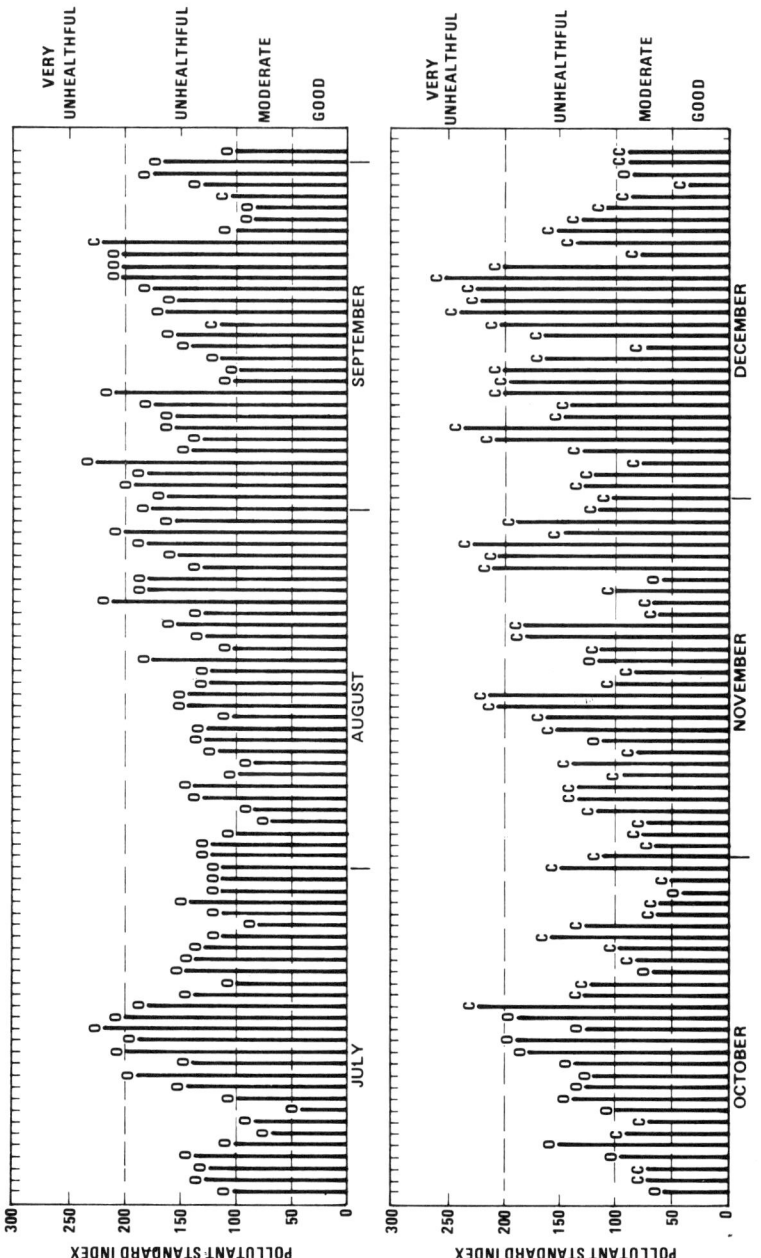

Figure 21. Time series plot of PSI in downtown Los Angeles, based on 1974 data.[38]

the smallest range was 152.2 in Seattle, and the largest range was 286.7 in Lennox (Table XXIX). For these eight data sets, the annual arithmetic mean of PSI ranged from 44.0 in Seattle to 119.2 in downtown Los Angeles, with standard deviations ranging from 21.3 in Seattle to 71.2 in Lennox. The coefficient of variation (ratio of the standard deviation to the mean) remained somewhat more constant, ranging from 0.421 in downtown Los Angeles to 0.677 in Houston. This result suggests that the dispersion or "spread" of the PSI frequency distribution, as measured by its standard deviation, may be, to some degree, a function of its mean. The calculated coefficient of skewness $\sqrt{b_1}$ was always positive, ranging from 0.02 in Chicago to 2.18 in Camden. The coefficient of skewness is a measure of the lack of symmetry of a distribution. The small $\sqrt{b_1}$ in Chicago suggests a nearly symmetrical distribution of observed index values, while the distributions for the other cities have a significant right "tail"—that is, they are skewed to the right. The coefficient of kurtosis, which gives a measure of the "peakedness" of the shape of the distribution (relationship of height to width), ranges from 2.32 in downtown Los Angeles to 12.75 in Camden, while Chicago has a value of 3.08. For the normal distribution, the theoretical value for $\sqrt{b_1}$ and b_2 are 0.0 and 3.0, respectively, suggesting that index values for Chicago tend more toward a normal distribution than those for the other cities.[42]

Table XXIX. Statistical Characteristics of PSI When Applied to 8 U. S. Cities[42]

City	n	$\bar{x}$	s	cov	$\sqrt{b_1}$	b_2	Lowest	Highest	Range
Los Angeles, CA	364	119.2	50.1	0.421	0.37	2.32	25.0	257.1	232.1
Lennox, CA	365	112.8	71.2	0.631	0.86	2.43	20.0	306.7	286.7
Anaheim, CA	365	81.5	43.0	0.528	0.97	3.73	12.5	255.0	242.5
Chicago, IL	364	98.5	42.0	0.426	0.02	3.08	6.3	224.1	217.8
Newark, NJ	353	65.3[a]	37.6	0.575	1.34	5.29	0.0[a]	212.4	212.4
Camden, NJ	356	50.7[a]	28.2	0.556	2.18	12.75	0.0[a]	232.4	232.4
Houston, TX	363	59.3[a]	40.2	0.677	0.83	3.75	0.0[a]	208.2	208.2
Seattle, WA	357	44.0[a]	21.3	0.485	1.32	6.68	0.0[a]	152.2	152.2

Explanation of Terms

n = number of days for which data were available
$\bar{x}$ = arithmetic mean
s = arithmetic standard deviation
cov = coefficient of variation, $s/\bar{x}$
$\sqrt{b_1}$ = coefficient of skewness
b_2 = coefficient of kurtosis

[a]In these computer runs, the index was set to zero if data for all five air pollutants were missing; thus, the lowest value and mean are underestimated.

AIR POLLUTION INDICES 175

The cumulative frequencies for the index data can be plotted readily on logarithmic probability paper (Figure 22). On this paper, lack of curvature, or "straightness" of the lines, usually is interpreted to mean that the distribution tends toward lognormality. These lines illustrate a problem described earlier as the "ranking paradox" (Chapter I, page 20). Notice that the lines for two of the cities, downtown Los Angeles and Lennox, intersect. If the cities are ranked from most severe to least severe using either the medians (50 percentiles) or arithmetic means, the highest index values would be in downtown Los Angeles, followed by Lennox, Anaheim, Newark and Camden. If, however, rankings are based on the highest index values observed (for example, values greater than PSI = 200), Lennox and downtown Los Angeles are reversed. These two orderings again illustrate two separate, important characteristics which must be considered when comparing environmental measurements: (1) frequency of adverse conditions and (2) degree of severity of adverse conditions. This ranking paradox is not unique to PSI but will arise for most other index structures as well.

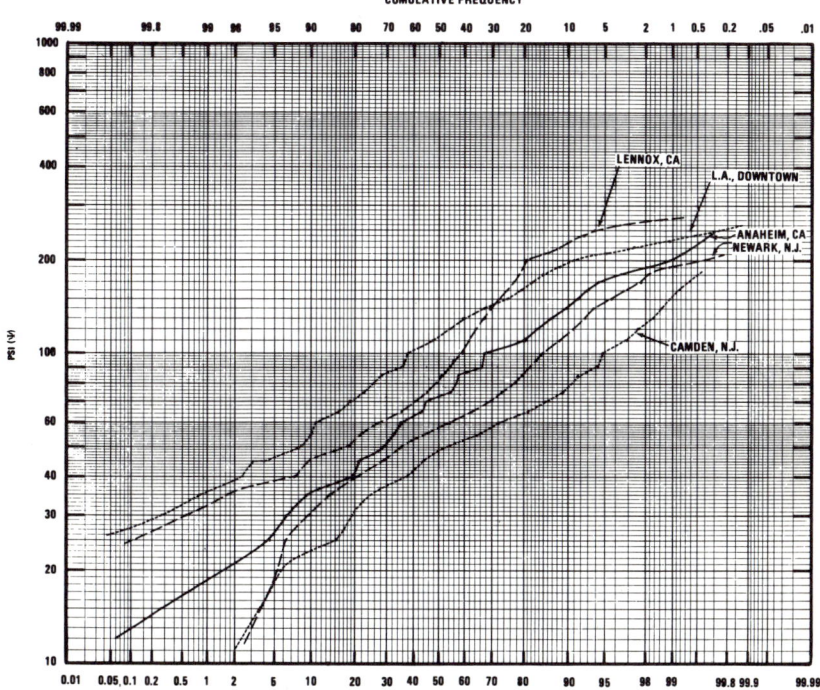

Figure 22. Logarithmic probability plots of PSI for five U.S. cities.[42]

Comparison of Cities

The ranking paradox, or frequency/severity issue, is not the only problem that must be considered when comparing air quality in different cities. Other problems arise from nonuniformity in quality control practices at particular air monitoring stations and from nonstandardization of station sites. As discussed by Ott,[46] variation in monitoring station sites tends to bias measured concentrations, particularly for carbon monoxide, making it difficult to generalize about the representativeness of a given station or to compare cities readily using data from these stations.

The PSI guidelines document by Hunt et al.[41] stresses that rankings of the air quality of different cities should include additional factors besides the index:

> PSI is designed for the daily reporting of air quality to advise the public of potentially acute, but not chronic health effects. To properly rank the air pollution problems in different cities, one should rely not just on air quality data, but should include all data on population characteristics, daily population mobility, transportation patterns, industrial composition, emission inventories, meteorological factors, and the spatial representativeness of air monitoring sites. A correct ranking should also consider the number of people actually exposed to various concentrations, as well as the frequency and duration of their exposure.

Although a really meaningful ranking should include these factors, some of this information currently is unavailable. Thus, while keeping in mind the limitations of the data, it is of interest to apply PSI to different cities and examine the results. Presumably, as quality control practices improve and station sites become more standardized, more meaningful comparisons will ultimately be possible.

In the computer evaluation of PSI, Ott and Hunt[42] plotted the distributions of descriptor words for all five cities on one figure (Figure 23). These distributions differed strikingly. In Anaheim (top, left), the 365 days of the year were divided almost equally among three descriptor words: "good," "moderate," and "unhealthful," with PSI above 200 occurring only 1% of the time. In Lennox, PSI was above 200 more often, causing 20% of the days to be designated "unhealthful." Downtown Los Angeles gave a pattern that was different from either Anaheim or Lennox: the majority of the descriptor words (53%) fell into a single category, "very unhealthful." Of the five cities, Camden had the lowest concentrations, with 52% of the days reported as "good," and only 5% above the NAAQS (PSI > 100). In Newark, the majority of the PSI values were in the "good" or "moderate" descriptor categories, and just 16% fell into the range above the NAAQS. In Los Angeles, by contrast, 62% were in the range above the NAAQS. For all

AIR POLLUTION INDICES 177

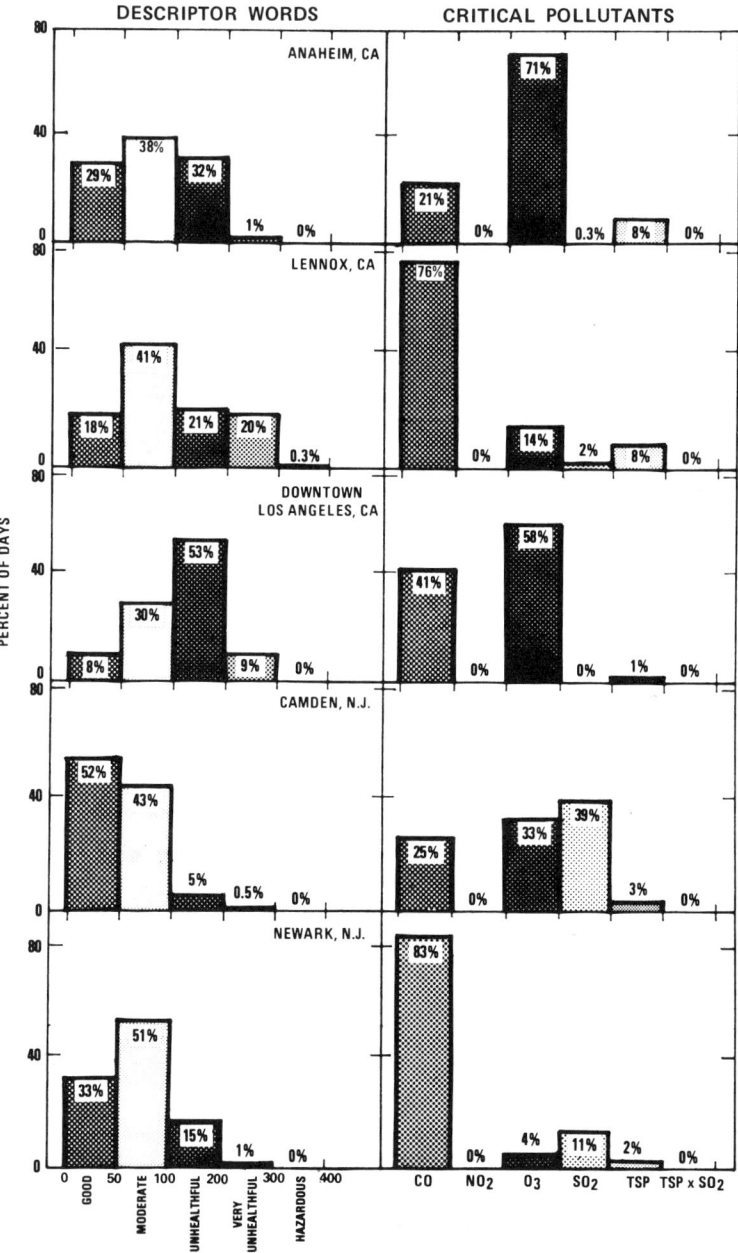

Figure 23. Distributions of PSI descriptor words and critical pollutants in five U.S. cities; based on 1974 data.[42]

five cities, PSI values above 300 were very rare: there was only one instance, in Lennox, of a "hazardous" descriptor word.

The distributions of critical pollutants also showed considerable variation from city to city, as shown by the histograms on the right side of Figure 23. In the California cities, the most common critical pollutants were CO and oxidants, both of which were attributable to the motor vehicle. Taken together, they accounted for 89% of the critical pollutant days in Lennox, 92% of the days in Anaheim, and 99% of the days in downtown Los Angeles. Although CO and oxidants were the most frequently occurring critical pollutants, the proportion of CO days relative to oxidant days varied from city to city. In Anaheim, photochemical oxidants was the critical pollutant for 71% of the days. In Lennox, the situation was reversed: CO showed up as the critical pollutant for 75% of the days, and oxidants showed up only 14% of the time. In the three California cities, SO_2 accounted for a very small proportion of the days—never more than 2%—while TSP varied from 1% in downtown Los Angeles to 8% in Lennox and Anaheim.[42]

In the two New Jersey cities, Newark and Camden, SO_2 showed up as the critical pollutant more frequently than in the California cities. In Camden, it was the most commonly occurring critical pollutant, accounting for 39% of the days, while in Newark it was the second most common critical pollutant, accounting for 11% of the days. Although SO_2 showed up more frequently in the two New Jersey cities, carbon monoxide and oxidants together still accounted for more than half of the days—58% in Camden and 87% in Newark. In the two eastern cities, TSP accounted for a very small share of the total critical pollutant days—3% in Camden and 2% in Newark. The low incidence of TSP days in all cities is due, in part, to the smaller amount of TSP data available from the high-volume sampler, which usually gives a reading once every few days. The product TSP x SO_2 never appeared as the critical pollutant, nor did NO_2.[42]

The article by Ott and Hunt[42] summarizes the results of applying PSI to data from five U.S. cities as follows:

- The dominant pollutants were CO and oxidants, together accounting for more than 89% of the days in the California cities and more than 50% of the days in the New Jersey cities.
- Sometimes CO occurred more frequently than oxidants; sometimes the reverse occurred.
- In addition to CO and oxidants, SO_2 was a common critical pollutant in the New Jersey cities, alone accounting for 39% of the days in Camden and 11% of the days in Newark.
- NO_2 never appeared as the critical pollutant.
- TSP x SO_2 never appeared as the critical pollutant.

AIR POLLUTION INDICES 179

- The most severe descriptor word, "hazardous," occurred only once.
- The ranking of cities is a complex task, not easily accomplished with this or with other indices.

The authors conclude that PSI has a variety of possible uses:

> Our evaluation of the PSI air pollution index recommended by EPA suggests that the index reflects day-to-day variations of air pollutant concentrations in a reasonable fashion. The index appears to be a useful tool for presenting and interpreting air quality data. It also can be used in trend studies to display annual air quality levels at monitoring stations in U.S. cities. Its mathematical properties and the simplicity of the data it generates, as revealed by our computer programs, make the index an effective means for compactly representing a year of air quality data for any air monitoring station or city. Although these displays permit different cities to be compared in a very approximate fashion, such comparisons are complex and must be carried out with great caution due to variations in air monitoring practices.[42]

As reported in the *Compendium*[8] of air pollution indices, air pollution control agencies in the Steubenville-Pittsburgh-Wheeling (tri-state) area reported air quality data to the public using three different approaches. Prior to adopting PSI, the Allegheny County Air Pollution Control Advisory Committee wanted to discover how PSI would perform in comparison with the indices then in use. At the Committee's request, Rubin et al.[47] carried out an evaluation similar to the one performed by Ott and Hunt,[42] except that PSI was compared with the indices used in the area. Because the tri-state air monitoring network measures COH routinely, they used an arbitrary conversion factor of 1 COH unit = 150 $\mu g/m^3$ in order to compute the PSI subindex for TSP.

For seven air monitoring stations, they observed arithmetic mean values of PSI ranging from 27.8 to 63.9, with standard deviations ranging from 14.7 to 39.6. Although the means and standard deviations were generally lower than those obtained for the cities studied by Ott and Hunt,[42] the coefficients of variation were similar, ranging from 0.451 to 0.703. Rubin et al.[47] plotted PSI values from three air monitoring stations in the tri-state metropolitan area for a month of data (Figure 24). Each station was in a different state but was within the metropolitan area. They felt that the results showed that PSI, if adopted by all three agencies, would give the public a greater understanding of air pollution levels in the area:

> .... The marked similarity of index values and trends during the one month studied reflects similar levels of air quality throughout the region. Adoption of the PSI would thus give tri-State area residents a much clearer and consistent picture of air quality. This, in contrast to the present situation where television viewers receiving both Ohio and Pennsylvania

180 ENVIRONMENTAL INDICES

broadcasts hear the same index value called "excellent" in Ohio and "unsatisfactory" in Allegheny County.[47]

After considering the analysis performed by Rubin et al.,[47] the Allegheny County Air Pollution Control Advisory Committee in January 1977 voted unanimously to have Allegheny County adopt PSI in lieu of its present system effective July 1, 1977. PSI is now being used routinely in the tri-state area.

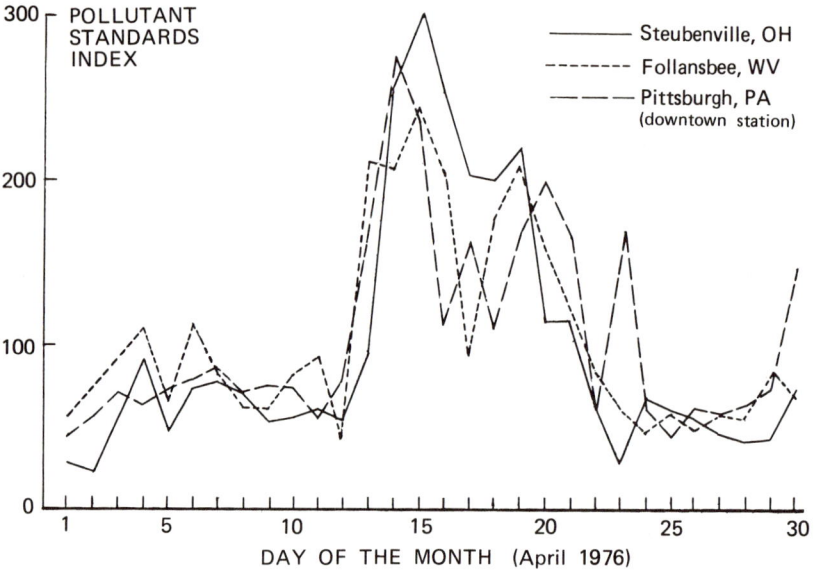

Figure 24. Comparison of PSI values at three air monitoring stations in the tri-state metropolitan area, from Rubin et al.[47]

PSI Bars

Although the results from the computer evaluation by Ott and Hunt[42] gave interesting statistics and visual presentations, these findings may not completely satisfy air pollution specialists who are interested in the behavior of the individual pollutants. Can a single diagram be devised which shows all of the PSI subindices for a year of data on a simple format? What is needed is a chart that provides sufficient detail to the expert while being understandable to the layman.

To solve this problem, INDEX.PLOT was modified[37] to print out the frequency of occurrence of descriptor categories by subindex. These results, in turn, were plotted on bar charts, which, for convenience, we shall call "PSI bars." PSI bars were generated for this book using the same data employed in the Ott and Hunt[42] evaluation, giving an interesting visual display of air pollution levels in the five cities (Figure 25). Each horizontal bar denotes a different pollutant variable, and the shaded area depicts the proportion of the year that the subindex lies within each descriptor category. The values for n listed at the right of the diagram denote the number of observations (number of days) on which the PSI bar was based.

The ranking paradox is easily seen in these diagrams. The area of the "unhealthful" region (second level of shading) in Los Angeles exceeds that in Lennox. By contrast, the area of the "very unhealthful" region (third level of shading) in Lennox exceeds that in Los Angeles. The higher severity of levels in Lennox can be attributed mainly to CO, while the higher frequency of "unhealthful" levels in Los Angeles can be attributed to CO and ozone.

PSI bars are designed to provide the specialist with the information required to compare individual pollutants in each city in a meaningful fashion. In addition, the layman, by scanning the diagram with his eye, should be able to gain an impression of the severity of the air pollution problem in each city. Ideally, the observer's mind performs the aggregation, evaluating the air pollution problem in each city according to the darkness of the shading and the area it occupies. By considering these diagrams, the observer should be able to "compare" air pollution problems in different cities. For each city in Figure 25, the bottom bar shows the values calculated using PSI.

The PSI bar approach also can be applied to data sets with missing values, such as the data for the three cities originally considered in the Ott and Hunt[42] computer evaluation (Figure 26). Generally, the observer should pay attention to the values for n listed at the right side of each bar. Notice, for example, that the TSP PSI bar for Houston gives the impression that over half the year is in the "moderate" range. Yet it is based on only 25 TSP observations, which may or may not be representative of the year. The PSI bar for oxidants (O_3) is based on data for 327 days and therefore is the most representative of the five PSI bars for Houston. Oxidants tend to dominate PSI for this city, probably because of the missing values for CO, NO_2, SO_2, and TSP. Hopefully, most air quality reports will have fewer missing data than the examples in Figure 26.

The author believes that PSI bars may offer a powerful tool for presenting air quality data from different locations to both the specialist and the layman. If air pollution problems in different cities are to be compared, PSI bars appear to offer an effective graphical technique for doing so. If the

182 ENVIRONMENTAL INDICES

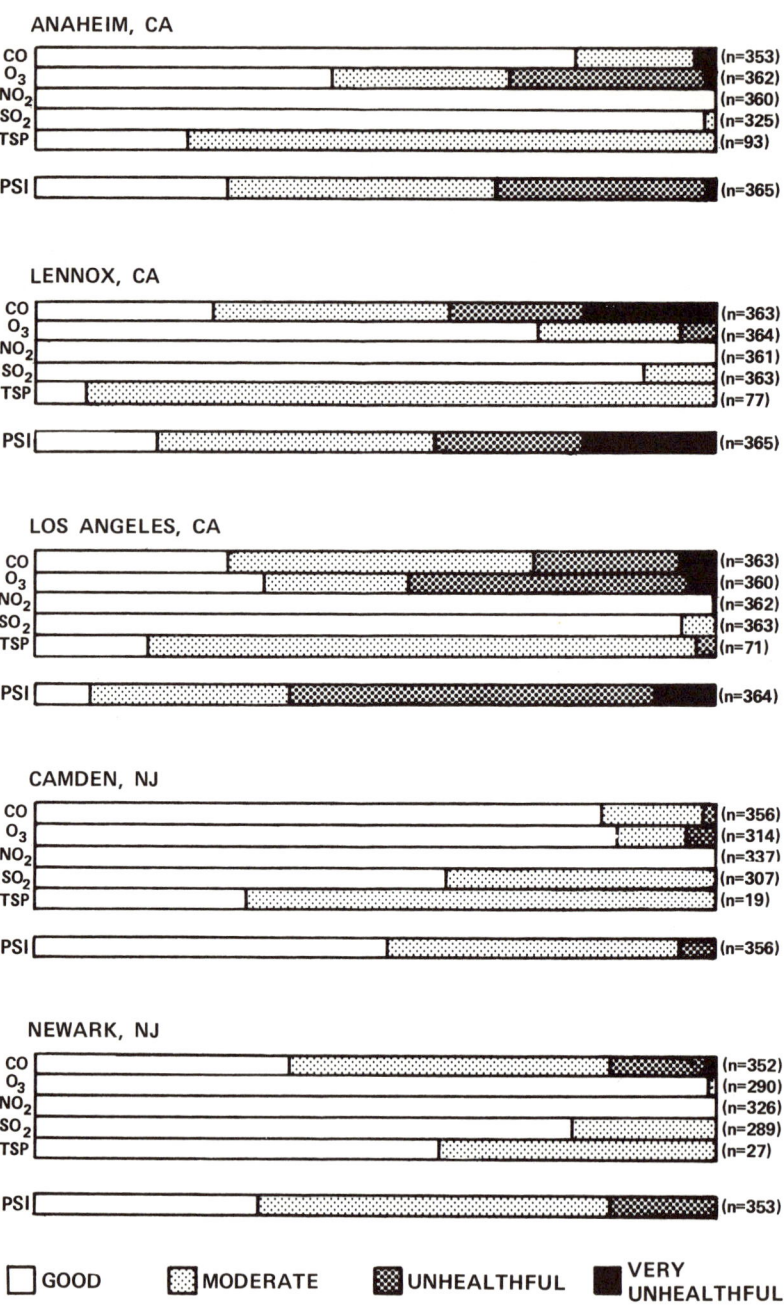

Figure 25. PSI bars calculated by the author for five U.S. cities; from 1974 data.

shaded areas of the PSI bars were color-coded using successive colors of the spectrum, as suggested in the discussion of Figure 20, they would probably become an even more effective data presentation tool. Federal, state and local agencies charged with routinely summarizing and comparing air quality data from different monitoring stations may wish to consider the use of PSI bars. This technique also should be useful to research investigators wishing a simple but accurate way for presenting monitoring data. Each set of PSI bars represents, in effect, a condensation of thousands of hourly measurements. Most of the information that is meaningful for decision-making, however, is retained. With suitable graphical experimentation, it should be possible to present a dozen or more cities on a single page using this

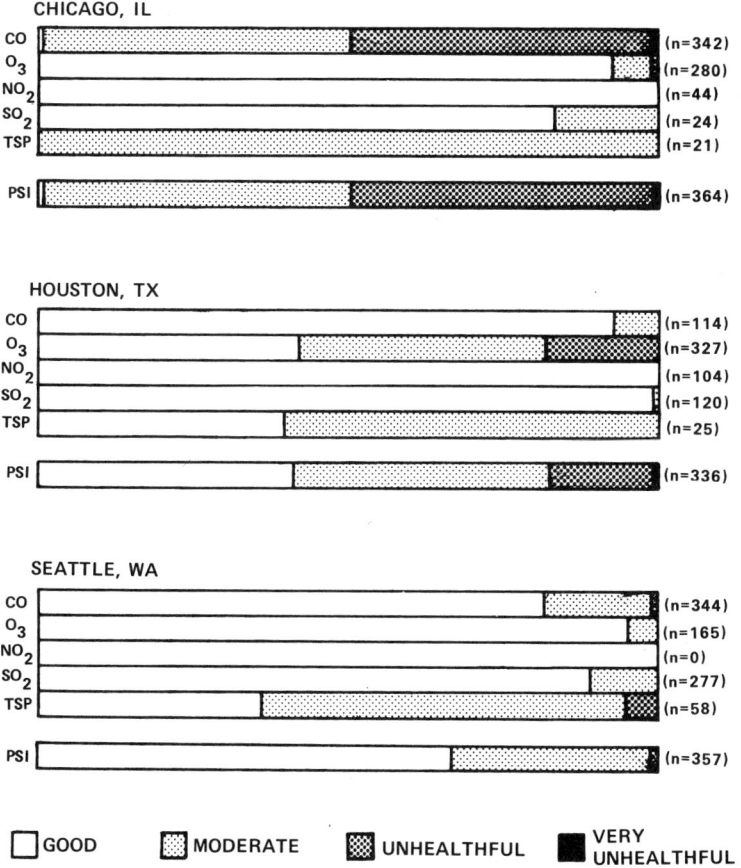

Figure 26. PSI bars calculated by the author for three U.S. cities; from 1974 data with many missing values.

Adoption of PSI

When published in September 1976, PSI was intended only as a suggested possible index structure, not as a mandatory approach for reporting air quality data, as stated in the *Federal Register*: "This guideline suggests the use of the Pollutant Standards Index (PSI) for those local and state air pollution control agencies wishing to report an air quality index on a daily basis."[39] Thus, the decision of whether or not to use PSI was left up to state and local air pollution control agencies.

Because of the many desirable features of PSI and the obvious advantages of national uniformity, most members of the Working Group that developed PSI felt that it would automatically be adopted on a gradual basis over the next five years. Because PSI embodied the most common characteristics of the indices then in use, the Working Group also felt that adoption would not present serious obstacles for most air pollution agencies. In general, adoption has proceeded at a rapid pace. However, new legislation passed by Congress has altered the original philosophy of voluntary adoption and may bring about more rapid implementation of PSI.

U.S. Adoption

In July 1977, 11 months after publication of PSI, Hunt[48,49] found that 9 state and 21 local air pollution agencies either had adopted PSI or were planning to adopt it (Table XXX). PSI was under active consideration by 7 additional state and 11 additional local air pollution control agencies.

Table XXX. U. S. Air Pollution Control Agencies Which Have Adopted, or Plan To Adopt, PSI, as of July 1977[48,49]

State Agencies	Local Agencies	
Alabama	Mobile County, AL	Jefferson County, KY
Colorado	Fresno County, CA	Northern Ohio Valley
Connecticut	Broward County, FL	Air Authority, OH
Kentucky	Dade County, FL	Allegheny County, PA
Massachusetts	Duval County, FL	Lucerne-Lackawanna
Mississippi	Hillsborough County, FL	County, PA
Oregon	Palm Beach County, FL	Knox County, TN
Pennsylvania	Sarasota County, FL	Memphis-Shelby County,
Virginia	Fulton County, GA	TN
	Cedar Rapids, IA	Houston, TX
	Wichita-Sedgwick	Wheeling, WV
	County, KS	Milwaukee, WI
		Racine, WI

In most instances, PSI has remained intact. However, some agencies are using the PSI concept but are reporting fewer than the original five pollutants. Others have made minor modifications to the descriptor words. Hunt concludes: "In general, however, we are much closer to having a uniform air quality index across the nation than ever before."[48]

In August 1977, Congress passed the Clean Air Act Amendments of 1977,[50] which amend the original 1970 law. Section 309 (which amends Section 319) contains language stating that EPA must issue regulations establishing a nationally uniform air quality monitoring system that incorporates a uniform index:

> Not later than one year after the date of enactment of the Clean Air Act Amendments of 1977 and after notice and opportunity for public hearing, the Administrator [of EPA] shall promulgate regulations establishing an air quality monitoring system throughout the United States which—(1) utilizes uniform air quality monitoring criteria and methodology and measures such air quality according to a uniform air quality index, (2) provides for air quality monitoring stations in major urban areas and other appropriate areas throughout the United States to provide monitoring such as will supplement (but not duplicate) air quality monitoring carried out by the States required under any applicable implementation plan, (3) provides for daily analysis and reporting of air quality based upon such uniform air quality index, and (4) provides for recordkeeping with respect to such monitoring data and for periodic analysis and reporting to the general public by the Administrator with respect to air quality based upon such data.[50]

Examination of Congressional testimony on this topic reveals that these legislative changes were made in response to findings of the 1976 *Compendium* of air pollution indices by Thom and Ott.[8] The legislation calls for nationwide implementation of a uniform air pollution index. In response to this legislation, EPA has drafted regulations that make the use of PSI mandatory in urban areas with populations greater than 250,000.

International Adoption

As PSI evolved, there was considerable communication between the U.S. and Canada. In 1975, the Canadian Federal-Provincial Committee on Air Pollution established a subcommittee to examine the need for an air quality index, and if the need were indicated, to recommend an air pollution index that could be used in Canadian cities. Although the Canadian index was developed independently, it was based on many of the concepts in the *Compendium*[8] and on the design criteria proposed by Ott and Thom.[29] Thus, it closely resembles PSI.

As described by Young et al.,[51] the proposed index for Canada is based on the National Ambient Air Quality Objectives which "... have been determined cooperatively by the federal and provincial governments to provide a uniform scale for assessing the quality of air in all parts of Canada." The Objectives are similar to the U.S. NAAQS and Alert levels. Like PSI, the Canadian index uses segmented linear subindex functions and reports only the maximum subindex ("maximum mode"). It includes five pollutant variables: CO, NO_2, oxidants (O_3), COH and SO_2. Its developers have examined the performance of the index using data from downtown air monitoring stations in Toronto, Hamilton and Windsor.[51]

Like Canada, other countries throughout the world have adopted ambient air quality standards.[52] However, air pollution indices have not yet proliferated on an international level to the same degree that they did in the U.S. Hopefully, an international proliferation of indices can be prevented. Various organizations, such as the United Nations and the World Health Organization,[53,54] currently are developing air quality criteria which eventually may lead to international air quality standards. In the future, further international activity in air pollution control and the dissemination of air pollution information may enhance the need for an international index. To help meet this need, and to prevent the proliferation of disparate indices which occurred in the U.S., Thom and Ott[55] have proposed the UNiform International Pollution IndEX (UNIPEX) system, a uniform international approach for designing indices that is based on the 10 criteria for a uniform index.[29]

Indices designed using the UNIPEX system must employ segmented linear subindex functions and aggregate the subindices by means of the maximum operator. Like PSI, each UNIPEX subindex uses two breakpoints that have a scientific basis: a short-term health standard (UNIPEX = 100) and a level reflecting significant harm to health (UNIPEX = 500). However, other countries do not necessarily use the same administrative limits embodied in the U.S. episode criteria (Alert, Warning and Emergency levels). Thus, a simpler approach than PSI is used between 100 and 500. In this range, the relationship is linear, and the scale is divided into four equal segments, giving breakpoints A, B and C. Breakpoints A, B and C, and the level reflecting significant harm to health (breakpoint D) are described by Thom and Ott[55] as follows:

 A: Air quality beginning to degrade; initial emission control actions taken; at this level, general health effects include the significant aggravation of symptoms.

 B: Air quality continuing to degrade; additional emission control actions taken; at this level, general health effects include the premature onset of certain diseases.

AIR POLLUTION INDICES 187

C: Air quality is continuing to degrade toward a level of significant harm to the health of persons; the most stringent emission control actions are taken to prevent concentrations from reaching the emergency episode level; at this level, the general health effects include the premature death of the ill and elderly.

D: Air pollution levels which could cause significant harm to the health of the general population.

The concentrations to be selected for the short-term health standard and breakpoint D are left to the individual country that is implementing UNIPEX. Because these two breakpoints are based on scientific facts, however, it is hoped that the concentrations selected would be identical to (or very close to) the U.S. NAAQS and Significant Harm level. However, demographic and health characteristics peculiar to each country may dictate that different numbers be selected in some cases. Because UNIPEX is linear between 100 and 500, the concentrations for breakpoints A, B and C are determined automatically once the other two breakpoint concentrations are selected. In this sense, UNIPEX may be an improvement over the PSI functions.

Structurally, the UNIPEX system may be viewed as a "linearized" version of PSI between 100 and 500. In Figure 27, the PSI subindex function for ozone is plotted alongside the UNIPEX subindex function. If PSI were modified to coincide with the UNIPEX concept, what changes in breakpoints would be necessary? The changes for all of the breakpoints affected can be calculated easily (Table XXXI). All changes are less than 30%; 14 are less than 20%, 9 are less than 10% and 5 are less than 5%.

UNIPEX has no official standing and is just a conceptual approach proposed in the scientific literature by Thom and Ott.[55] Because the administrative limits for Alert, Warning and Emergency have been widely adopted in the U.S., it would not seem desirable at this late date to change them to conform to the UNIPEX system. However, many other countries are more flexible, because they have not yet adopted episode criteria.

The Mexican government has designed a candidate air pollution index for Mexico City that is based on the UNIPEX system. For each subindex, the first health-related breakpoint (I = 100) is based on proposed Mexican ambient air quality standards (analogous to the U.S. NAAQS) and on scientific information on the health and economic effects of air pollution in Mexico. The second health-related breakpoint (I = 500) is essentially the same as the U.S. Significant Harm levels. The subindices are aggregated using the maximum operator. In 1978, the index was implemented routinely using data from the extensive air monitoring network operated in Mexico City.[56]

Thom and Ott[55] believe that worldwide adoption of a uniform air pollution index system would have a number of benefits:

188 ENVIRONMENTAL INDICES

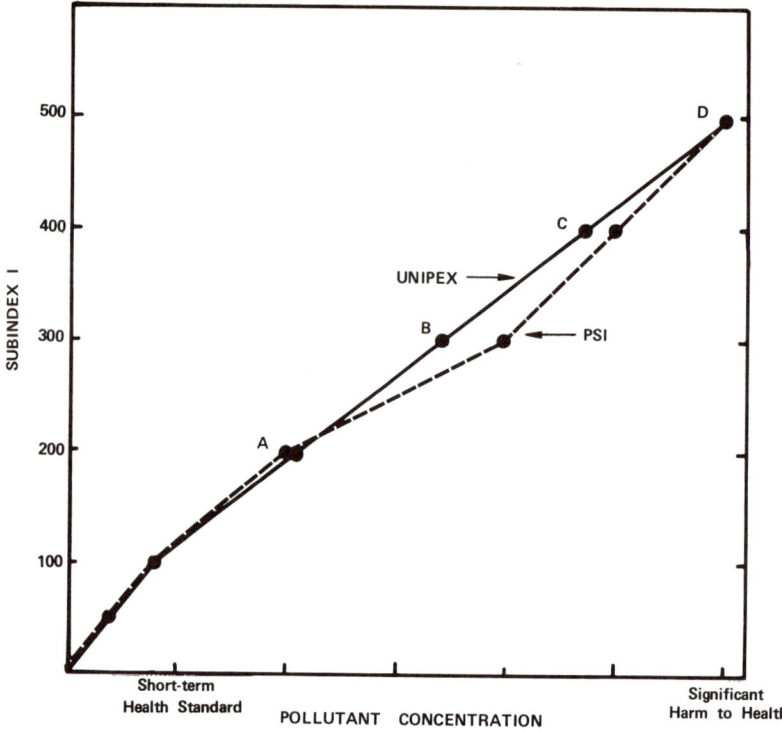

Figure 27. Comparison of PSI subindex function for oxodants (ozone) with UNIPEX subindex function.[55]

In conclusion, the use of the ten criteria to develop air pollution indices in other countries could lead to a uniform international air pollution index which is similar structurally, but simpler than the U.S. Pollutant Standards Index. The development of a uniform international index would (1) help prevent the proliferation of indices which occurred in the United States, (2) provide officials with a uniform measure for evaluating the impact of international regulatory actions and agreements, (3) simplify the understanding of air pollution levels by foreign travellers, and (4) enable air pollution levels to be more readily compared, thus opening new lines of international communication.

Table XXXI. Changes in Breakpoints if PSI is Linearized as in UNIPEX System

Pollutant	(A) Alert			(B) Warning			(C) Emergency		
	Old ($\mu g/m^3$)	New ($\mu g/m^3$)	Change	Old ($\mu g/m^3$)	New ($\mu g/m^3$)	Change	Old ($\mu g/m^3$)	New ($\mu g/m^3$)	Change
CO	17[a]	21.9[a]	+28.8%	34[a]	33.8[a]	-0.6%	46[a]	45.6[a]	-0.9%
NO_2	-	-	-	2260	2003	-11.4%	3000	2877	-4.1%
O_3	400	420	+5%	800	680	-15%	1000	940	-6%
TSP	375	445	+18.7%	625	630	+0.8%	875	815	-6.9%
SO_2	800	929	+16.1%	1600	1493	-6.7%	2100	2056	-2.1%
TSP x SO_2	261[b]	207[b]	-20.7%	393[b]	348[b]	-11.5%	393[b]	490[b]	+24.7%

[a]Expressed as mg/m^3;
[b]Expressed as $(\mu g/m^3)^2$ in thousands.

CONCLUSION

At present, the air pollution field appears to be at the initial stages of an entirely new concept: uniform reporting of air quality data through indices. Adoption of PSI, although in its infancy, promises to revolutionize the manner in which air quality data are reported, giving both the scientist and the layman a quantitative tool for gaining greater insight into the quality of the atmosphere. Its values can be treated in a statistical fashion, allowing analysis of trends and comparison of air quality at different locations, and its daily reports can be presented to millions of people through radio, TV and newspapers. Future events will determine whether a uniform index, both nationally and internationally, can bring about increased communication, greater use of air quality data, and increased understanding of environmental problems.

PROBLEMS FOR STUDY

1. Hydrocarbons is a pollutant variable that is monitored routinely in most U.S. cities, but it is rarely included in air pollution indices. Discuss the possible reasons for this.

2. After examining the discussion of each of the major air pollutants given in Chapter I, write a short narrative paragraph that could be used on radio or TV to explain the sources and effects of each pollutant in PSI.

3. Federal episode criteria and a Significant Harm level have been published for the product COH x SO_2, where SO_2 is expressed in volumetric units. (a) Using the same philosophy that was used in developing PSI's other subindex functions, plot a PSI subindex function for COH x SO_2, with SO_2 expressed in ppm, and list the breakpoints. [Answers: 0.2, 0.8, 1.2, 1.5] (b) Using the conversion factor for SO_2 given in this chapter, plot a similar COH x SO_2 subindex function in gravimetric units and list the breakpoints. [Answers: 524, 2095, 3143, 3929] (c) Using Equation 22, calculate the slopes for the segments of the subindex functions in (a) and (b). [Answers: 166.7, 250.0, 333.3; 0.064, 0.096, 0.127]

4. Suppose that the following concentrations for four pollutants are measured during the first 10 days of the month at a given air monitoring station. For each 24-hour period, the table lists the following values: CO, highest 8-hour average; O_3, highest 1-hour average; SO_2 and TSP, 24-hour average. Calculate PSI for each day and list the critical pollutant. [Answers: 70, CO; 80, CO; 129, CO; 150, O_3; 125, O_3; 100, O_3; 92, SO_2; 90, SO_2; 86, TSP; 85, TSP]

Pollutant Concentrations Measured at an Urban Air Monitoring Station

Day	CO (mg/m^3)	O$_3$ (μg/m^3)	SO$_2$ (μg/m^3)	TSP (μg/m^3)
1	7	40	80	60
2	8	64	140	76
3	12	184	200	96
4	11	280	220	112
5	9	220	230	160
6	9	160	240	212
7	7	136	320	204
8	9	112	310	212
9	6	80	250	208
10	5	56	240	204

5. Suppose that a particular country wishes to develop a national air pollution index using the UNIPEX system. After evaluating conditions in the country and scientific evidence on the effects of CO on health, officials have set the health-related standard for CO at 14 ppm. They have selected a value for significant harm to health identical to the U.S. Significant Harm level. Using the UNIPEX system, plot the CO subindex function in volumetric units and list the five breakpoints. [Answers: 14, 23, 32, 41, 50]

6. Using the UNIPEX system with breakpoints based on the U.S. NAAQS and Significant Harm levels, plot the individual subindex functions for CO, oxidants, NO$_2$, TSP and SO$_2$ in gravimetric units. Discuss the important differences between this index and PSI.

7. Can you think of any way to handle the ranking paradox which arises when one attempts to compare air quality levels at two different locations?

REFERENCES

1. "National Primary and Secondary Ambient Air Quality Standards," *Code of Federal Regulations,* Title 40, "Protection of Environment," Part 50, revised as of July 1, 1976.
2. "Clean Air Act Amendments of 1970," *Public Law 91-604.*
3. "Federal Air Quality Control Regions," U.S. Environmental Protection Agency, Rockville, MD, Publication No. AP-102 (Jaunary 1972).
4. Blacker, Stanley M., Wayne R. Ott and Thomas W. Stanley. "Measurement and the Law: Monitoring for Compliance with the Clean Air Act Amendments of 1970," *International Journal of Environmental Studies* 11:169-185 (1977).
5. "Example Regulations for Prevention of Air Pollution Emergency Episodes," *Code of Federal Regulations,* Title 40, "Protection of Environment," Appendix L to Part 51, revised as of July 1, 1976.

6. "Requirements for Preparation, Adoption and Submittal of Implementation Plans," *Federal Register*, 36(158):15486-15506, Part II, Washington, DC (August 14, 1971).
7. "Prevention of Air Pollution Emergency Episodes," *Code of Federal Regulations*, Title 40, "Protection of Environment," Part 51.16, revised as of July 1, 1976.
8. Thom, Gary C. and Wayne R. Ott. *Air Pollution Indices: a Compendium and Assessment of Indices Used in the United States and Canada* (Ann Arbor, MI: Ann Arbor Science Publishers, Inc., 1976).
9. Green, Marvin H. "An Air Pollution Index Based on Sulfur Dioxide and Smoke Shade," *J. Air Poll. Control Assoc.* 11(12):703-706 (December 1966).
10. Rich, T. A. "Air Pollution Studies Aided by Overall Air Pollution Index," *Environ. Sci. and Technol.* 1(10):796-800 (October 1967).
11. "M.U.R.C. Index Tells Detroiters How Dirty Air Is," *Air Engineering*, 10(6):28 (June 1968).
12. Fensterstock, Jack C., Kenneth Goodman, George M. Duggan and William S. Baker. "The Development and Utilization of an Air Quality Index," Paper No. 69-73, presented at the 62nd Annual Meeting of the Air Pollution Control Association, New York, NY, June 1969.
13. Miller, M. E., and George C. Holzworth. "An Atmospheric Diffusion Model for Metropolitan Areas," *J. Air Poll. Control Assoc.* 17(1):46-50 (January 1976).
14. Shenfeld, L. "Note on Ontario's Air Pollution Index and Alert System," *J. Air Poll. Control Assoc.* 20(9):622 (September 1970).
15. Babcock, Lyndon R., Jr. "A Combined Pollution Index for Measurement of Total Air Pollution," *J. Air Poll. Control Assoc.* 20(10):653-659 (October 1970).
16. Thomas, W. A., L. R. Babcock and W. B. Schults. "Oak Ridge Air Quality Index," Report No. ORNL-NSF-EP-8, Oak Ridge National Laboratory, Oak Ridge, TN (September 1971).
17. Larsen, Ralph I. "A New Mathematical Model of Air Pollutant Concentration Averaging Time and Frequency," *J. Air Poll. Control Assoc.* 19(1):24-30 (January 1969).
18. McGuire, Terry and Kenneth E. Noll. "Relationship Between Concentrations of Atmospheric Pollutants and Averaging Times," *Atmos. Environ.* 5(5):291-298 (May 1971).
19. Babcock, Lyndon R., Jr., and Niren L. Nagda. "Indices of Air Quality," *Indicators of Environmental Quality*, edited by William A. Thomas, (New York: Plenum Press, 1972).
20. Thomas, William A. "Public Acceptance of an Air Quality Index," *J. Environ. Educ.* 6(4):18-24 (1975).
21. Bisselle, C. C., S. H. Lubore and R. P. Pikul. "National Environmental Indices: Air Quality and Outdoor Recreation," Report No. MTR-6159, MITRE Corporation, McLean, VA (April 1972).
22. "Environmental Quality: The Third Annual Report of the Council on Environmental Quality," U.S. Government Printing Office, Washington, DC (August 1972).
23. Miller, Terry L. "Short Time Averaging Relationships to Air Quality Standards (STARAQS)—A Predictive Air Quality Index Model for Use

by Air Pollution Agencies," Paper No. 73-351, presented at the 66th Annual Meeting of the Air Pollution Control Association, Chicago, IL, June 24, 1973.
24. Inhaber, H. "Environmental Quality: Outline for a National Index for Canada," *Science* 186(29):798-805 (November 1974).
25. Inhaber, H. "A Set of Suggested Air Quality Indices for Canada," *Atmos. Environ.* 9:353-364 (1974).
26. Marchesani, Vincent J., Joseph N. DePierro, Terrence A. Juchnowski, Ralph J. Pfannenstiel, John H. Serkies and Althea S. Thornton. "New Jersey Air Quality Index," *Atmos. Environ.* 9:683-686 (1974).
27. Cullen, James J., Theodore V. Flaherty, Jr., and Stanley M. Barnett. "Indices for Dissemination of Ambient Air Quality Information to the Public," Paper No. 74-221, presented at the 67th Annual Meeting of the Air Pollution Control Association, Denver, CO, June 9, 1974.
28. Reidy, M. and C. Dziewulski. "Homogeneous Dissemination of Ambient Air Quality Levels to the News Media and the Public or the Need to Eschew Obfuscation," Paper No. 73-352, presented at the 66th Annual Meeting of the Air Pollution Control Association, Chicago, IL, June 24, 1973.
29. Ott, Wayne R. and Gary C. Thom. "A Critical Review of Air Pollution Index Systems in the United States and Canada," *J. Air Poll. Control Assoc.* 26(5):460-470 (1976).
30. "1973-1974 Directory of Governmental Air Pollution Agencies," Air Pollution Control Association, Pittsburgh, PA (1973).
31. Ott, Wayne R. and Gary C. Thom. "Air Pollution Indices in the United States and Canada—The Present Picture," presented at the 171st National Meeting of the American Chemical Society, New York, NY, April 8, 1976.
32. National Weather Service, Operations Manual, Air Pollution Weather Forecasts, WSOM Issuance 75-15, Part C, Chapter 30, July 16, 1975.
33. Hunt, William F., William M. Cox, Wayne R. Ott and Gary C. Thom. "A Common Air Quality Reporting Format, Precursor to an Air Quality Index," *Proceedings of the Fifth Annual Environmental Engineering and Science Conference,* Louisville, KY, March 3-4, 1975, pp. 99-121.
34. Thom, Gary C. and Wayne R. Ott. "Progress Toward a Uniform Air Pollution Index," *J. Air Poll. Control Assoc.* 25(11):1157-1158 (November 1975).
35. Thom, Gary C. and Wayne R. Ott. "A Proposed Uniform Air Pollution Index," *Atmos. Environ.* 10:261-264 (1976).
36. Thom, Gary C., Wayne R. Ott, William F. Hunt and John B. Moran. "A Recommended Standard Air Pollution Index," presented at the 171st National Meeting of the American Chemical Society, New York, NY, April 7, 1976.
37. Ott, Wayne R. "A FORTRAN Program for Computing the Pollutant Standards Index (PSI)," U.S. Environmental Protection Agency, Washington, DC, EPA-600/4-78-001 (May 1978).
38. Hunt, William F., Jr., and Wayne R. Ott. "Pollutant Standards Index (PSI) Evaluation Study," U.S. Environmental Protection Agency, internal report, Research Triangle Park, NC (April 1976).

39. "Guideline for Public Reporting of Daily Air Quality," *Federal Register* 41(174):37660-37671, Washington, DC (September 7, 1976).
40. "A Recommended Air Pollution Index," report prepared by the Federal Interagency Task Force on Air Quality Indicators, Council on Environmental Quality, Washington, DC (September 1976).
41. Hunt, William F., Jr., Wayne R. Ott, John Moran, Raymond Smith, Gary Thom, Neil Berg and Barry Korb. "Guideline for Public Reporting of Daily Air Quality—Pollutant Standards Index (PSI)," U.S. Environmental Protection Agency, Research Triangle Park, NC, EPA-450/2-76-013 (August 1976).
42. Ott, Wayne R. and William F. Hunt, Jr. "A Quantitative Evaluation of the Pollutant Standards Index," *J. Air Poll. Control Assoc.* 26(11): 1050-1054 (November 1976).
43. Wallace, Lance A. and Wayne R. Ott. "Rapid Techniques for Calculating the Pollutant Standards Index (PSI)," U.S. Environmental Protection Agency, Washington, DC, EPA-600/4-78-002 (March 1978).
44. "EPA's New Air Pollution Index," U.S. Environmental Protection Agency, Office of Public Awareness, Washington, DC 20460.
45. McAdie, H.G. and D. K. A. Gillies. "The Operational Forecasting of Undesirable Pollution Levels Based on a Combined Pollution Index," *J. Air Poll. Control Assoc.* 23(11):941-944 (November 1973).
46. Ott, Wayne R. "Development of Criteria for Siting Air Monitoring Stations," *J. Air Poll. Control Assoc.* 27(6):543-547 (June 1977).
47. Rubin, Edward S., Richard W. Walters, Paul J. Grogan and Sara M. Friedman. "Comparative Analysis of Present and Proposed Air Pollution Indices for Allegheny County," presented at the 70th Annual Meeting of the Air Pollution Control Association, Toronto, Ontario, Canada, June 20-24, 1977.
48. Hunt, William F., Jr. "Pollutant Standards Index," letter to the editor, *J. Air Poll. Control Assoc.* 27(7):620 (July 1977).
49. Hunt, William F., Jr., Raymond Smith, Wayne R. Ott and Wilson B. Riggan. "The Pollutant Standards Index (PSI)—An Early Warning System for Air Pollution," presented at the Eighth International Scientific Meeting of the International Epidemiological Association, Las Croabas, Puerto Rico, September 18-23, 1977.
50. "Clean Air Act Amendments of 1977," *Public Law 95-95* amending *Public Law 91-604* (August 3, 1977).
51. Young, J. W. S., R. J. Fry, J. M. Hewings, D. G. Kelley, J. C. Lack, R. E. Munn, H. J. Roireau, R. W. Shaw and L. Shenfeld. "A Proposed Air Quality Index for Canada," presented at the 4th International Clean Air Congress, Tokyo, Japan, May 16-20, 1977.
52. Martin, W. and A. C. Stern. "The World's Air Quality Management Standards," Volume I, "The Air Quality Management Standards of the World, Including United States Federal Standards," U.S. Environmental Protection Agency, Washington, DC, EPA 650/9-75-001-a (October 1974).
53. World Health Organization. "Air Quality Criteria and Guides for Urban Air Pollutants," WHO Technical Report Series No. 506 (1972).
54. Yanagisawa, Saburo. "Air Quality Standards National and International," *J. Air Poll. Control Assoc.* 23(11):945-948 (November 1973).

55. Thom, Gary C. and Wayne R. Ott. "Criteria for a Uniform International Air Pollution Index," presented at the 4th International Clean Air Congress, Tokyo, Japan, May 16-20, 1977.
56. "An Air Quality Index for Mexico City," Direccion General de Saneamiento Atmosferico, Subsecretaria de Mejoramiento del Ambiente, Mexico, D. F., draft report (October 6, 1977).

CHAPTER IV

WATER POLLUTION INDICES

In this chapter, primary emphasis is upon indices that can be used with data available from today's water quality monitoring activities (see Chapter I, pages 39 to 44). Although a significant number of biological indices of water quality (systems that are based on the presence or absence of certain species of organisms) also have been developed, most of these require data on variables other than those routinely monitored by U.S. water pollution control agencies. Therefore, biological indices are discussed only briefly.

As indicated in Chapter II (page 49), some water quality index developers refer to increasing scale indices as "water pollution" indices and decreasing scale indices as "water quality" indices. In this book, the terms "water pollution" and "water quality" are used interchangeably and do not denote a particular type of scale.

INDICES IN THE LITERATURE

Several comprehensive reviews of published water pollution indices have been undertaken. The first of these appeared in 1974 in a doctoral thesis by Landwehr.[1] In cooperation with other federal agencies, the Council on Environmental Quality (CEQ) has supported a survey and evaluation of water quality indices through contracts with Mathtech, Inc., and Energy Resources Co., Inc. The literature reviews appear in the contractors' final reports by Rosen et al.[2] and Orlando and Wrightington[3] and in a paper by Orlando, Wrightington and Maxim.[4]

Attempts were made in Germany as early as 1848 to relate the level of water purity and pollution to the occurrence of certain biological organisms.[1] Over the last 130 years, various European countries have developed and applied different systems to classify the quality of the waters within their boundaries. These water classification systems usually were of two types:

198 ENVIRONMENTAL INDICES

those concerned with the amount of pollution present and those concerned with living communities of macro- and microscopic organisms (for example, fish, benthic organisms and plants). Rather than assigning a numerical value to represent water quality, these classification systems categorized water bodies into one of several pollution classes or levels. By contrast, indices that use a numerical scale to represent gradations in water quality levels are a recent phenomenon, beginning with Horton's index[5] in 1965.

To present the many physical and chemical indices found in the literature in an orderly fashion, this book has classified them into four general categories:

- General Water Quality Indices
- Specific-Use Indices
- Planning Indices
- Statistical Approaches

The biological approaches are discussed as a separate class of techniques.

General Water Quality Indices

As discussed in Chapter I (page 40), water has a variety of different uses: for example, supply of public drinking water, crop irrigation, recreation, and maintenance of fish and wildlife habitats. Water quality requirements vary, depending on the intended use. Some indices, however, are based on the assumption that "water quality" is a general attribute of surface waters, irrespective of the use to which the water is put. We shall call these "general water quality indices." This section discusses five indices designed for general water quality use:

- Horton's Quality Index
- National Sanitation Foundation's Water Quality Index
- Prati's Implicit Index of Pollution
- McDuffie's River Pollution Index
- Dinius' Social Accounting System

Horton's Quality Index

Horton,[5] who proposed the first formal water quality index in the literature in 1965, saw indices as tools for evaluating abatement programs and for giving public information:

> A quality index system offers a means for measuring pollution abatement progress since stream conditions at any time can be compared with conditions that are desired or planned for the future. In this way the rating system can be useful for administrative purposes and for meaningful communication with the public.

Horton argued that "water quality" and "pollution" are relative terms, and presenting water quality in the "black and white" terms required by "stream classification" systems is misleading because it does not allow for gradations in conditions. Furthermore, lack of agreement among different agencies on standards and criteria makes such classification systems unworkable:

> The difficulty with these systems, unfortunately, is that different administrative agencies do not always agree on the same set of values. For example, one agency might establish Class "A" water as water with a coliform count of less than 1,000 MPN (Most Probable Number)/100 ml. Another agency might rule that a count of 2,000 is satisfactory for its Class "A" water. Now, both agencies might agree that water with 1,000 coliforms/100 ml is of better quality than water with 2,000 coliforms/100 ml, but because of the rigidity with which the systems are employed, there is no basis for comparing water quality as designated under one system with that as designated under the other.
>
> In addition, the inflexibility of the system may lead to irrational situations. In the same example, both agencies might designate Class "A" water as suitable for recreational purposes. The result would be that water with 2,000 coliforms/100 ml could be used for recreation in one area but not in another.[5]

To avoid these problems, Horton concluded that there is need for a "... system whereby water quality can be rated on a comparative basis ..."; this system would "... avoid any connotation of absolute values, such as pollution and non-pollution, or suitability and non-suitability." Thus, a properly designed index should permit the user to compare different locations and different points in time in terms of gradations in water quality.

In selecting the variables to be included in the index, Horton[5] imposed the following criteria:

- The number of variables should be limited to avoid making the index unwieldy.
- The variables should be of significance in most parts of the country.
- The variables should reflect the availability of the data.

Although the 10 variables selected for Horton's example index (Tables I and II) resulted from his work with the Ohio River Water Sanitation Commission (ORSANCO), they are applicable to other parts of the U.S. Dissolved oxygen (DO), pH, coliforms, specific conductance, alkalinity, and chloride content are widely measured water pollution variables (see page 44). Specific conductance was intended to serve as an approximate measure of total dissolved solids (TDS), and carbon chloroform extract (CCE) was included to reflect the influence of organic matter. One of the variables, "sewage treatment" (percentage of the population served), is designed to reflect the effectiveness of abatement activities on the premise that "chemical and biological measures of quality are of little significance until substantial progress has been made

Table I. Weights for Horton's Water Quality Index[5]

Variable	w
Dissolved Oxygen	4
Sewage Treatment	4
pH	4
Coliforms	2
Specific Conductance	1
Carbon Chloroform Extract	1
Alkalinity	1
Chloride	1

Table II. Breakpoints for Horton's Water Quality Index[5]

I	Dissolved Oxygen (%)	Coliforms (MPN/100 ml)[a]	Carbon Chloroform Extract (10^{-3} mg/l)
100	>70	<1,000	0-100
80	50-70	1,000- 5,000	100-200
60	30-50	5,000-10,000	200-300
30	10-30	10,000-20,000	300-400
0	<10	>20,000	>400

I	pH	Specific Conductance (μmho)	Alkalinity (mg/l)	Chlorides (mg/l)
100	6-8	0- 750	20-100	0-100
80	5-6; 8-9	750-1,500	5-20; 100-200	100-175
40	4-5; 9-10	1,500-2,500	0-5; >200	175-250
0	<4; >10	>2,500	Acid	>250

I	Sewage Treatment (% population served)	Coefficients
100	95-100	If temperature is above critical level, $M_1 = 1/2$; otherwise, $M_1 = 1$.
80	80- 95	
60	70- 80	
40	60- 70	
20	50- 60	If "obvious pollution" is present, $M_2 = 1/2$; otherwise, $M_2 = 1$.
0	< 50	

[a]MPN = most probable number.

in eliminating discharges of raw sewage." The index weights range from 1 to 4 (Table I), and the breakpoints (Table II) give staircase step function subindices (see page 60). Horton's index does not include toxic substances, because he felt that toxic substances were not eligible for index rating since ". . . under no circumstances should streams contain substances that are injurious to humans, animals, or aquatic life."

Horton's Quality Index (QI) uses a linear sum aggregation function. It consists of the weighted sum of the subindices divided by the sum of the weights and multiplied by two coefficients, M_1 and M_2, which reflect temperature and "obvious pollution," respectively:

$$QI = \frac{\sum_{i=1}^{n} w_i I_i}{\sum_{i=1}^{n} w_i} M_1 M_2 \qquad (1)$$

If all 10 pollutant variables are available, the equation is written more simply:

$$QI = \frac{M_1 M_2}{18} \sum_{i=1}^{8} w_i I_i \qquad (2)$$

Each of the coefficients, M_1 and M_2, takes on either of two values, 1 or 1/2, depending on temperature conditions and the extent of obvious pollution. For example, in the Ohio River Basin, a temperature below 93°F (34°C) was considered acceptable ($M_1 = 1$), but a temperature above this level was considered unacceptable ($M_1 = 1/2$). Obvious pollution, M_2, is treated dichotomously like temperature. It reflects offensive conditions, such as formation of sludge deposits; presence of oil, debris, foam, scum or other floating materials; and discharge of wastes that create a color or odor nuisance.

Horton's index has the advantage that it is relatively easy to apply, although the coefficients M_1 and M_2 require some tailoring to fit individual situations. The index structure and its weights and rating scales were considered preliminary and were based on the judgment of the author and his associates:

> The ratings shown are presented for illustrative purposes. In this case they represent the preliminary judgment of the ORSANCO staff. Hopefully, they will offer a basis for discussion through which a consensus might be reached.[5]

Horton's index is an important example of an early index design that influenced later index structures. Like many of the water quality indices which have followed, it is a decreasing scale index.

National Sanitation Foundation's Water Quality Index

In 1970, Brown, McClelland, Deininger and Tozer[6] presented a water quality index similar in structure to Horton's index. This effort was supported by the National Sanitation Foundation (NSF), and the resulting index is known as the National Sanitation Foundation Water Quality Index (NSF WQI). The NSF WQI was developed using a formal procedure based on the Rand Corporation's Delphi technique[7,8] to combine the opinions of a large panel of water experts from throughout the U.S. In this approach, members of the panel were polled by mail using questionnaires. The results from these questionnaires were tabulated and reported to each member, enabling him to see how his own responses compared with those of the group as a whole. Then each member was polled again to arrive at a final consensus. The NSF WQI developers felt that such a procedure helped minimize problems associated with the arbitrary judgment incorporated into previous indices:

> In an attempt to alleviate the limitations of previous efforts to derive a WQI–due to the "subjective" establishment of selected parameters, rating scales, and parameter weightings–systematic opinion research technology was utilized to incorporate the judgment of a large and diverse panel. The research procedure was designed to minimize formation of a judgment based on any professional viewpoint or individual geographic area.[6]

In carrying out this procedure, the investigators assembled a panel of 142 persons from throughout the U.S. with expertise in various aspects of water quality management (Table III). As described by Brown et al.[6] and McClelland,[9] a series of three questionnaires was mailed to the members of this panel.

In questionnaire No. 1, the respondents were asked to consider 35 water pollutant variables for possible inclusion in a water quality index (Table IV). Respondents were permitted to add any variables to the list which they felt should be included in the NSF WQI. They were asked to designate each variable as follows: "do not include," "undecided," or "include." Respondents also were asked to rate each variable marked "include" according to its significance as a contributor to overall water quality. This rating was done on a scale of "1" (highest relative significance) to "5" (lowest relative significance).[6,9]

When respondents returned questionnaire No. 1, the results were tabulated on a computer printout. Of the original panel of 142 members, 102 respondents (72%) completed and returned this questionnaire. However, 8 of the

Table III. Professional Fields of Participants in the NSF WQI Panel[6]

Regulatory Officials (federal, interstate, state, territorial and regional)	101
Managers of Local Public Utilities	5
Consulting Engineers	6
Academicians	26
Others (industrial waste control engineers and representatives of professional organizations)	4
Total	142

Table IV. 35 Candidate Variables Considered for the NSF WQI in Questionnaire No. 1[9]

Dissolved Oxygen	Oil and Grease
Fecal Coliforms	Turbidity
pH	Chlorides
Biochemical Oxygen Demand (5-Day)	Alkalinity
	Iron
Coliform Organisms	Color
Herbicides	Manganese
Temperature	Fluorides
Pesticides	Copper
Phosphates	Sulfates
Nitrates	Calcium
Dissolved Solids	Hardness
Radioactivity	Sodium and Potassium
Phenols	Acidity
Chemical Oxygen Demand	Bicarbonates
Carbon Chloroform Extract	Magnesium
Ammonia	Aluminum
Total Solids	Silica

102 respondents did not return their questionnaires in sufficient time for inclusion in the tabulation, and the printout therefore summarized the responses from 94 panel members (Figure 1). The printout also listed the panel member's own replies alongside the group response. When questionnaire No. 2 was mailed, the printout was included, giving each member both a summary of the group response and his own replies.

In questionnaire No. 2, each member was asked to review his original ratings and to modify his response if he desired. He was instructed to note his replies for each variable and to compare them with those of the entire

204 ENVIRONMENTAL INDICES

WATER QUALITY INDEX
QUESTIONNAIRE 1

THIS IS A STATISTICAL SUMMARY OF ALL RESPONSES

YOUR RESPONSE WAS THE FOLLOWING

VARIABLE	NO.	A NO ANSWER	B DON'T INC.	C UNDECIDED	D INCLUDE	SIGNIFICANCE					AVG.	A	B	C	D	SIGNIFICANCE				
						1	2	3	4	5						1	2	3	4	5
DISSOLVED OXYG	1	1	2	1	90	75	7	5	1	2	1.31				X	1				
BOD	2	2	9	9	74	48	11	11	1	3	1.65				X					
COD	3	1	18	20	55	20	15	15	3	2	2.13				X			3		
DISSOLV SOLIDS	4	1	6	10	77	27	24	20	4	3	2.09		X					3		
TOTAL SOLIDS	5	2	8	12	71	21	23	18	6	3	2.25				X					
COLOR	6	2	9	13	78	14	31	15	6	6	2.55			X						
P H	7	2	1	2	89	56	19	8	4	2	1.62				X					
CARB CHLOR EXT	8	1	14	23	56	16	21	19	5	4	2.18			X				3		
HARDNESS	9	5	10	15	64	12	11	28	5	4	2.72			X						
ALKALINITY	10	3	10	12	69	15	17	32	3	2	2.42			X						
ACIDITY	11	5	22	18	49	5	13	23	5	2	2.76			X						
ALUMINUM	12	4	48	26	16	0	6	7	3	0	2.81			X						
AMMONIA	13	2	15	17	60	14	24	17	3	2	2.25			X						
BICARBONATE	14	2	34	21	37	4	11	15	4	3	2.76			X						
CALCIUM	15	3	31	18	42	9	11	14	3	5	2.62			X						
CHLORIDES	16	2	4	11	77	16	27	25	6	3	2.39				X					
COPPER	17	5	22	22	45	8	13	17	3	4	2.60			X						
FLUORIDES	18	3	16	17	58	14	15	17	5	7	2.59			X			2			
IRON	19	3	9	16	66	15	19	17	12	3	2.53			X						
MANGANESE	20	2	14	17	61	14	16	17	10	4	2.57			X						
MAGNESIUM	21	3	35	16	40	4	12	17	2	5	2.80			X						
NITRATE+NITRITE	22	2	3	7	82	26	35	14	4	3	2.06				X		2	3		
PHOSPHATES	23	1	2	9	82	29	32	15	4	3	2.01				X					
SILICA	24	5	47	25	17	0	3	6	5	3	3.41				X	1	2			
SODIUM+POTASS	25	4	31	17	42	5	14	14	6	2	2.74			X		1	2			
SULFATES	26	2	17	15	60	10	13	29	6	2	2.62		X							
TEMPERATURE	27	1	1	5	87	49	15	13	4	6	1.89				X					
PESTICIDES	28	3	6	11	74	32	23	12	6	1	1.93				X					
COLIFORM ORG	29	2	13	10	69	39	15	10	2	3	1.77				X					
FECAL COLIFORM	30	0	1	10	83	68	6	6	1	2	1.35				X					
RADIOACTIVITY	31	2	5	15	72	32	12	19	5	4	2.12				X					
PHENOLS	32	5	11	10	68	23	21	18	4	2	2.13				X		2	3		
OIL+GREASE	33	3	16	12	65	19	19	19	7	2	2.26				X		2			
HERBICIDES	34	3	12	23	56	28	13	11	2	1	1.87	1			X		2			
TURBIDITY	35	0	5	4	85	26	27	15	12	5	2.33			X			2			

Figure 1. Example of results of computer printout of questionnaire No. 1.[6]

group. The purpose of this feedback process was to obtain greater convergence of opinion regarding the significance of each variable for overall water quality. However, the investigators report that there was little change in the significance ratings expressed in questionnaire No. 2 when compared with those in questionnaire No. 1.[6]

A list of candidate variables was sent along with questionnaire No. 2. The list included the original 35 variables, ranked in order of decreasing importance as determined by the average significance ratings of the group from the first questionnaire, and 9 new variables. The new variables had been added to the first questionnaire by several respondents, and they were now introduced for group consideration: chromium (hexavalent), total organic carbon, cyanides, specific conductance, lead, arsenic, cadmium, selenium, and zinc. From the total of 44 variables, panelists were asked to designate not more than 15 variables which they considered to be the "most important" for inclusion in a water quality index.

Of the 94 respondents receiving the second questionnaire, 77 completed and returned it, giving an 82% response. Using expert opinion derived from the initial stages of this study, the investigators identified 9 individual variables and 2 grouped variables of greatest importance. The individual variables were DO, fecal coliforms, pH, 5-day biochemical oxygen demand (BOD_5), nitrates, phosphates, temperature, turbidity, and total solids (TS). The grouped variables were toxic substances and pesticides.

In questionnaire No. 3, the respondents were asked to develop a rating curve for each of the 9 individual variables. This was accomplished by providing blank graphs to each respondent. Levels of "water quality" from 0 to 100 were indicated on the ordinate of each graph, while various levels (or strengths) of the particular variable were arranged along the abscissa. The respondent was asked to draw a curve on each graph which, in his judgment, represented the variation of water quality produced by the various quantities of each pollutant variable.

The investigators subsequently averaged the curves from the respondents to produce a set of "average curves," one for each pollutant variable.[9] The resulting curves are shown in Figures 2 through 10. In each figure, the solid line represents the arithmetic mean of all respondents' curves, while the dotted lines bounding the shaded area represent the 80% confidence limits. Approximately 80% of the respondents' curves lie within the shaded zone. A narrow band of shading, such as the one for DO (Figure 2), denotes greater agreement among respondents than a wide band, such as the one for turbidity (Figure 9).

Special procedures were used to deal with the two grouped variables, pesticides and toxic substances. For pesticides, respondents were asked to consider an approach in which the NSF WQI would automatically be set

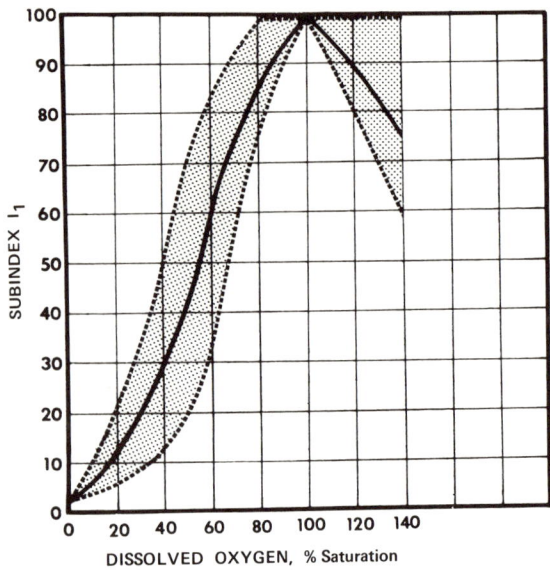

Figure 2. Subindex function for DO in the NSF WQI.[9] (For DO $> 140\%$, $I_1 = 50$.)

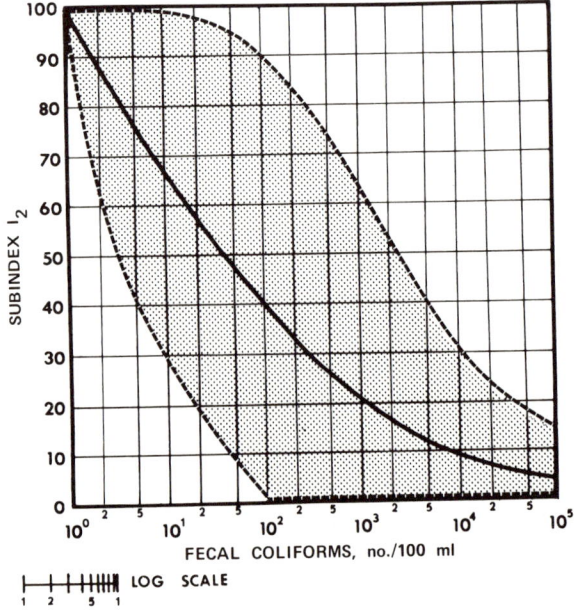

Figure 3. Subindex function for fecal coliforms (average number of organisms per 100 ml) in the NSF WQI.[9] (For fecal coliforms $> 10^5/100$ ml, $I_2 = 2$.)

WATER POLLUTION INDICES 207

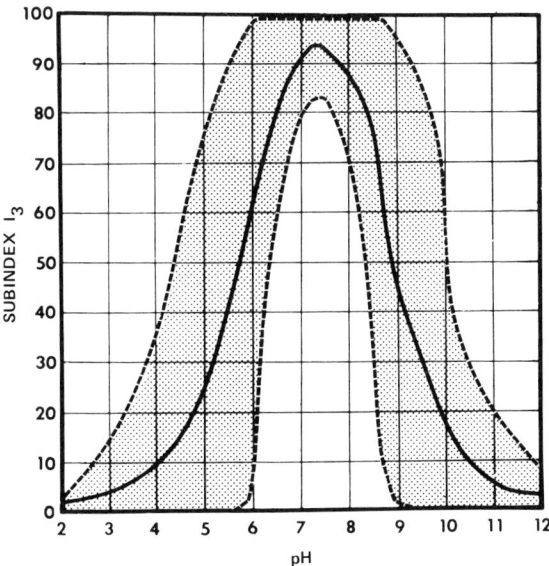

Figure 4. Subindex function for pH in the NSF WQI.[9]

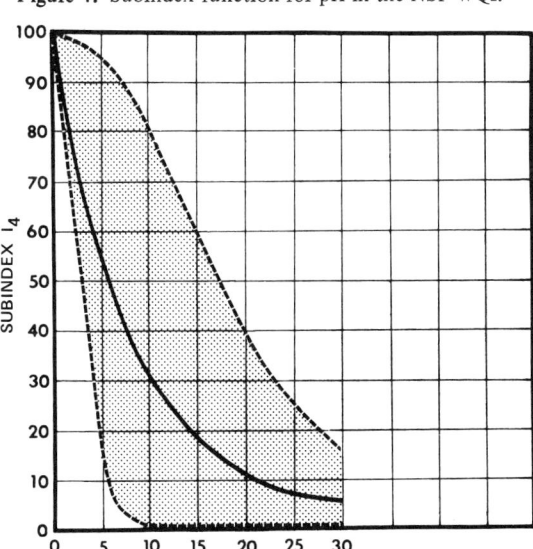

Figure 5. Subindex function for BOD_5 in the NSF WQI.[9] (For $BOD_5 > 30$ mg/l, $I_4 = 2$.)

208 ENVIRONMENTAL INDICES

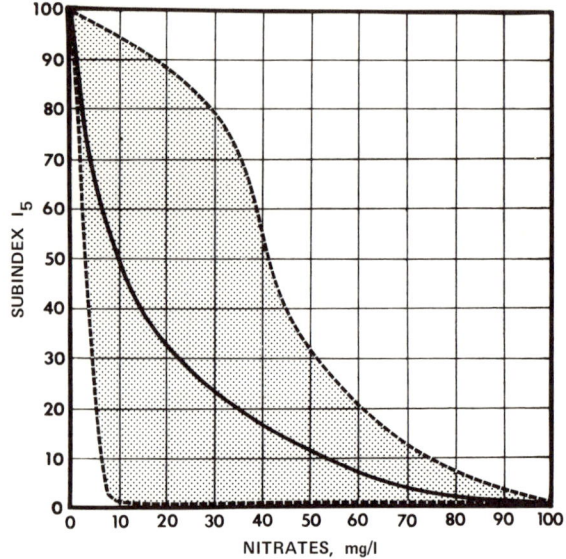

Figure 6. Subindex function for nitrates in the NSF WQI.[9] (For nitrates > 100 mg/l, $I_5 = 1$.)

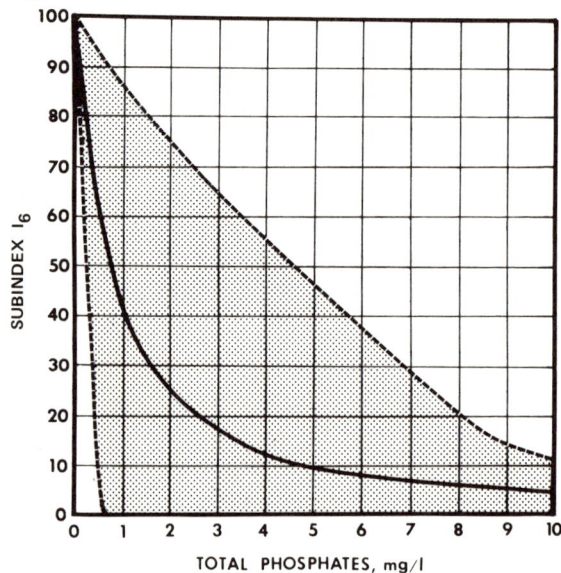

Figure 7. Subindex function for total phosphates in the NSF WQI.[9] (For total phosphates > 10 mg/l, $I_6 = 2$.)

WATER POLLUTION INDICES 209

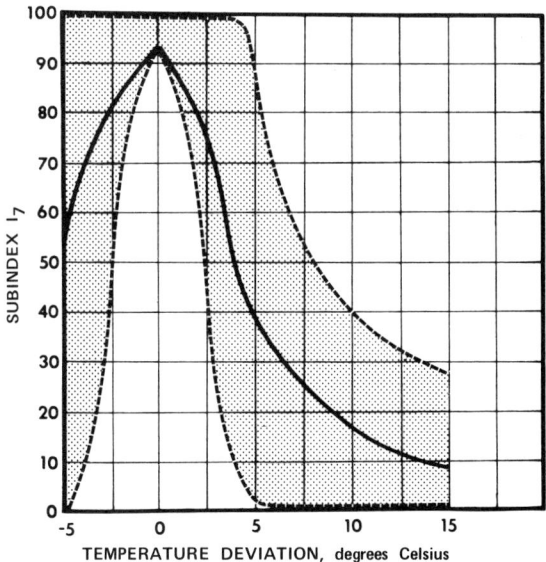

Figure 8. Subindex function for temperature deviation from equilibrium (ΔT) in the NSF WQI.[9] (For $\Delta T > 15°C$, $I_7 = 5$.)

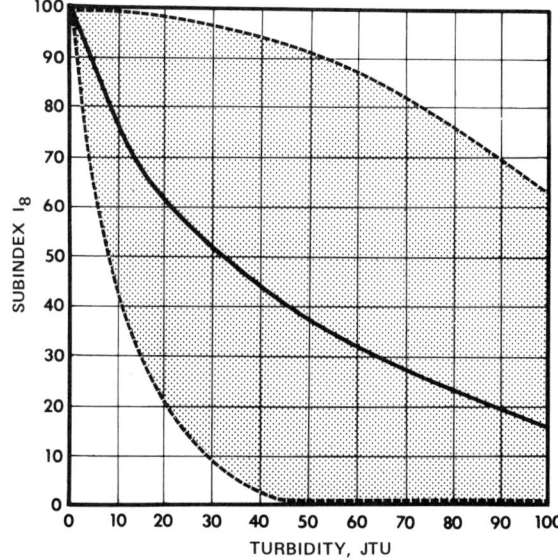

Figure 9. Subindex function for turbidity (Jackson Turbidity Units) in the NSF WQI.[9] (For turbidity > 100 JTU, $I_8 = 5$.)

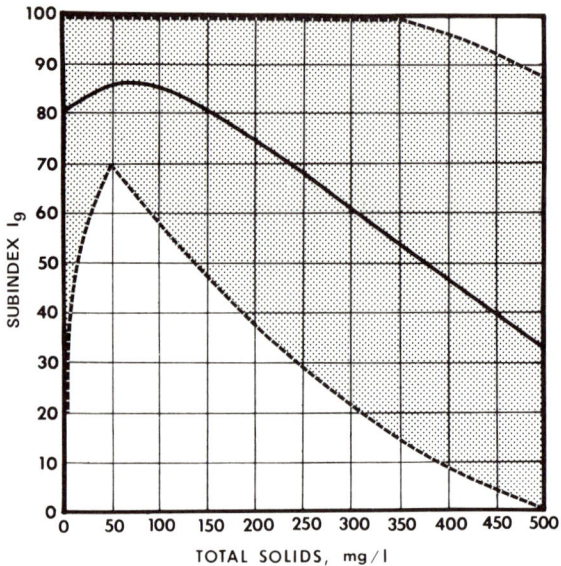

Figure 10. Subindex function for total solids in the NSF WQI.[9] (For total solids > 500 mg/l, I_9 = 20.)

to zero, the lowest point on the water quality scale, if the concentration of detectable pesticides (of all types) exceeds 0.1 mg/l (100 parts per billion). The proposed procedure for toxic substances was similar. If any toxic substance exceeded its assigned upper limit, the NSF WQI would be set to zero. The maximum permissible upper limits for toxic substances came from published drinking water standards. The panelists agreed with these procedures for handling pesticides and toxic substances.

The investigators sought to derive a set of weights for the index which would sum to 1.0 but which would reflect the significance ratings assigned to the variables by the panelists. The following procedure was used. The arithmetic means of the significance ratings were calculated for all variables rated in questionnaire No. 3 (Table V). "Temporary weights" then were derived by dividing the significance rating of each variable into the rating for the variable with the highest significance rating, DO (1.4). Finally, each temporary weight was divided by the sum of the temporary weights, giving the subindex weights (last column of Table V). Thus, the subindex weight for DO is 0.17 and that for turbidity is 0.08, giving a ratio of 0.17/0.08 = 2.1. This is the reciprocal of the original ratings of 1.4 and 2.9, since 1.4/2.9 = 1/2.1. Note that the ratios are only approximately preserved due to error in rounding the final weights to two significant figures.

Table V. Significance Ratings and Weights for
Nine Pollutant Variables Included in the NSF WQI[6]

Variable	Mean of all Significance Ratings Returned by Respondents	Temporary Weights	Final Weights[a]
Dissolved Oxygen	1.4	1.0	0.17
Fecal Coliforms	1.5	0.9	0.15
pH	2.1	0.7	0.12
5-Day Biochemical Oxygen Demand	2.3	0.6	0.10
Nitrates	2.4	0.6	0.10
Phosphates	2.4	0.6	0.10
Temperature	2.4	0.6	0.10
Turbidity	2.9	0.5	0.08
Total Solids	3.2	0.4	0.08
Total			1.00

[a]A computer program developed later for the NSF WQI incorporates weights with slightly different numerical values: fecal coliforms, 0.16; pH, 0.11; BOD_5, 0.11; total solids, 0.07. The program is available from the National Sanitation Foundation, P.O. Box 1468, Ann Arbor, MI 48106.

The final curves (Figures 2 through 10) are the subindex functions of the index, one for each pollutant variable. To calculate the index, one reads the subindex value I_i from the appropriate rating curve for pollutant variable i. In the original structure proposed by Brown et al.,[6] the index, NSF WQI_a, is the weighted linear sum of the subindices:

$$\text{NSF WQI}_a = \sum_{i=1}^{n} w_i I_i \qquad (3)$$

Although the additive form of the index has been widely field-tested and applied, an alternate multiplicative form, NSF WQI_m, was proposed subsequently to overcome the eclipsing which occurs when a single pollutant variable shows extremely poor water quality.[1,9-11] In the multiplicative form, which is equivalent to the weighted product aggregation function discussed on pages 82 and 83, the same weights become powers of the subindices:

$$\text{NSF WQI}_m = \prod_{i=1}^{n} I_i^{w_i} \qquad (4)$$

If any one subindex approaches zero, the overall index approaches zero.

The temperature pollutant variable is defined as the deviation from equilibrium temperature (degrees Celsius). Equilibrium temperature is that which occurs without the influence of a heated or cooled discharge. In field applications, two temperatures are taken: one at the sampling site and one at some point upstream where a heated or cooled discharge is known to be absent.

Because the subindex functions are not represented by equations but by empirical curves, we classify the subindices as implicit nonlinear functions. In computer programs available for this index,[11] the curves usually are divided into segments which are entered into the program as data, and linear interpolation is undertaken between the points.

McClelland[9] regards the NSF WQI as ". . . a management and general administrative tool intended for use in communicating water quality information to the lay public and to legislative decision makers . . . .", but " . . . not a complex predictive model for technical and scientific application." The NSF WQI developers view the index as a means for making the public aware of water quality levels:

> A variety of public communication and education usages for the index are envisioned. For example, monthly published indices for selected bodies of water would be a useful device for informing the public of pollution abatement progress by municipalities, industries, etc. The positive citizen awareness and action programs which have been initiated in a number of cities through the use of air pollution indices might well be duplicated for water quality.[6]

They also suggest a way of reporting the NSF WQI which relates the index values to five descriptor words and to colors of the spectrum (Table VI). In this system, "bad" water quality is represented by orange (26-50), "medium" water quality by yellow (51-70), and "good" water quality by green (71-90).

Table VI. Descriptor Words and Colors Suggested for Reporting the NSF WQI[11]

Descriptor Words	Numerical Range	Color
Very Bad	0- 25	Red
Bad	26- 50	Orange
Medium	51- 70	Yellow
Good	71- 90	Green
Excellent	91-100	Blue

The NSF WQI has been widely field-tested and applied to data from a number of different geographical areas. In 1973, it was applied by Brown, McClelland, Deininger and Landwehr[10] to data from California, Colorado, Maryland, Michigan, Ohio, Pennsylvania and Tennessee. In 1974, it was applied by McClelland[9] to the Kansas River Basin. Some applications[10] have employed three-dimensional graphs to display water quality profiles showing the NSF WQI values of a stream versus time and distance (Figure 11). Its developers view this index as an effective technique for reporting water quality data, examining trends, and evaluating the effectiveness of water pollution control programs:

> The NSF water quality index has been shown by field validation to be responsive to changes in water quality conditions. It is a useful tool for aggregating and summarizing stream quality data, illustrating trends, and comprehensibly reporting success—or lack of success—in federal, interstate, state, local, or private water quality improvement programs.[10]

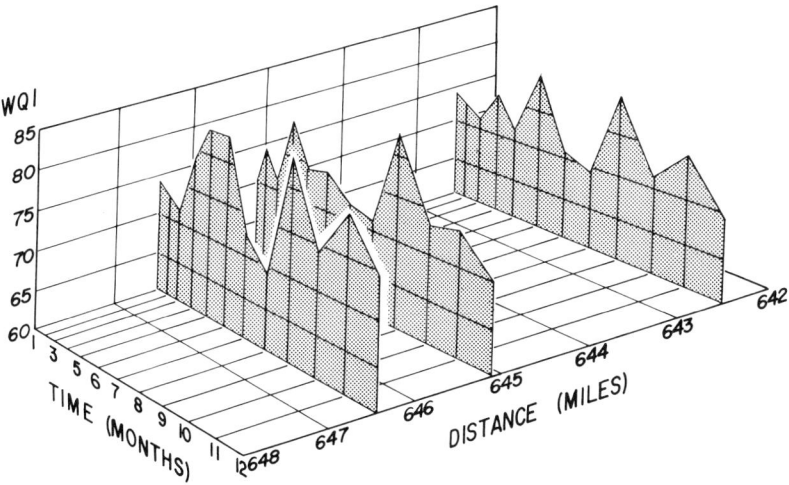

Figure 11. Water quality "profiles" of the Tennessee River, Tennessee Valley Authority, using the NSF WQI (1971 data).[10]

As we shall see, the NSF WQI is the most widely used of all existing water quality indices. In a 1977 national survey of water pollution control agencies, Ott[12] found that 12 out of 60 state and interstate agencies were using a water quality index. Of the 12 index users, 7 had selected the NSF WQI. Although the index is widely known, some water quality specialists have been reluctant to adopt it, citing various technical reasons. However, the NSF WQI developers have suggested that this reluctance may stem from "psychological rather than technical" reasons.[13]

Prati's Implicit Index of Pollution

In 1971, Prati, Pavanello and Pesarin[14] proposed an index for surface waters based on the water quality classification systems used in a number of different countries. These investigators saw their index as a possible tool for establishing a comparative inventory of the quality of water resources in a region or country, but they did not believe it should be used to make wastewater treatment decisions:

> Although this method lacks somewhat in objectivity, we think it may serve a useful purpose in securing comparability as regards the degree of pollution present in rivers in various countries. There is hardly need to point out that the method is not applicable to surveys or investigations aimed directly at determining the degree of treatment to be applied to specific wastewaters or the siting of treatment plants to obtain a given river water quality.[14]

In developing this index, the Implicit Index of Pollution, the authors first reviewed the water quality classification systems that have been adopted in England, Germany, the Soviet Union, Czechoslovakia, New Zealand, Poland, and some states in the U.S. From this information, they developed their own classification system involving 13 pollutant variables (Table VII). The system has five different water quality classes, I-V, and subindex ranges are assigned to each class. The upper limits of the first four ranges are 1, 2, 4 and 8, which correspond to a geometric progression in which 2 is raised to successive

Table VII. Prati's Classification System for Surface Water Quality[14]

Condition:	Excellent	Acceptable	Slightly Polluted	Polluted	Heavily Polluted
Index of Quality:	1	2	4	8	>8
pH	6.5-8.0	6.0-8.4	5.0-9.0	3.9-10.1	<3.9 to >10.1
Dissolved Oxygen (%)	88-112	75-125	50-150	20-200	<20 to >200
5-Day Biochemical Oxygen Demand (ppm)	1.5	3.0	6.0	12.0	>12.0
Chemical Oxygen Demand (ppm)	10	20	40	80	>80
Permanganate (mg/l O_2; Kubel Test)	2.5	5.0	10.0	20.0	>20.0
Suspended Solids (ppm)	20	40	100	278	>278
Ammonia (ppm)	0.1	0.3	0.9	2.7	>2.7
Nitrates (ppm)	4	12	36	108	>108
Chlorides (ppm)	50	150	300	620	>620
Iron (ppm)	0.1	0.3	0.9	2.7	>2.7
Manganese (ppm)	0.05	0.17	0.5	1.0	>1.0
Alkly Benzene Sulfonates (ppm)	0.09	1.0	3.5	8.5	>8.5
Carbon Chloroform Extract (ppm)	1.0	2.0	4.0	8.0	>8.0

integer powers (0,1,2,3). Toxic substances were not included, because the investigators felt that concentrations above a threshold level for any toxic substance should automatically result in the index being classified in the highest category (heavily polluted). Note that this classification system is an increasing scale type in which the values increase with increasing degree of pollution.

For each subindex, the investigators developed explicit mathematical functions (Table VIII) consistent with the classification ranges. These functions were based on their own judgment of the severity of pollution effects within each range. Linear interpolation was used in the subindices for BOD_5, chemical oxygen demand (COD), and permanganate (Kubel test), giving simple linear functions for each of these pollutant variables (Figure 12). A linear function also was used for CCE, one of the variables included in Horton's index. Nonlinear functions were used for four of the pollutant variables (suspended solids, ammonia, nitrates and iron). For each of the five remaining pollutant variables, a set of different equations was used for different variable ranges, giving segmented nonlinear functions. In the subindex for pH, small "kinks" can be observed at the breakpoints, particularly where nonlinear segments join linear segments (Figure 15 of Chapter II, page 16). In the subindex for DO, the slopes of the two segments at DO = 50% are so similar that the location of the breakpoint cannot be detected easily (Figure 13). Because the index has an increasing scale, the pH subindex reaches a *minimum* value at X = 7.0. This is the opposite of a decreasing scale index such as the NSF WQI, for which the pH subindex has a *maximum* value at approximately the same pH value (X=7.3).

Prati's index is computed as the arithmetic mean of the 13 subindices:

$$I = \frac{1}{13} \sum_{i=1}^{13} I_i \qquad (5)$$

The index ranges from 0 to 14 (and above). This index was applied by the investigators to data on surface waters in the province of Ferrara, Italy. All the pollutant variables were not available for this pilot application, however, and no papers describing the subsequent fate of this index, or any more extensive applications, can be found in the literature.

McDuffie's River Pollution Index

In 1973, McDuffie and Haney[15] presented a relatively simple water quality index which they called the River Pollution Index (RPI). They felt that indices could be applied to river water data to facilitate a variety of analyses: "A valid index would provide a measurement and picture of water quality

Table VIII. Subindex Equations for Prati's Index[15]

Dissolved Oxygen (%)
$I = 0.00168X^2 - 0.249X + 12.25$ $\quad 0 \leqslant X < 50$
$I = -0.08X + 8$ $\quad 50 \leqslant X < 100$
$I = 0.08X - 8$ $\quad 100 \leqslant X$

pH (units)
$I = -0.4X^2 + 14$ $\quad 0 \leqslant X < 5$
$I = -2X + 14$ $\quad 5 \leqslant X < 7$
$I = X^2 - 14X + 49$ $\quad 7 \leqslant X < 9$
$I = -0.4X^2 + 11.2X - 64.4$ $\quad 9 \leqslant X < 14$

5-Day Biochemical Oxygen Demand (mg/l)
$I = 0.666667X$

Chemical Oxygen Demand (mg/l)
$I = 0.1X$

Permanganate, Kubel Test (mg/l O_2)
$I = 0.4X$

Suspended Solids (mg/l)
$I = 2^{[2.1 \log(0.1X - 1)]}$

Ammonia (mg/l)
$I = 2^{[2.1 \log(10X)]}$

Nitrates (mg/l)
$I = 2^{[2.1 \log(0.25X)]}$

Chlorides (mg/l)
$I = 0.000228X^2 + 0.0314X$ $\quad 0 \leqslant X < 50$
$I = 0.000132X^2 + 0.0074X + 0.6$ $\quad 50 \leqslant X < 300$
$I = 3.75 (0.02X - 5.2)^{0.5}$ $\quad 300 \leqslant X$

Iron (mg/l)
$I = 2^{[2.1 \log(10X)]}$

Manganese (mg/l)
$I = 2.5X + 3.9\sqrt{X}$ $\quad 0 \leqslant X < 0.5$
$I = 5.25X^2 + 2.75$ $\quad 0.5 \leqslant X$

Alkyl Benzene Sulfonates (mg/l)
$I = -1.2X + 3.2\sqrt{X}$ $\quad 0 \leqslant X < 1$
$I = 0.8X + 1.2$ $\quad 1 \leqslant X$

Carbon Chloroform Extract (mg/l)
$I = X$

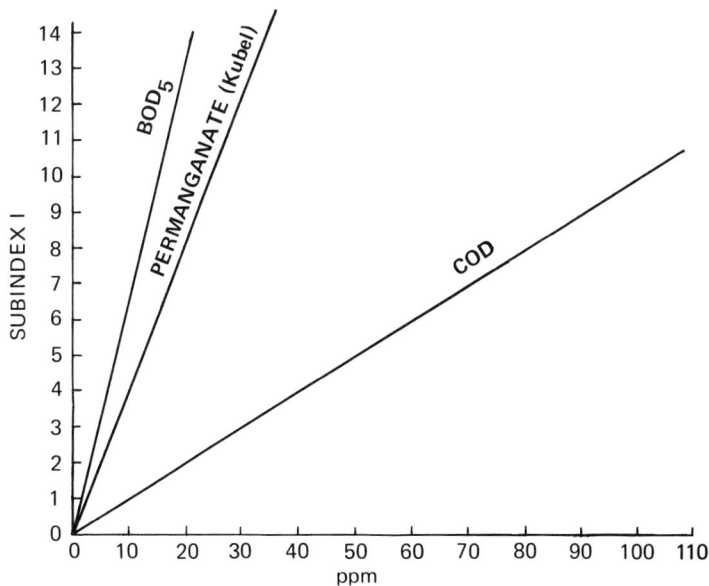

Figure 12. Prati's (linear) subindex functions for BOD$_5$, permanganate and COD.[14]

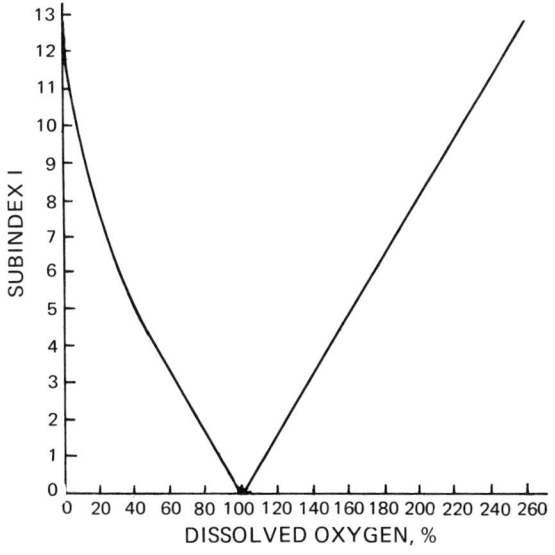

Figure 13. Prati's (segmented nonlinear) subindex function for DO.[14]

at any instant, and a way to compare different rivers as well as trends over the years for a particular river."[15]

Although eight pollutant variables are discussed in their paper, either fewer or more than eight variables can be included in the index, depending on the availability of data. The subindex for each pollutant variable is based on the ratio of the measured value to its "natural" level. The natural level is the normal value of the pollutant variable in "good" or "unpolluted" water. To make the subindices vary from 10 ("natural" levels) to 100 ("highly polluted" levels), the ratio of the observed to natural level is multiplied by a scaling factor (usually 10). Thus, many of the subindices are of the general linear form:

$$I_i = 10\left(\frac{X}{X_N}\right)_i \qquad (6)$$

where I_i = subindex for the ith pollutant variable
X = observed value of the pollutant variable
X_N = natural level of the pollutant variable

Six of the eight subindices described by McDuffie and Haney[15] are explicit linear functions, and two (coliform count and temperature) are explicit nonlinear functions (Table IX). In this table, we have simplified McDuffie and Haney's original equations, and the scaling factors and recommended limits are incorporated into the resulting equations. The subindices

Table IX. Subindex Functions for McDuffie's Index[15]

Percent Oxygen Deficit:	$I = 100 - X$	$X = DO\ (\%)$
"Biodegradable" Organic Matter:	$I = 10X$	$X = BOD_5\ (ppm)$
"Refractory" Organic Matter:	$I = 5(X - Y)$	$X = COD\ (ppm)$ $Y = BOD_5\ (ppm)$
Coliform Count (no./100 ml):	$I = 10\left(\frac{\log X}{\log 3}\right)$	
Nonvolatile Suspended Solids (ppm):	$I = X$	
Average Nutrient Excess:	$I = 5\left(\frac{X}{0.2} + \frac{Y}{0.1}\right)$	$X = $ Total N (ppm) $Y = $ Total PO_4 (ppm)
Dissolved Salts:	$I = 0.25X$	$X = $ Specific Conductance (μmho/cm)
Temperature (°C)	$I = \frac{X^2}{6} - 65$	

include "oxygen deficit," which is the inverse of DO and provides a measure of the degree to which a particular water sample has been degraded. This subindex can become negative when the water is supersaturated with oxygen. "Refractory" organic matter includes those forms of organic matter (for example, cellulose) which are oxidized by potassium dichromate in the COD test but are not very active in the BOD test. Nonvolatile suspended solids represents the amount of inorganic insoluble matter, an indication of the extent of erosion or inorganic pollution upstream, which causes undesirable turbidity or muddiness in surface waters and objectionable bottom deposits. The "average nutrient excess" subindex assumes that the "normal" concentration (natural level) of total nitrogen is 0.2 ppm and that the "normal" concentration of total phosphate is 0.1 ppm. The temperature subindex was empirically derived from New York State water quality data, and it is a parabolic function in which I = 0 corresponds to approximately 20°C. The index does not include pH or toxic substances, but McDuffie and Haney indicated that these might be added at a later time.

The index is computed as the sum of n subindices times a scaling factor $10/(n + 1)$:

$$RPI = \frac{10}{n+1} \sum_{i=1}^{n} I_i \qquad (7)$$

The purpose of the scaling factor is to make the index, which has an increasing scale, vary from approximately 100 ("natural" levels) to approximately 1,000 ("highly polluted" levels). However, the index can go below 100, and, theoretically, it can approach zero. Thus, the theoretical range is from 0 to above 1,000.

The RPI was applied on a test basis using data from New York State's Water Quality Surveillance Network and from other sources. It was calculated for stations located on the Susquehanna, Genesee, Delaware, Mohawk and Hudson Rivers. The index tended to be high at times of low flow (late summer), when dilution of pollution was minimal, and low at times of high flow (spring runoff), when dilution was greatest. On the Susquehanna River, the index reflected the impact of the sewage outfall at Binghamton, New York, by showing distinct upstream and downstream differences. The eventual fate of the RPI is unknown, and no further applications of this index appear in the literature. As will be discussed in future sections of this chapter, New York State has chosen the NSF WQI for displaying its water quality data.

Dinius' Social Accounting System

Dinius[16] has proposed a water quality index as a first step toward designing a "rudimentary social accounting system" which would measure the costs and impact of pollution control efforts. The conceptual framework was patterned after the balance sheets used by accountants to describe the assets and liabilities of a firm. Dinius felt that the social accounting system would facilitate the reporting of environmental quality data to the citizen and administrator, thereby improving the monitoring of public expenditures for regulatory actions:

> Such a reporting and measurement system should (1) specifically inform the administration and the public of the quantity and location of pollution and (2) indicate the results of money and community effort expended for the control of natural resources.[16]
>
> Hopefully the described evaluating and systemizing techniques may lay the groundwork for the development of a simple universal social accounting system out of which may grow a system so refined and so perfected that the government and the public will have the means to recognize, assess, and thereby use their dollars and their time most effectively to control all classes of pollution.[16]

Although Dinius' water quality index was viewed as an initial step toward the development of the larger system, she suggested that it could be used as a possible water quality reporting system for the State of Alabama.

This water quality index includes 11 pollutant variables. Like Horton's index and the NSF WQI, it has a decreasing scale, with values expressed as a percentage of "perfect water quality," which corresponds to 100%:

> The measurement unit, called the quality unit Q, is the degree of pollution existing in the water at the testing date, expressed as a percent of "perfect" water. Thus highly polluted water would have a quality unit approaching 0%.[16]

Like Prati's index and McDuffie and Haney's index, the subindices are represented by explicit mathematical functions. Each subindex was developed from a review of the published scientific literature. Dinius examined the water quality ascribed by various authorities to different levels of pollutant variables, and from this information she generated 11 subindex equations (Table X).

The index is calculated as the weighted sum of the subindices, like Horton's index and the additive version of the NSF WQI:

$$I = \frac{1}{21} \sum_{i=1}^{11} w_i I_i \qquad (8)$$

The weights range from 0.5 to 5 on a "basic scale of importance." On this scale, 1, 2, 3, 4 and 5 denote, respectively, "very little," "little," "average," "great" and "very great" importance. Like the subindex functions, the weights are based on the investigator's evaluation of the importance of each pollutant variable and her review of the literature (Table XI). The weights

Table X. Subindex Functions for Dinius' Water Quality Index[16]

Dissolved Oxygen (%):	$I = X$	
5-Day Biochemical Oxygen Demand (mg/l):	$I = 107X^{-0.642}$	
Total Coliforms (MPN/100 ml):[a]	$I = 100X^{-0.30}$	
Fecal Coliforms (MPN/100 ml):[a]	$I = 100(5X)^{-0.30}$	
Specific Conductance (μmho/cm at 25°C):	$I = 535X^{-0.3565}$	
Chlorides (mg/l):	$I = 125.8X^{-0.207}$	
Hardness ($CaCO_3$, ppm):	$I = 10^{1.974 - 0.00132X}$	
Alkalinity ($CaCO_3$, ppm):	$I = 108X^{-0.178}$	
pH:	$I = 10^{0.2335X + 0.440}$	for $X < 6.7$
	$I = 100$	$6.7 \leqslant X \leqslant 7.58$[b]
	$I = 10^{4.22 - 0.293X}$	for $X > 7.58$[b]
Temperature (°C):	$I = -4(X_a - X_s) + 112$	X_a = actual temperature X_s = standard temperature
Color (C units):	$I = 128X^{-0.288}$	X is in C units measured after all suspended matter, evaluated by turbidity, is removed

[a]MPN = most probable number.
[b]Differs slightly from the value originally published.

Table XI. Weights for Dinius' Water Quality Index[16]

Variable	w
Dissolved Oxygen (%)	5
5-Day Biochemical Oxygen Demand	2
Fecal Coliforms	4
Total Coliforms	3
Specific Conductance	1
Chlorides	0.5
Hardness	1
Alkalinity	0.5
pH	1
Temperature	2
Color	1

sum to 21, which is the denominator in the index equation. The index is defined over the range from 0 to 100, although limits must be placed on the range of each variable to avoid values over 100.

Although this is a general water quality index, Dinius suggested that specific water uses could be accommodated by interpreting the index numbers differently for each water use. She proposed descriptor language for each of six specific water uses (Figure 14). Except for the terms "not acceptable" and "acceptable," which appear in several columns in the figure, the descriptor language differs strikingly for the various water uses. This language illustrates the diverse ways in which water quality can be interpreted for different uses.

To use this index in her social accounting system, Dinius treats water quality, or the "worth" of the water, as capital is treated by accountants. Pure water (100% quality) is treated as the original asset. Pure water (assets) minus the percentage of pollution (liabilities) equals the worth of the water (capital). The capital is weighted by the length of all streams, expressed in shoreline miles:

> Since Q is the capital, or value of the water when 100 is "perfect," liabilities are calculated by subtracting Q from 100. The liability, or 100% - Q, is also quantified in terms of equivalent miles by multiplying the above liability percent times the shoreline miles.[16]

One goal of this system is to produce an overall water quality "balance sheet" on a statewide basis.

The index was appllied by Dinius on an illustrative basis to data on several streams in Alabama. However, the literature does not contain any subsequent applications of this index. As we shall see in later sections of this chapter, the State of Alabama currently does not use a water quality index to report or analyze its water quality data.

Specific-Use Water Quality Indices

The descriptive language proposed by Dinius to enable a single index to be applied to different water uses illustrates the contrast in meaning of the different water quality variables when uses are taken into account. Walski and Parker[17] view the problem of different water uses as the most significant challenge facing index developers:

> The most significant problem facing the creation of water quality indices is that the uses for water are manifold and the quality of water demanded for each purpose varies tremendously. A high value of a certain parameter [variable] may be desirable in one instance and indifferent or even detrimental in another. For example, a high dissolved oxygen concentration is essential if good fishing is to be found in a body of water, but is only of marginal value in a drinking water supply, while it is highly undesirable in

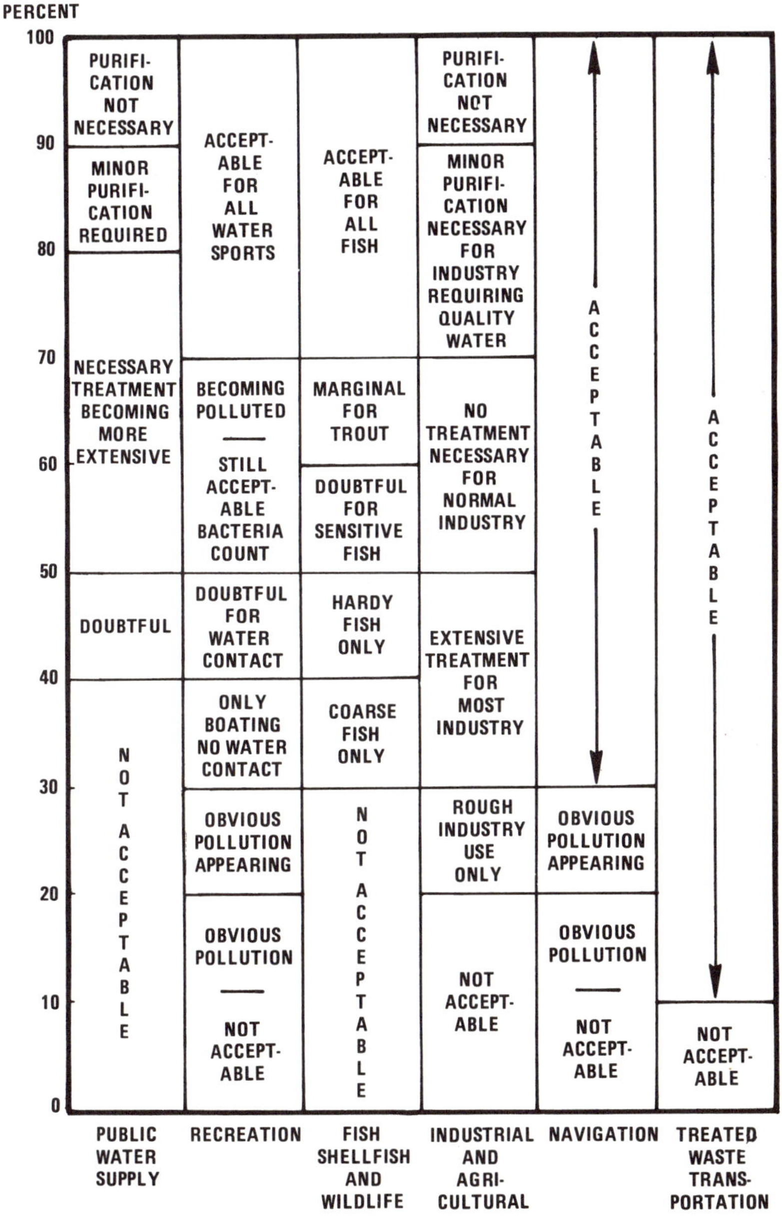

PERCENT	PUBLIC WATER SUPPLY	RECREATION	FISH SHELLFISH AND WILDLIFE	INDUSTRIAL AND AGRICULTURAL	NAVIGATION	TREATED WASTE TRANSPORTATION
100–90	PURIFICATION NOT NECESSARY	ACCEPTABLE FOR ALL WATER SPORTS	ACCEPTABLE FOR ALL FISH	PURIFICATION NOT NECESSARY	ACCEPTABLE	ACCEPTABLE
90–80	MINOR PURIFICATION REQUIRED			MINOR PURIFICATION NECESSARY FOR INDUSTRY REQUIRING QUALITY WATER		
70–60	NECESSARY TREATMENT BECOMING MORE EXTENSIVE	BECOMING POLLUTED — STILL ACCEPTABLE BACTERIA COUNT	MARGINAL FOR TROUT	NO TREATMENT NECESSARY FOR NORMAL INDUSTRY		
60–50			DOUBTFUL FOR SENSITIVE FISH			
50–40	DOUBTFUL	DOUBTFUL FOR WATER CONTACT	HARDY FISH ONLY	EXTENSIVE TREATMENT FOR MOST INDUSTRY		
40–30		ONLY BOATING NO WATER CONTACT	COARSE FISH ONLY			
30–20	NOT ACCEPTABLE	OBVIOUS POLLUTION APPEARING	NOT ACCEPTABLE	ROUGH INDUSTRY USE ONLY	OBVIOUS POLLUTION APPEARING	
20–10		OBVIOUS POLLUTION —		NOT ACCEPTABLE	OBVIOUS POLLUTION —	
10–0		NOT ACCEPTABLE			NOT ACCEPTABLE	NOT ACCEPTABLE

Figure 14. Descriptive language suggested by Dinius[16] to enable a single water quality index to be applied to different water uses. (Reprinted with permission, copyright by the American Geophysical Union.)

boiler feed water. Therefore, should a water quality index include dissolved oxygen as a parameter, and, if so, should it positively or negatively affect the index and what weight should be assigned to it?

Even within one use category, such as recreation, different variables have different importance:

> Even when recreation is taken as the use of water to be considered, there remain discrepancies between the uses that come under this heading. Boating, for example, is unaffected by dissolved oxygen concentration and coliform count, while swimming is drastically affected by the coliform count and fishing is affected by both.[17]

Some water quality specialists who do not accept the concept of general water quality believe that each index should be designed for a specific water use. A number of "specific-use" water quality indices have been proposed:

- O'Connor's indices (fish and wildlife, public water supply)
- Deininger and Landwehr's PWS index (public water supply)
- Walski and Parker's index (recreation)
- Stoner's index (public water supply, irrigation)
- Nemerow and Sumitomo's index (human contact, indirect contact, remote contact)

O'Connor[18] developed two specific-use indices and compared their performance with a general water quality index, the NSF WQI. Deininger and Landwehr[19] developed a public water supply index, but they concluded that its structure and performance are not sufficiently different from those of a general water quality index to warrant its use. Stoner[20] proposed an index which, with appropriate changes in coefficients, can be used either for public water supply or for irrigation. Nemerow and Sumitomo[21] proposed indices for three separate water uses which, when combined, give a general water quality index.

O'Connor's Indices

How badly do general water quality indices, such as the NSF WQI, fail when they are applied to specific water uses? Can a successful specific-use index be designed? Is it necessary to develop a system of water quality indices, one for each water use?

As an important step toward answering these questions, O'Connor[18] developed two water quality indices for specific, but very different, water uses. He applied the indices to water quality data to compare their performance with each other and with that of the NSF WQI. If water use is important enough to require different indices for different uses, he reasoned that these two indices, each designed for very different uses, should give very different results when applied to actual data. (This hypothesis assumes, of course, that each index accurately reflects the water use for which it was designed.)

O'Connor's first index, the Fish and Wildlife (FAWL) index, was intended to describe the quality of a surface body of raw water used to sustain a population of fish and wildlife. His second index, the Public Water Supply (PWS) index, was intended to describe the quality of a surface body of raw water which will be treated as necessary and used for public water supplies. Both indices were developed using an approach similar to the Delphi technique[7,8] employed by Brown et al.[6] but somewhat more interactive. It employed face-to-face interviews between the investigator and eight water quality experts distributed over various parts of the U.S. After two visits to each of the experts and a series of three written questionnaires, satisfactory agreement about the structure of the two indices was achieved.

In the first questionnaire, each respondent was asked about his feelings regarding the concept of general water quality: "Do you (or did you) feel comfortable with the concept of overall water quality or would you prefer to make judgments with a specific use in mind?"[18] The respondent also was asked whether he felt that the variables, weights, and subindex curves in an index would be different for different water uses. He was asked to rate the relative importance of 36 pollutant variables for each of the two uses on a scale from 0 to 100, ranging from "irrelevant" to "extremely important," and to evaluate the reliability and difficulty of measuring each pollutant variable.

The replies from the first questionnaire indicated that only three of the eight respondents felt comfortable with the concept of general water quality. All respondents felt that the index variables, weights and subindex rating scales would vary with use. Following the questionnaire, an interview lasting from 3 to 4 hours was conducted with each respondent, and the respondent was asked to reevaluate his importance ratings from the original questionnaire. He also was asked to select variables for inclusion in the FAWL and the PWS, keeping in mind both the difficulty of measuring each variable and the need for each variable to reflect the maximum amount of nonredundant water quality information. Each respondent also was asked to plot a subindex curve by establishing a point of maximum quality on a blank chart and drawing a continuous curve from 0 to 100. This approach was similar to the one used by Brown et al.[6] in developing subindex curves for the NSF WQI. Toxic substances also were handled in a way similar to that for the NSF WQI. The experts agreed that, if any toxic substance exceeded its recommended limit, the index would be set to zero.

The number of variables initially selected by the eight experts for inclusion in the FAWL index ranged from 6 to 14, and the number selected for the PWS ranged from 7 to 22. As the study progressed, the investigator was able to obtain greater consensus among the respondents regarding parameters, weights and subindex curves. Some variables were eliminated because they had been mentioned by only one respondent, and others were dropped because they appeared redundant. In a second interview, each respondent was allowed to see the distribution of importance weights and the average rating curves of the entire group, which he then could compare with his own responses. He could change his own replies if he so chose. This process resulted in a final list of 9 pollutant variables for the FAWL and 13 pollutant variables for the PWS. If the resulting variables and their weights are compared with those of the NSF WQI (Table XII), it is seen that six variables are common to the FAWL and PWS indices, and four variables (DO, pH, nitrates and turbidity) are common to all three indices. The relative weighting of these variables tended to differ in the three indices. For example, DO received the largest weighting in the NSF WQI and FAWL indices but received relatively small weighting in the PWS index.

Table XII. Comparison of Weights Used in Three Water Quality Indices

Pollutant Variable	NSF WQI[6]	O'Connor's Indices[18]	
		FAWL	PWS
Dissolved Oxygen	0.17	0.206	0.056
Fecal Coliforms	0.15		0.171
pH	0.12	0.142	0.079
5-Day Biochemical Oxygen Demand	0.10		
Nitrates	0.10	0.074	0.070
Phosphates	0.10	0.064	
Temperature[a]	0.10	0.169	
Turbidity	0.08	0.088	0.058
Total Solids	0.08		
Dissolved Solids		0.074	0.084
Phenols		0.099	0.104
Ammonia		0.084	
Fluorides			0.079
Hardness			0.077
Chlorides			0.060
Alkalinity			0.058
Color			0.054
Sulfates			0.050
Total	1.00	1.000	1.000

[a]Departure from equilibrium temperature.

The overall FAWL and PWS indices are computed as the weighted sum of the subindices times a factor which takes into account pesticides and toxic substances:

$$I_{FAWL} = \delta \sum_{i=1}^{9} w_i I_i \tag{9}$$

$$I_{PWS} = \delta \sum_{i=1}^{11} w_i I_i \qquad (10)$$

where δ = 0 if pesticides or toxic substances exceed recommended limits
       δ = 1 otherwise

As a final step in his study, O'Connor compared these specific-use indices with each other and with the NSF WQI by examining correlations among the indices using several data sets. Data sets 1 and 2 were artificially created for testing purposes. Data sets 3 and 4 were obtained from actual measurements on a stream, and data set 5 consists of data sets 3 and 4 pooled together. For the five data sets, the correlations between the FAWL and PWS indices were generally low, ranging from 0.501 to 0.733 (Table XIII).

Table XIII. Correlations between FAWL, PWS and NSF WQI Indices for Five Data Sets, as Reported by O'Connor[18]

	Data Set				
	1	2	3	4	5 (3+4)
FAWL Index	0.670	0.667	PWS Index 0.501	0.733	0.682
NSF WQI	0.879	0.854	FAWL Index 0.792	0.863	0.840
NSF WQI	0.684	0.744	PWS Index 0.652	0.860	0.809

Correlations between the NSF WQI and the FAWL index were much higher, ranging from 0.792 to 0.879, and correlations between the NSF WQI and the PWS index ranged from 0.652 to 0.860. Thus, each of the two specific-use indices correlated better with the NSF WQI than they did with each other. The investigator concludes that water use is an important concept to be considered in index development:

> Can a single index of *overall* water quality be used to describe the quality of water to the public, or must the specific use to which the water is put be considered? The low correlations between the PWS and FAWL indices ... clearly indicate that indices developed for different uses will assign very different quality values to water samples, and this indicates that specific use must be considered. One should note, however, that overall quality is a specific use if one chooses to define it as such. Specifying a use is choosing a subset of all uses and all users. In making valid judgments about overall quality, one would have to consider all uses and all users, and this seems difficult if not impossible.[18]

After evaluating his findings, O'Connor concludes that the higher correlations found between either of the two specific-use indices and the NSF WQI suggest the possibility that the NSF WQI may be described as a linear combination of the FAWL and PWS indices. Thus, the FAWL and PWS indices may be reporting a subset of the information contained in the NSF WQI, and the NSF WQI may depict a kind of "weighted average" of all possible specific-use indices. O'Connor's findings apparently strengthen the case for specific-use indices and the case for a general water quality index as well. He concludes that he would develop both types of indices if he were to undertake his study again:

> The experimenter in this study has concluded that if he were starting this project with enough time and resources, he would develop indices for each of the major uses for water. Then, in describing the quality of a body of water, he would report all of them along with an index of overall quality which was a weighted linear average of the indices for respective uses. As indicated, the WQI is very likely a good approximation to that average.[18]

An interesting part of O'Connor's study sought to examine the importance of the shapes of the quality rating curves. To assess the way in which the shapes of the curves affected the overall results, he created straight-line approximations of each subindex curve. This was done by drawing straight lines from the point on each curve where the subindex reached 100 to the point at which it reached 0. Then the correlations between these crude straight-line approximations and the original indices were examined. For the five data sets, the correlations between the two versions of the FAWL index were moderately high, but they varied with the data set: 0.988, 0.985, 0.885, 0.709 and 0.670. The correlations for the two versions of the PWS index were surprisingly high for all five data sets: 0.994, 0.995, 0.952, 0.968 and 0.978. O'Connor concludes that the actual shapes of rating curves may not be as important as the other factors which must be considered in developing an index, and, therefore, ". . . more time should be spent in identifying critical parameters than in ascertaining small differences between curves."[18]

Deininger and Landwehr's PWS Index

In 1971, Deininger and Landwehr[19] presented a specific-use index intended for water used for public water supply (PWS). In their study, several different forms of the index structure were developed and compared. The index was developed by sending a questionnaire to 12 members of the original panel of 142 water quality specialists who participated in the development of the NSF WQI. The overall approach was similar to that used by Brown et al.[6] The 12 respondents were teachers or professors of sanitary, civil, or public health engineering (5 respondents); consulting engineers (4 respondents); and administrative and public health officials (3 respondents). On the questionnaire, it was emphasized that each respondent should keep "in the back of his mind a free-flowing stream which will serve as a source of raw water for a public water supply."[19]

Respondents were given a list of 16 candidate variables and were asked to designate each one as "include" or "don't include" in the public water supply index. They also were allowed to add variables which did not appear on the original list, and space was provided on the form for writing in these new variables. For each variable designated "include," the respondent was asked to rate the importance of the variable for public water supply on a scale from 0 to 100 and to indicate the rating on a visual scale (Figure 15). This visual approach was very similar to the one used by O'Connor[18] in his questionnaire. To develop the subindex curves, respondents were given blank graphs (Figure 16) similar to those used by Brown et al.[6] The respondent was asked to draw a subindex quality curve on this graph, keeping in mind the use of the water:

> You are asked to draw a graph, illustrating your feelings as to how the raw water quality of public water supply is affected by different concentrations of these parameters. That is, the X-axis for each graph represents a measure of concentration for each parameter. The Y-axis represents your opinion as to the quality of the water at varying concentrations, with "0" indicating "lowest" quality, and "100", "highest."[19]

A total of 24 pollutant variables were added by respondents to the original list of 16 variables. The new variables included chlorides, ammonia, radioactivity, pesticides, and various toxic substances, such as the heavy metals. To select the variables to be used in the index, the investigators used the following criterion: if 9 of the 12 respondents (75%) had voted to include a variable, it would be included in the public water supply index. A total of 13 variables satisfied this criterion (Table XIV). However, the

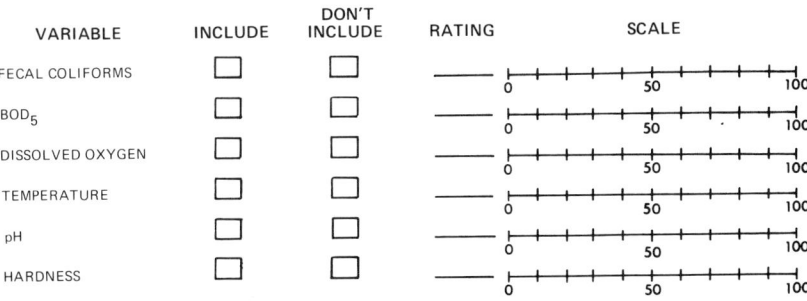

Figure 15. Example from questionnaire by Deininger and Landwehr[19] given to respondents to select and rate pollutant variables.

230 ENVIRONMENTAL INDICES

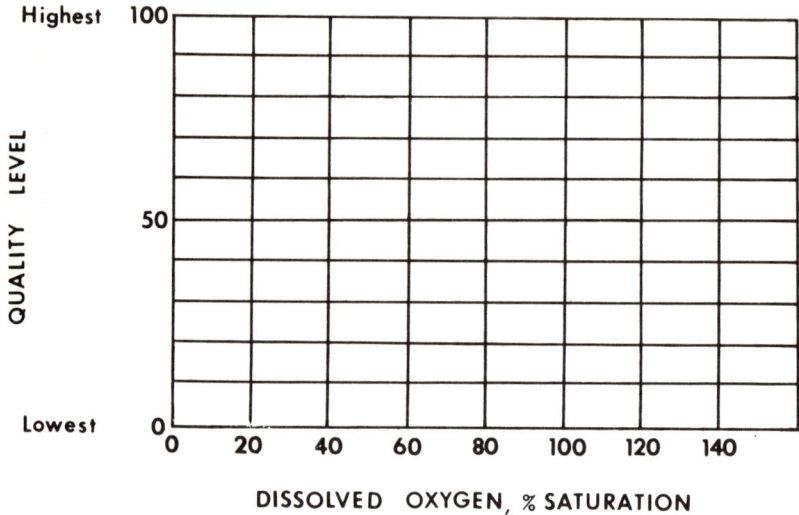

Figure 16. Example from questionnaire by Deininger and Landwehr[19] showing graph used for drawing individual subindex curves.

Table XIV. Variables Selected for Inclusion in Public Water Supply Index of Deininger and Landwehr[19]

Variable	No. of Respondents Voting to Include[a]	Average Importance Rating
Dissolved Oxygen	11	41
Fecal Coliforms	12	94
pH	11	56
5-Day Biochemical Oxygen Demand	9	61
Nitrates	12	68
Temperature	10	44
Turbidity	11	61
Dissolved Solids	10	64
Phenols	10	64
Color	10	65
Hardness	11	55
Fluorides[b]	11	57
Iron[b]	10	57

[a]From a total of 12 respondents.
[b]Not included in the 11-variable version of the index.

investigators were reluctant to include two of these variables, iron and fluorides, because ". . . they usually arise as a problem in well-water situations, rather than free-flowing streams, and so are not pertinent to the conditions specified."[19] Furthermore, since fluorides are reflected in dissolved solids, it was felt that this variable might introduce redundancy. To deal with these problems, the investigators decided to create two public water supply indices, an 11-variable version (without iron and fluorides) and a 13-variable version. The importance ratings were used to develop weights for each of the two versions. The subindex curves were averaged to give mean subindex functions for each of the 13 pollutant variables.

Two aggregation functions were considered: an additive form and a geometric mean. The 11-variable and 13-variable versions of the index were computed for each aggregation function:

Additive

$$PWS_{11} = \sum_{i=1}^{11} w_i I_i \qquad (11)$$

$$PWS_{13} = \sum_{i=1}^{13} w_i I_i \qquad (12)$$

Geometric Mean

$$PWS_{11} = \left[\prod_{i=1}^{11} I_i^{w_i} \right]^{1/11} \qquad (13)$$

$$PWS_{13} = \left[\prod_{i=1}^{13} I_i^{w_i} \right]^{1/13} \qquad (14)$$

For comparison purposes, the investigators also formulated a special geometric mean version of the NSF WQI.

When the weights of the additive forms of PWS_{11} and PWS_{13} are compared, they are seen to be very close (Table XV). The weights of the two versions of the geometric mean indices also were very close to each other (not shown). Seven variables were common to the NSF WQI and the PWS_{11} and PWS_{13} indices: DO, fecal coliforms, pH, BOD_5, nitrates, temperature and turbidity. The investigators note that the common variables suggest that the same information is contained in the different indices:

> This indicates that a good deal more than 50% (at least 7/9, 7/11, and 7/13, respectively, although probably much more) of the information about the water quality conditions is duplicated within each index. The non-specific-use-oriented index is practically embedded within the use-oriented indices.[19]

Table XV. Comparison of Weights in the NSF WQI and the
Two (Additive) Water Supply Indices

Pollutant Variable	NSF WQI[6]	Deininger and Landwehr[19]	
		PWS_{11}	PWS_{13}
Dissolved Oxygen	0.17	0.06	0.05
Fecal Coliforms	0.15	0.14	0.12
pH	0.12	0.08	0.07
5-Day Biochemical Oxygen Demand	0.10	0.09	0.08
Nitrates	0.10	0.10	0.09
Phosphates	0.10		
Temperature[a]	0.10	0.07	0.06
Turbidity	0.08	0.09	0.08
Total Solids	0.08		
Dissolved Solids		0.10	0.08
Phenols		0.10	0.08
Color		0.10	0.08
Hardness		0.08	0.07
Fluorides			0.07
Iron			0.07
Total	1.00	1.01	1.00

[a]Departure from equilibrium temperature.

After examining the weights for the two forms of the PWS_{11} and PWS_{13} indices and the NSF WQI, the investigators conclude that the additive and geometric weights are "usually quite close in value" for all six schemes. They also found that the subindex curves were relatively close to the original NSF WQI curves (Figure 17).

The investigators also examined the performance of the six indices using four sets of simulated water quality data. Although these indices differ in concept and form, the resulting values were generally close. Among the six indices, the maximum difference using any data set was 20 points, and the largest difference among the additive or geometric indices themselves did not exceed 11 points.

After reviewing their findings, the investigators conclude that the development of specific-use water quality indices does not appear warranted:

> The comparisons show that this index developed with a specific use-orientation does not seem to rate water quality levels in a manner markedly different from the rating made by a general, non-specific use oriented index. Thus, it is possible to argue that water of a certain quality retains that relative quality rating regardless of the use for which it is being

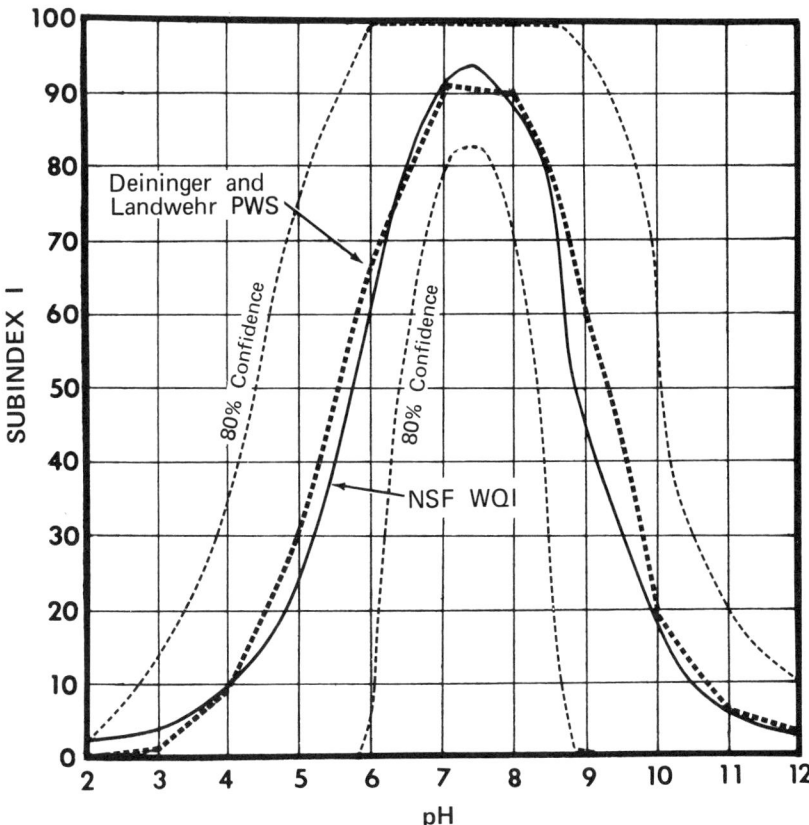

Figure 17. Comparison of pH subindices from the Deininger and Landwehr[19] PWS index (based on 12 respondents) and the NSF WQI.[9]

considered. Hence, waters of different streams can certainly be compared with regard to changes in quality levels, using one uniformly applied rating scheme.[19]

Instead of developing a number of indices for the many water uses, it appears to be more meaningful to further develop and refine a sensitive and general water quality index.[19]

Like O'Connor's study, this investigation was designed to examine the characteristics of specific-use indices and not to develop indices for routine application. Their conclusion, if correct, implies that a general water quality index can do the job of a specific-use index.

Walski and Parker's Index

In 1974, Walski and Parker[17] published a water quality index specifically intended for the recreational use of water, such as swimming and fishing. These investigators felt that most people are apt to judge surface waters on the basis of their recreational uses:

> For the purpose of this study, the use of water for recreation was treated as the primary consideration. This is not to imply that recreation is the most important use of water, but, rather, that people are most likely to judge the water quality by whether a water body is fit for fishing or swimming. People generally do not expect surface water to be of drinking quality, and purposes such as navigation and industrial cooling are too far removed from the populace to make them crucial considerations. The population may also be aware that the water quality requirements for navigation and cooling are less severe than for recreation.[17]

The variables selected for inclusion in the index were based on the investigators' evaluation of 65 commonly measured variables. In making this selection, they sought to represent water quality by as few variables as possible. They also tried to pick variables which can be measured rapidly in order to "...communicate the index to the populace quickly so that it can give the present actual water quality in the water body rather than yesterday's or last week's water quality." The investigators introduced four general categories of variables: (1) those which affect aquatic life (*e.g.*, DO, pH and temperature), (2) those which affect health (*e.g.*, coliforms), (3) those which affect taste and odor (*e.g.*, threshold odor number), and (4) those which affect the appearance of the water (*e.g.*, turbidity, grease and color). A total of 12 different pollutant variables were used in the index.

The subindices consist of nonlinear and segmented nonlinear explicit functions (Table XVI). Except for the two unimodal variables, pH and temperature, all subindices are represented by negative exponential equations (base e raised to negative powers). The pH and temperature subindices are represented by parabolic equations. Two subindices were used for temperature: one for actual temperature and another for departure from equilibrium temperature. In the subindex for actual temperature, temperatures around 20°C were considered desirable for American freshwater fish, with maintenance of a viable fishery considered to be increasingly difficult at lower or higher temperatures (Figure 18).

The coefficients and factors in the subindex equations were based on the investigators' judgment and on their review of the literature. For each subindex, values of the pollutant variable were identified which correspond to "intolerable" (I = 0.01), "poor" (I = 0.1), "good" (I = 0.9), and "perfect" (I = 1.0) water quality. An example of the approach can be seen in the development of the turbidity subindex function, or "sensitivity" function:

Table XVI. Subindex Functions for Index Proposed by Walski and Parker[17]

Pollutant Variable	Equation	Range
Dissolved Oxygen (mg/l)	$I = e^{[0.3(X - 8)]}$	$0 < X \leq 8$
	$I = 0$	$8 < X$
pH (standard units)	$I = 0$	$X < 2$
	$I = 0.04[25 - (X - 7)^2]$	$2 \leq X \leq 12$
	$I = 0$	$12 < X$
Total Coliforms (no./100 ml)	$I = e^{-0.0002X}$	
Temperature (°C)		
Actual:	$I = 0.0025[1 - (X - 20)^2]$	$0 \leq X \leq 40$
Deviation:	$I = 0$	$\Delta X < -10$
	$I = 0.01(100 - \Delta X^2)$	$-10 \leq \Delta X \leq 10$
	$I = 0$	$10 < \Delta X$
Phosphates (mg/l)	$I = e^{-2.5X}$	
Nitrates (mg/l)	$I = e^{-0.16X}$	
Suspended Solids (mg/l)	$I = e^{-0.02X}$	
Turbidity (JTU)[a]	$I = e^{-0.001X}$	
Color (C units)	$I = e^{0.002X}$	
Grease		
Thickness (μ):	$I = e^{-0.35X}$	
Concentration (mg/l):	$I = e^{-0.016X}$	
Odor (threshold odor number)	$I = e^{-0.1X}$	
Secchi Disk Transparency (m)	$I = \log(X + 1)$	$X \leq 9$
	$I = 1$	$9 < X$

[a]JTU = Jackson Turbidity Units.

To determine the sensitivity function for turbidity, two facts were noted. The first is that the upper limit placed by some states on the turbidity of a water to be employed as a source for a water supply is less than 250 units as reported by McKee and Wolf. Therefore, a turbidity of 100 units was assigned a value of 0.9. The second point is that Kemp reported that turbidities of 3,000 units are considered dangerous to fish over long exposures. Therefore, a turbidity of 2,300 units was assigned a value of 0.1.[17]

The resulting turbidity subindex function (Figure 19) is as follows:

$$I = e^{-0.001X} \tag{15}$$

236 ENVIRONMENTAL INDICES

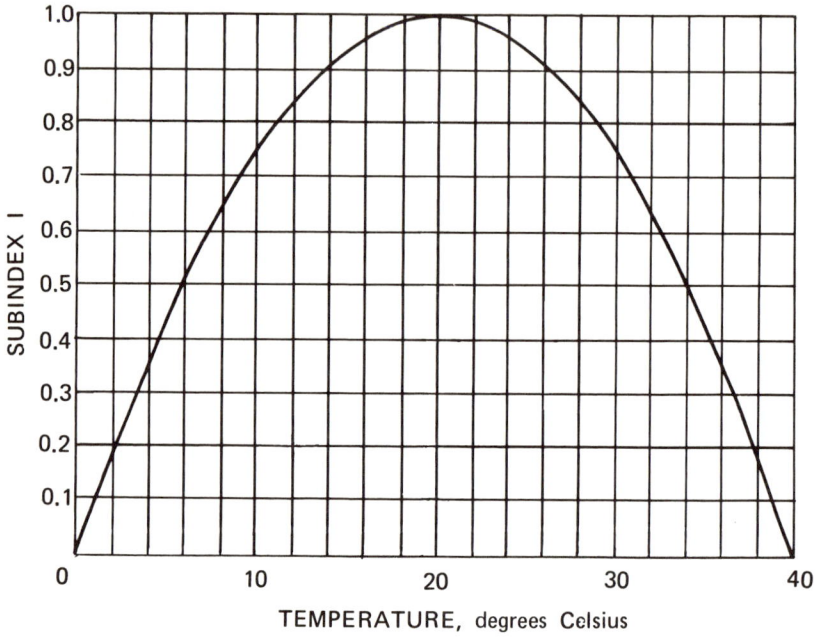

Figure 18. Parabolic function for water temperature used in the index of Walski and Parker.[17]

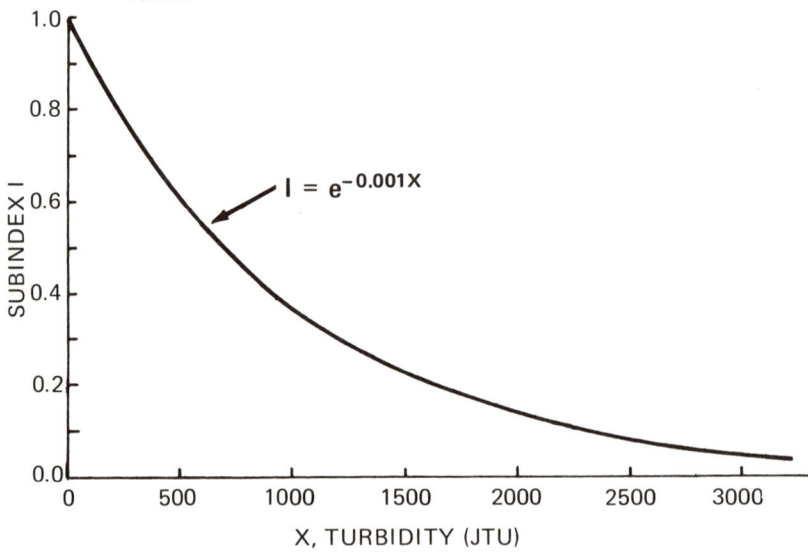

Figure 19. Exponential subindex function for turbidity proposed by Walski and Parker.[17]

This equation gives I = 0.1 for X = 2,300, I = 0.9 for X = 100, and I = 1.0 for X = 0 Jackson Turbidity Units. These turbidity levels are considerably higher than those found in most of the other water quality indices. Because the subindices used by Walski and Parker usually contain only one constant, fitting each equation to the desired values is necessarily approximate.

To aggregate subindices, Walski and Parker chose a geometric mean over an arithmetic mean, because they felt it can better handle the problem caused when just one subindex indicates poor water quality and the corresponding subindices indicate good water quality; that is, the problem of eclipsing:

> An arithmetic mean is the obvious first choice, but it has the shortcoming of failing to go to zero when the value of one of the parameters becomes execrable to the point of rendering the water useless for recreation. For example, a river may contain no color, odor, grease, or coliform bacteria, but may have a pH of one, and still have a good value for its water quality index if an arithmetic mean is used. Therefore, one would wish an average where an extremely low value, zero, would cause the index to go to zero. A geometric mean has this property and is better since it is less affected by extreme values.[17]

Their aggregation function is as follows:

$$I = \left[\prod_{i=1}^{12} I_i^{w_i} \right]^{1/12} \quad (16)$$

The published article on this index does not give the values of the weights.

This index was applied on a pilot basis to data from the Stones River near Nashville, Tennessee, and to data from sampling points along the Cumberland River upstream and downstream from the Metro Nashville Wastewater Treatment Plant. Because the index is designed for water intended for recreational use, the investigators suggest that it would be desirable to make the water quality measurements at beaches and other areas where people would be planning to fish or swim. Although the index was intended for routine application, it does not appear to have been implemented routinely by any existing water pollution control agencies.

Stoner's Index

Suppose one wishes to develop a single index which can accommodate two different water uses by changing subindex equations and weights. What would such an index look like? Stoner[20] has proposed a specific-use water quality index designed for two water uses: public water supply and irrigation. This index employs a single aggregation function which selects from two sets of recommended limits and subindex equations. Stoner views this approach as a general structure designed to accommodate any water use:

With the method to be described, a water quality index for any water use, whether broad or narrow in scope, can be developed if certain information can be provided. The minimum information items needed are (1) a set of limits for each water quality characteristic to be considered, (2) a rationale for the establishment of the limits, and (3) some information on, or appraisal of, the relationship of various concentrations of each water quality characteristic to the specific use for which the index is being developed.[20]

Although Stoner applies the index to just two water uses, it could be adapted to additional water uses as well.

Two general types of variables are used in Stoner's index:

Type 1: Variables normally considered toxic (for example, lead, chlordane and radium-226)

Type 2: Variables which affect health or aesthetic characteristics (for example, chlorides, sulfur, color, taste and odor)

The Type I pollutant variables are treated in a dichotomous manner, giving subindex step functions. Each Type I subindex is assigned the value of zero if the concentration is less than or equal to the recommended limit and the value -100 if the recommended limit is exceeded (Figure 20). The recommended limits are based on water quality criteria such as those published by the National Academy of Sciences.[22] An important reason for treating Type I variables as step functions was the difficulty of relating very low concentrations of a toxic substance to various levels of water quality:

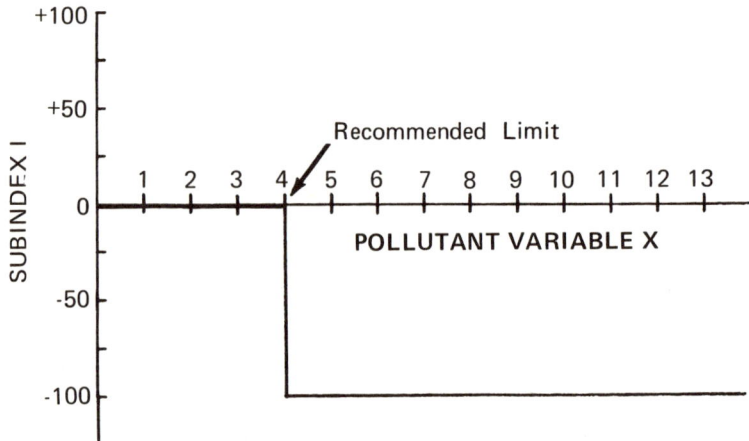

Figure 20. Step functions used in Stoner's index[20] for Type I pollutant variables.

These parameters were similar, in that their limits were in the very low concentration range. Examples of these are lead at 0.05 milligrams per litre (mg/l), chlordane at 0.003 mg/l, and radium 226 at 20 picocuries per litre (pC/l). The information available for these parameters indicated that the probability of assigning a mathematical statement to describe the effect of concentration with a reasonable degree of validity would not be practical. In other words, in practice, it would be difficult to judge the fitness of a water with 0.001 mg/l chlordane against a water with 0.002 mg/l chlordane. One other point of similarity was that when prescribed limits were exceeded by any of these parameters, it would indicate a significant health hazard.[20]

A total of 26 Type I pollutant variables are used in the public water supply version of the index, and 5 Type I variables are used in the irrigation version (Table XVII).

The Type II pollutant variables in Stoner's index are represented, on the other hand, by explicit mathematical functions. For these subindices, Stoner chose the simple linear function $a + bX$, the parabolic function $a + bX^2$, and the combination of the two, $a + bX + cX^2$. He felt that the Type II subindices should employ the simplest possible functions that can be supported by existing knowledge: "Because there was no evidence to indicate that a complicated function would be any more valid than a simple function, the simplest functions that described the effect of the constituents were used."[20] A total of 13 Type II pollutant variables are included in the public water supply version of the index (Table XVIII) and 16 Type II variables are included in the irrigation version (Table XIX). All equations are based on the recommended limits in published water quality criteria. The constants in each subindex equation are selected so that $I = 0$ when a recommended limit is reached, and $I = 100$ when the "ideal" value of that pollutant variable is attained. For example, the public water supply nitrite subindex (Figure 21) gives $I = 0$ when the recommended limit of 1.0 mg/l nitrite as nitrogen (NO_2-N) is reached and $I = 100$ when the ideal limit of 0.0 mg/l is attained. Because the safe concentration level of fluorides is related to air temperature, the fluoride subindex is computed by first computing the ratio of the observed fluoride concentration X (mg/l) to the maximum recommended fluoride concentration X_m at specified temperature ranges (Table XX); that is, by computing $R = X/X_m$. The public water supply version of the index includes methylene blue active substances (MBAS) because of "...their tendency to produce undesirable aesthetic effects, foaming, dispersion of insoluble or sorbed substances, and because of their tendency to interfere with the removal of substances by coagulation, sedimentation, and/or filtration."[20] As shown in Tables XVII and XIX, the Type II pollutant variables are classified into different groups, with weights specified for each group.

The overall index is computed by combining the unweighted Type I subindices with the weighted Type II subindices:

Table XVII. Recommended Limits for Type I Variables
(Toxic Substances) Used in Stoner's Index[20]

Pollutant Variable	Recommended Limit	
Public Water Supply		
Arsenic	0.1	mg/l
Barium	1.0	mg/l
Cadmium	0.01	mg/l
Chromium	0.05	mg/l
Lead	0.05	mg/l
Mercury	0.002	mg/l
Selenium	0.01	mg/l
Cyanide	0.2	mg/l
Aldrin	0.001	mg/l
Chlordane	0.003	mg/l
DDT	0.05	mg/l
Dieldrin	0.0005	mg/l
Endrin	0.0001	mg/l
Heptachlor	0.0001	mg/l
Heptachlor Epoxide	0.0001	mg/l
Lindane	0.005	mg/l
Methoxychlor	1.0	mg/l
Toxaphene	0.005	mg/l
Organophosphates, Carbamates	0.1	mg/l
2,4-D (2,4-Dichlorophenoxyacetic acid)	0.02	mg/l
Silvex	0.03	mg/l
2,4,5-T (2,4,5-Trichlorophenoxyacetic acid)	0.002	mg/l
Radium-226	20	pCi/l
Iodine-131	100	pCi/l
Strontium-90	200	pCi/l
Strontium-89	2,000	pCi/l
Irrigation		
Molybdenum	0.01	mg/l
Selenium	0.02	mg/l
Dalapon	0.2	µg/l
TCA (trichloroacetic acid)	0.2	µg/l
2,4-D	0.1	µg/l

$$I = \sum_{i=1}^{n} T_i + \sum_{j=1}^{m} w_j I_j \tag{17}$$

where T_i = subindex for the ith Type I pollutant variable
w_j = weight for the jth Type II pollutant variable
I_j = subindex for the jth Type II pollutant variable

Table XVIII. Subindex Functions for Stoner's Public Water Supply Index[20]

Variable	Subindex Function
Group A (w = 0.134)	
Ammonia-Nitrogen (mg/l)	$100 - 200X$
Nitrite-Nitrogen (mg/l)	$100 - 100X^2$
Fecal Coliforms (no./100 ml)	$100 - 0.000025X^2$
Group B (w = 0.089)	
pH (standard units)	$-1{,}125 + 350X - 25X^2$
Fluorides (see text)	$98.8 + 24.7X - 123X^2$
Group C (w = 0.067)	
Chlorides (mg/l)	$100 - 0.4X$
Sulfates (mg/l)	$100 - 0.4X$
Group D (w = 0.053)	
Phenols (μg/l)	$100 - 100X$
Methylene Blue Active Substances	$100 - 200X$
Group E (w = 0.045)	
Copper (mg/l)	$100 - 100X^2$
Iron (mg/l)	$100 - 333X$
Zinc (mg/l)	$100 - 20X$
Color (Pt-Co units)	$100 - 0.0178X^2$

Table XIX. Subindex Functions for Stoner's Irrigation Index[20]

Variable	Subindex Function
Group A (w = 0.111)	
Sodium Absorption Ratio	$100 - X^2$
Specific Conductance (μmho)	$100 - 0.0002X^2$
Fecal Coliforms (no./100 ml)	$100 - 0.0001X^2$
Group B (w = 0.074)	
Arsenic (mg/l)	$100 - 1{,}000X$
Boron (mg/l)	$100 - 100X^2$
Cadmium (mg/l)	$100 - 10^6 X^2$
Group C (w = 0.0555)	
Aluminum (mg/l)	$100 - 4X^2$
Beryllium (mg/l)	$100 - 10^4 X^2$
Chromium (mg/l)	$100 - 10^4 X^2$
Cobalt (mg/l)	$100 - 2{,}000X$
Manganese (mg/l)	$100 - 500X$
Vanadium (mg/l)	$100 - 1{,}000X$
Group D (w = 0.028)	
Copper (mg/l)	$100 - 2{,}500X^2$
Fluorides (mg/l)	$100 - 100X^2$
Nickel (mg/l)	$100 - 2{,}500X^2$
Zinc (mg/l)	$100 - 25X^2$

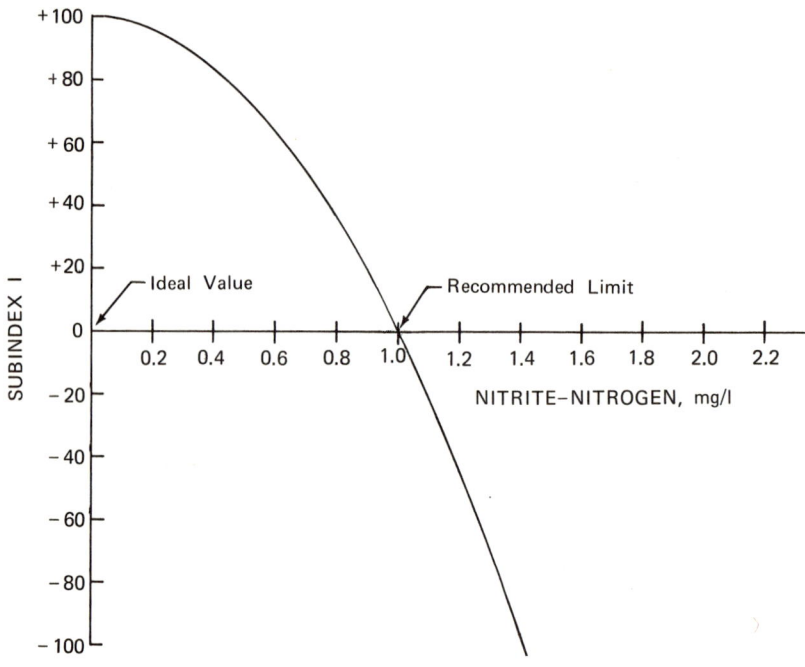

Figure 21. Example of a parabolic subindex function for a Type II variable, nitrite-nitrogen, from Stoner's index.[20]

Table XX. Recommended Limits for Fluoride Concentration as a Function of Temperature, from Stoner's Index[20]

Annual Average of Daily Maximum Temperatures (°C)	Maximum Fluoride Concentration, X_m (mg/l)
26.5-32.8	1.4
21.9-26.4	1.6
18.1-21.8	1.8
14.7-18.0	2.0
12.5-14.6	2.2
9.7-12.4	2.4

The right-hand term in Equation 17 can never exceed 100. When one Type I variable exceeds its recommended limit, the left-hand term becomes -100, making the overall index zero or less. Thus, the index becomes negative even if only one Type I pollutant variable exceeds its recommended limit. When a Type II pollutant variable exceeds its recommended limit sufficiently to make the water unfit for its intended use, the right-hand term also may become negative. Thus, Stoner's index has a decreasing scale which ranges from I = 100 (best possible water quality) to a large negative number (worst water quality). Stoner suggests that negative values of the index may be interpreted as a measure of water treatment costs:

> One can argue that, once the concentration of a constituent reaches a concentration that renders the water unfit, any further increase in concentration does not make the water any more unfit. This is probably true; for example, a water with 20,000 mg/l chloride is probably no more unfit for drinking than a water containing 10,000 mg/l chloride. The WQI is, however, designed in part to provide managers and planners with information they can use in decision-making processes. In general, the greater the negative number the more the need for treatment increases.[20]

Stoner applied the index to several waters in Texas and one in Florida (Table XXI) to illustrate how it can be used to select which stream is suitable for public water supply. The index ranged from I = -8,560 (Station 5, Buffalo Bayou) to I = 87.5 (Station 3, Little Wehiva River). Stoner[20] draws the following conclusions from these results:

1. Of the first three sources, the Little Wehiva River is the most suitable and the Cedar Bayou is the least suitable.
2. Any of the first three sources could be used for public water supplies.
3. The Cedar Bayou, although adequate, should be examined further to determine the reason for its relatively low score. (In this case, it turned out to be color, phenols and fecal coliforms.)
4. The adequacy of the Buffalo Bayou is dependent upon water discharge and, generally, is unfit for use without further treatment.
5. The Buffalo Bayou at low-water discharge can be extremely hazardous for use as a public water supply.

Stoner also applied the index to waters intended for irrigation use, but he found it difficult to find water quality data with all the irrigation variables represented, and missing values had to be estimated.

Stoner[20] concludes that specific-use indices can meet a variety of needs:

- They provide directly comparable numbers such that various waters could be judged for use in specific categories.
- They allow for comparison over time, *i.e.*, time trends.
- They provide numbers that managers and other nontechnical personnel can use more easily.

244 ENVIRONMENTAL INDICES

Table XXI. Results of Applying Stoner's Public Water Supply
Index to Selected Waters[20]

Monitoring Station	Index Value
1. Cedar Bayou near Crosby, Texas (November 8, 1972)	57.1
2. Chocolate Bayou near Alvin, Texas (May 16, 1973)	81.5
3. Little Wehiva River near Altamonte Springs, Florida (May 17, 1973)	87.5
4. Buffalo Bayou at Piney Point, Texas Discharge = 400 cfs (November 15, 1972)	61.3
5. Buffalo Bayou at Piney Point, Texas Discharge = 59 cfs (March 12, 1973)	-8,560.0
6. Buffalo Bayou at Piney Point, Texas Discharge = 155 cfs (May 14, 1973)	28.3
7. Buffalo Bayou at Piney Point, Texas Discharge = weighted average of Stations 4 and 6	-44.1

- They indicate waters of "good" and "bad" water quality for specific-use categories.

Stoner's index is interesting, as it shows that the complexity of an index greatly increases when it is used to reflect different water uses. If additional water uses such as recreation or maintenance of fishery and wildlife habitats were considered in the index, it would require additional variables, weights and subindex functions.

Nemerow and Sumitomo's Pollution Index

Nemerow and Sumitomo[21] have proposed an increasing scale water quality index consisting of three specific-use indices. The three separate water uses are denoted by j = 1, 2 and 3:

- Human Contact Use (j = 1)
- Indirect Contact Use (j = 2)
- Remote Contact Use (j = 3)

Human contact use includes uses in which humans come into direct contact with the water, such as drinking (including waters used for beverage manufacturing) and swimming. Indirect contact use includes uses in which humans have less direct contact with waters, such as fishing, food processing and agriculture. Finally, remote contact use includes uses in which human contact is very indirect, such as in navigation, industrial cooling, and some recreational activities (aesthetics, picnicking, hiking, and visits to the area).

Each specific-use index includes pollutant variables represented by linear or segmented linear subindex functions. The linear subindex functions are constructed so that I = 0 when a pollutant variable is at its desired value, $X = X_o$, and I = 1 when a pollutant variable is at its recommended limit, $X = X_s$. Assuming $X > X_o$, a typical linear subindex is shown in Figure 22; it has the following equation:

$$I = \alpha X + \beta \tag{18}$$

where $\alpha = \dfrac{1}{X_s - X_o}$

$\beta = -\dfrac{X_o}{X_s - X_o}$

For cases in which a subindex function has a desired value of $X_o = 0$, Equation 18 reduces to the simple ratio of the pollutant variable to its recommended limit, $I = X/X_s$. For unimodal subindex functions, such as pH, two line segments are joined together, one with a negative slope and one with a positive slope (Figure 23):

$$I = \dfrac{X_o - X}{X_o - X_a} \quad \text{for } 0 \leqslant X < X_o \tag{19}$$

$$I = \dfrac{X - X_o}{X_s - X_o} \quad \text{for } X_o \leqslant X \tag{20}$$

where: X = pollutant variable
X_a = lower recommended limit
X_s = upper recommended limit
X_o = desired level

Actually, Equation 20 is simply another way to write Equation 18.

To reduce eclipsing problems, the subindices are aggregated in a unique manner. For each specific use j, the maximum subindex is combined with the arithmetic mean of n subindices in a root-mean-square operation:

$$I_j = \sqrt{\dfrac{\left[\max_{\text{all } i}\{I_{ij}\}\right]^2 + \left[\dfrac{1}{n}\sum_{i=1}^{n} I_{ij}\right]^2}{2}} \tag{21}$$

246 ENVIRONMENTAL INDICES

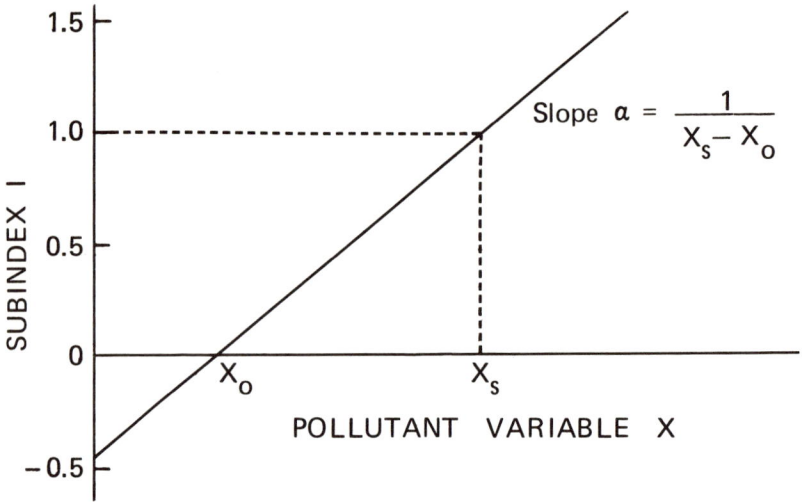

Figure 22. Linear subindex function from the water quality index of Nemerow and Sumitomo.[21]

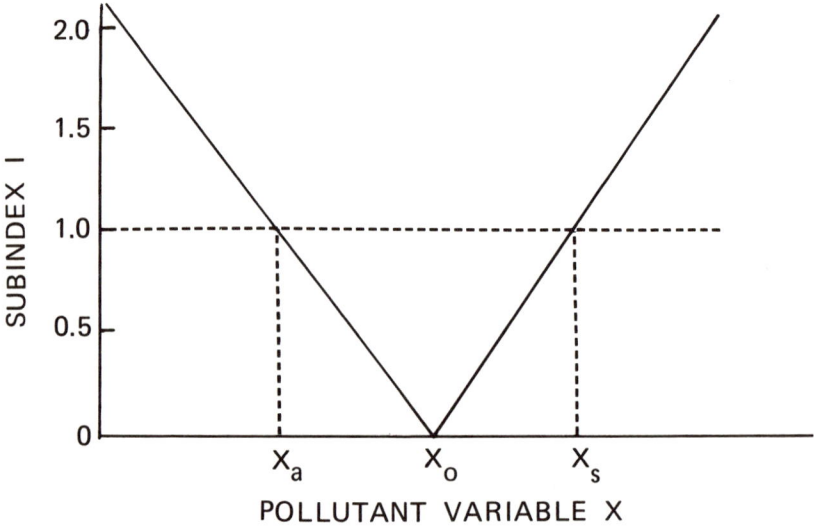

Figure 23. Subindex function for a unimodal pollutant variable such as pH from the water quality index of Nemerow and Sumitomo.[21]

Using this approach, each specific-use index reflects both the highest subindex (a measure of the extreme) and the average of all subindices (a measure of central tendency). The investigators recommend using 14 pollutant variables in the index (Table XXII).

Table XXII. Variables in the Index of Nemerow and Sumitomo[21]

Dissolved Oxygen	Dissolved Solids
Fecal Coliforms	Suspended Solids
pH	Color
Total Nitrogen	Hardness
Alkalinity	Chlorides
Temperature	Sulfates
Turbidity	Iron and Manganese

Finally, the general water quality index is computed as the weighted sum of the three specific-use indices:

$$I = \sum_{i=1}^{3} w_i I_i \qquad (22)$$

Nemerow and Sumitomo have applied this index to over 70 stations in New York State (creeks, rivers and lakes), computing the three specific-use indices as well as the general water quality index for each location.

In general, specific-use indices such as these illustrate the tendency for indices to become increasingly complex as greater realism is attempted. This is another example of the "classic dichotomy" discussed in Chapter I—the essential tradeoff between simplicity and accuracy.

Planning Indices

Another category of water quality indices, the "planning indices," is designed specifically for management decision-making. Unlike the general and specific-use indices, these indices usually do not depict ambient water quality or related conditions. Rather, they are "custom-designed" to assist the user in making specific decisions or in solving particular problems. Planning indices often incorporate variables other than those routinely measured by water pollution monitoring programs. For example, a planning index designed for allocating water pollution abatement funds might include the "cost of wastewater treatment facilities."

A great many planning indices have been proposed, and several examples will be discussed briefly:

- MITRE's Indices (PDI, NPPI and PAI)

- Dee's Environmental Evaluation System (EES)
- Inhaber's Canadian National Index
- Zoeteman's Pollution Potential Index
- Johanson and Johnson's Pollution Index

MITRE's Indices

As reported by Truett et al.[23] and Greeley et al.,[24] three water pollution indices were developed jointly by EPA and the MITRE Corporation. The first of these, the Prevalence Duration Intensity (PDI) index, closely resembles a general water quality index, except that it includes a variable designed to reflect the effects of water pollution on the environment. The PDI index is computed as the product of three variables divided by the total stream miles:

$$PDI = \frac{P \times D \times I}{M} \quad (23)$$

where   P = Prevalence: number of stream miles in violation of recommended limits within an area
D = Duration: number of quarter-year periods during which violations occur
I = Intensity: measure of the severity of the effects of violations
M = total stream miles

Each of the variables in the numerator can assume one of several fixed values. For example, duration can take on any of four different values (0.4, 0.6, 0.8 and 1.0). If violations occur only within a single quarter, D = 0.4, and if violations occur in all four quarters, D = 1.0. Intensity is computed as the sum of three other subindices: an ecological effects subindex, a utilitarian effects subindex, and an aesthetic effects subindex. The ecological effects subindex can take on any of four values (0.1, 0.3, 0.4 and 0.5); the utilitarian effects subindex can take on any of three values (0.1, 0.2 and 0.3) to reflect economic damages of water pollution; and the aesthetic effects subindex can take on one of two values (0.1, 0.2). Intensity can range from 0.3 to 1.

The PDI index was applied in 1970 and 1971 by various EPA regional offices to the nation's waters. The 1972 *CEQ Annual Report* evaluated the index favorably:

> The EPA PDI index has several advantages. It covers all U.S. surface waters. It considers the relationship of actual water quality to State standards of desirable water quality. And it allows for judgment as to the effects of the water pollution in any particular stream. It has proved a useful management tool, for planning, for directing resources to the most polluted areas and for suggesting improvements in monitoring coverage.[25]

However, a disadvantage is that the index depends on recommended administrative limits, such as state water quality standards, which generally vary from state to state. Thus, it is difficult to apply the PDI index to different states and make meaningful comparisons of water quality.

MITRE's National Planning Priorities Index (NPPI) is a planning index designed "... to assign priorities to each 'planning area' within the nation..." in order to "... ensure that funds are granted and used in a cost-effective manner for the planned waste treatment projects (plus other types of water quality control measures)."[23] It is computed as the weighted sum of 10 subindices (Table XXIII):

$$NPPI = \sum_{i=1}^{10} w_i I_i \tag{24}$$

Each subindex is computed using a segmented linear function (not shown). One subindex, which receives the greatest weight, is based on the PDI index itself. Notice that the NPPI includes variables not found in the ambient (general and specific-use) indices, such as "investment for fiscal year (FY) 1975-76," and "controllability," which is a measure of the difficulty of controlling the water pollution sources in the area.

The third MITRE index, the Priority Action Index (PAI), is similar to the NPPI but simpler in form. Like the NPPI, it was intended to assist decision-makers in allocating funds for wastewater treatment facilities. The PAI includes 4 of the 10 variables used in the NPPI: current area population (w = 0.17), downtown area population (w = 0.17), controllability (w = 0.26), and the PDI index (w = 0.40).

The three MITRE indices—PDI index, NPPI, and PAI—were computed for more than 1,000 planning areas covering the 48 contiguous U.S. states. The results were used to rank locations in terms of their priority for receiving assistance for water pollution control efforts. Although these indices occasionally appear in various reports, none of the three currently is being used routinely by decision-makers in the U.S.

Dee's Environmental Evaluation System

Dee et al.[26,27] proposed a general system for evaluating the environmental impact of large-scale water resource projects, such as those carried out by the Bureau of Reclamation. The Environmental Evaluation System (EES) is designed to assess environmental impact in four major categories: ecology, environmental pollution, aesthetics, and human interest. These four categories are represented by 18 components and, finally, by 78 individual variables. The 78 variables include factors such as soil erosion, sport fish, waterfowl, housing, land use, ethnic groups and noise. An important part of

Table XXIII. Variables Included in MITRE's NPPI[23]

Variable	Weight
Current Area Population	0.10
Downstream Affected Population	0.11
Investment in FY 1972	0.08
Investment in FY 1973-74	0.09
Investment in FY 1975-76	0.08
Controllability	0.16
Required Planning Level	0.06
Delta Planning Level	0.06
PDI Index	0.22
Per Capita Planning Cost	0.04
Total	1.00

this system is a water quality index, which is represented by 12 common water quality variables (DO, pH, turbidity, fecal coliforms, etc.) plus pesticides and toxic substances. The water quality variables use subindex functions similar to those in the NSF WQI.

Each of the 78 variables in the overall system is represented by subindices which range from 0 (extremely poor quality) to 1 (extremely good quality). The index is based on the weighted sum of the subindices, and the weights are chosen so that the index ranges from 0 to 1,000. Because the weights assigned to the water quality variables are relatively large, the water quality portion of the EES constitutes about one-third of the total index.

In practice, the user first calculates the index "without" the proposed water resource project. Then the calculation is repeated "with" the proposed project. The difference between the two scores is considered a measure of the environmental impact (EI) of the project:

$$EI = \sum_{i=1}^{78} w_i I_i \text{ [with]} - \sum_{i=1}^{78} w_i I_i \text{ [without]} \qquad (25)$$

The EES has been criticized by Andrews[28] because of the relative importance given to the variables by the choice of weights:

> One would doubt strongly, for instance, that a cross section of American society would attribute only 28 "social importance" points out of 1000 to impacts on ethnic and religious groups but 318 to water pollution. On the other hand, one might also question what professional basis might exist for the implied claim that upland game birds are equally as sensitive to disruption by water resource projects as Indians and twice as sensitive as other ethnic and religious groups.

He concludes that the EES developers tend to "... carry many of the shortcomings of others to such a level of elaboration and present their approach in so uncritical a fashion as to require detailed comment and criticism."[28] This problem is typical of systems which combine many diverse, unrelated variables, because the weights assigned to the variables are often arbitrary, and they embody the subjective values of the developers.

Inhaber's Canadian National Index

The Environmental Quality Index (EQI) suggested by Inhaber[29] in 1974 as a national index for Canada included an air quality index (Chapter III, pages 118 and 119), a water quality index, and a land quality index.

The water quality index[30] combines two subindices in a root-mean-square operation: an ambient water quality subindex and a pollutant source subindex based on effluents from point sources. The ambient water quality subindex is, in turn, comprised of three subindices: (1) a trace metals subindex based on cadmium, lithium, copper, zinc and the hardness of water; (2) a turbidity subindex; and (3) a commercial fish catch subindex based on the weight and mercury content of fish landed by Canadian ships. The pollutant source subindex is based on pollutant variables measured in effluents from five sources (municipal wastes and the petroleum-refining, chlor-alkali, fish-processing, and paper industries). The subindices are combined in successive root-mean-square operations. The rationale for the weights of the subindices and the manner in which they are aggregated is unclear. Although originally intended for nationwide use, there is no evidence that the index has been routinely applied in Canada.

Zoeteman's Pollution Potential Index

The Pollution Potential Index (PPI) developed by Zoeteman[31] in 1973 is a planning index based not on observed water quality variables but on indirect factors assumed to be responsible for pollution. It is based on the size of the population within a given drainage area, the degree of economic activity (measured by the per capita Gross National Product), and the average flow rate of the river:

$$PPI = \frac{NG}{Q} \times 10^{-6} \qquad (26)$$

where: N = number of people living in a drainage area
G = average per capita Gross National Product (GNP)
Q = yearly average flow rate (m^3/s)

Using historical records for the Rhine River in Germany, Zoeteman found that the PPI often correlated well with measured water quality variables (phosphates, nitrates, chlorides, BOD_5 and iron). He also applied the PPI to 160 river sites throughout the world, comparing PPI values with 11 pollutant variables for which more than 40 observations were available. The PPI ranged from 0.01 to 1,000 for these rivers. PPI values were strongly associated with coliform levels, organophosphate content, nitrate content, and hardness of the water. For PPI values above 10, a relationship also was found between the index and BOD_5, chlorides, fluorides, iron and zinc. The PPI was used to rank the pollution potential of the world's rivers:

> The highest potential pollution is found in relatively small rivers in the U.S.A. and Western Europe. Low PPI values are registered for rivers of different size in Asia, Africa and parts of Scandinavia. Among the larger rivers of the world with an average yearly discharge of more than 2000 m^3/s the Rhine proves to be potentially the most polluted one.[31]

By also considering the ratio of the drainage area of a river to its flow rate, Zoeteman classified the rivers into three groups (Figure 24) in terms of their potential for artificial pollution: "slightly polluted," "moderately polluted," and "strongly polluted."

Zoeteman[31] also applied the PPI in a detailed fashion to the Rhine River. He considered the index a tool for predicting future water pollution problems:

> ... the PPI concept offers a general matrix for water quality data of different origin, different kind and different time. From this matrix general trends can be derived to deduce river water quality in the past and to extrapolate quality trends into the future. Furthermore the PPI concept can be extended to a scientific yardstick in the process of decision-making concerning sanitation programming. The more data on water quality are available the more precise this yardstick will be.[31]

The PPI is very approximate; it does not, for example, take into account the impact of abatement activities in reducing pollution levels. The PPI illustrates the tradeoff between simplicity and accuracy: as an index becomes simpler, it generally becomes more approximate. Although very simple, the PPI is one of the few approaches that has been applied to waters throughout the world. Zoeteman[31] views it as a means for stimulating thought about water pollution problems:

> It is sincerely hoped that the PPI concept will stimulate people's imagination concerning the extent of future water quality problems and that it will contribute in this way to people's willingness to support the many activities that are needed now for the restoration and preservation of a healthy water quality in the rivers on earth.

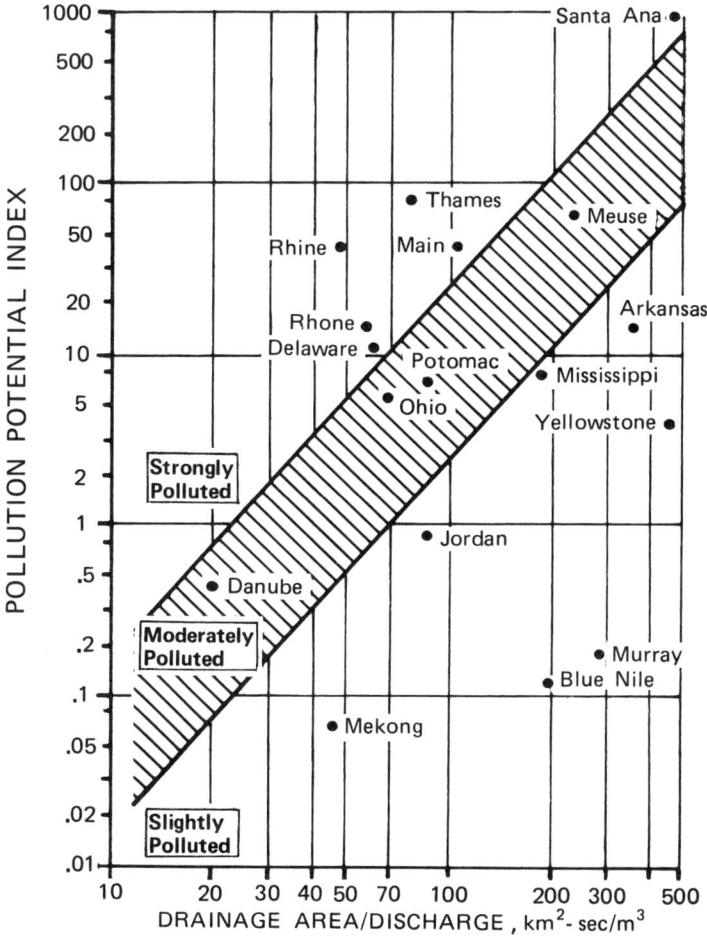

Figure 24. Zoeteman's classification of the world's rivers with respect to their potential for artificial pollution.[31]

Johanson and Johnson's Pollution Index

Over the years, toxic substances (for example, mercury, cadmium and arsenic) have accumulated in the sediments of U.S. ports, harbors and waterways. These pollutants come from wastewater outfalls, nonpoint sources, accidental spills, and disposal of dredge spoil material. Some of these pollutants are adsorbed or absorbed by small particles and travel considerable distances before settling out. When they settle, the result is a system of "in-place" pollutants which may be distributed over large areas of the waterway or accumulated in localized "hot spots."

Section 115 of the 1972 Federal Water Pollution Control Act Amendments[32] requires EPA "... to identify the location of in-place pollutants with emphasis on toxic pollutants in harbors and navigable waterways..." and to take steps to remove and dispose of them. To assist in the process of identifying candidate locations for the removal of these pollutants, Johanson and Johnson[33] developed a planning index and used it to screen 652 data sets from waterways across the nation. For each location, the Pollution Index (PI) was computed as follows:

$$PI = \sum_{i=1}^{n} w_i C_i \qquad (27)$$

where: w_i = weight for pollutant variable i
C_i = highest concentration of pollutant variable i reported in a location of interest

For each pollutant i, the weight was based on the reciprocal of the median of observed national concentrations. For example, the national median of mercury concentrations was 0.5 mg/kg and the weight for mercury, therefore, was set as $w = 1/0.5 = 2$. Using the index, it was possible to scan the data by computer and identify 23 locations receiving the highest priority for removal of in-place pollutants.

Statistical Approaches

Numerous statistical approaches also have been suggested for evaluating and interpreting water quality data. These approaches usually employ some standard statistical procedure already available in the literature, although the procedure often is adapted for use with water quality data. The statistical approaches have the advantage that they incorporate fewer subjective assumptions than the traditional indices; however, they are more complex and often more difficult to apply.

One class of statistical approaches, correlation techniques, examines the associations among variables to determine the importance of each as a determinant of water quality. Shoji, Yamamoto and Nakamura[34] applied "factor analysis" to the Yodo River System in Japan to examine the interrelationships among 20 pollutant variables. By comparing the correlation of each variable with every other variable and selecting combinations with the highest correlations, they identified three major factors: pollution, temperature and rainfall. With this procedure, the investigators determined the weights for an 18-variable Composite Pollution Index which they applied to data from each station on the Yodo River System.

In the U.S., Joung et al.[35] used factor analysis to develop water quality indices by examining water quality data from Carson Valley, Nevada. Ten pollutant variables were considered initially. By manipulating the matrix of correlation coefficients, they were able to identify linear combinations of the variables which best explain the variance but which have low correlations with each other. The approach retains the most important information in the raw data while eliminating redundant variables. They used the approach to identify the most significant variables and index weights for two water quality indices containing five variables each—the Index of Partial Nutrients and the Index of Total Nutrients. These indices then were applied to the Snake and Colorado River Basins in Nevada. Finally, the Index of Total Nutrients (with the variables DO, BOD_5, total phosphates, temperature and conductivity) was selected, and its performance was compared with that of the NSF WQI using water quality data from 20 locations throughout the U.S. Joung *et al.* felt that this approach offers a useful way for designing water quality indices and applying them to different geographical areas:

> The model estimated by factor analysis was shown to be unbiased, able to detect unique water quality relationships, and to have good potential as a water quality index. Such an index may enable geographic identification of problem areas contributing to water pollution and the development of appropriate water quality standards.[35]

In another correlation study, Coughlin *et al.*[36] examined the relationship between the NSF WQI and the uses of a stream made by nearby residents. They used "principal component analysis" to examine the relationships among individual NSF WQI variables and such factors as distance of residence from the stream, land values, and tendency for residents to walk along the stream or to wade or fish in it. They reported that increased water pollution was associated with reduced wading, fishing, picnicking, bird watching, walking and other activities.

Harkins' Index

In 1974, Harkins[37] presented a statistical approach for analyzing water quality data which is based on the rank order of observations. Harkins felt that absolute indices, such as the NSF WQI by Brown *et al.*,[6] lack objectivity:

> The index developed by Brown *et al.*[6] is not really objective because a panel of "experts" rates the water quality parameters to be used. Many users feel that different panels will give different ratings, thus destroying comparability and objectivity.[37]

> The experts frequently disagree on the relative importance of parameters, and panels of experts often give different weights to the same parameters. This failure to attain objectivity and consensus makes it difficult for an administrator to make a decision that cannot be challenged by another panel of experts.[37]

Harkins' index is an application of Kendall's[38] nonparametric classification procedure.

256 ENVIRONMENTAL INDICES

The approach begins by ranking the observations for each pollutant variable, including a "control value," which is usually a water quality standard or recommended limit. For each observation j of pollutant variable i, the transform z_{ij} is computed as the difference between the rank order of the observation and the rank order of the control value (R_{ic}), divided by the standard deviation of the ranks s_i:

$$z_{ij} = \frac{R_{ij} - R_{ic}}{s_i} \qquad (28)$$

where R_{ij} = rank of the jth observation of the ith variable
R_{ic} = rank of the control value for the ith variable
s_i = standard deviation of the ranks for the ith variable

Then the index is computed for each observation by adding the square of the transforms for n pollutant variables:

$$I_j = \sum_{i=1}^{n} z_{ij}^2 \qquad (29)$$

Because the transform z_{ij} includes the standard deviation of the ranks, it is first necessary to calculate the standard deviation s_i as follows:

$$s_i = \sqrt{\frac{m_i^2 - 1}{12}} \qquad (30)$$

where m_i = number of values (observations plus control value) for pollutant variable i

Since the variance of the ranks is the square of the standard deviation, it is sometimes more convenient to work with the quantity under the radical sign, $Var(R_i) = s_i^2 = (m_i^2 - 1)/12$. Within the observations, the same value often appears more than once; these repeated values (or "ties") reduce the variance and must be taken into account. When repeated values occur, the standard deviation s_i is calculated as follows:

$$s_i = \left[\frac{1}{12 m_i} [m_i^3 - m_i - \sum_{k=1}^{q_i} (t^3 - t)_k] \right]^{1/2} \qquad (31)$$

where m_i = number of values for each variable i
t = number of ties (repeated values)
q_i = number of separate occurrences of ties

Like Equation 30, the variance s_i^2 is the quantity inside the brackets. Actually, Equation 30 is a special case of Equation 31, and this can be seen by setting t = 0. In the ranking process, tied values are ranked by assigning each of them the average value of the ranking positions they occupy. For example, four tied values occupying ranks 2, 3, 4 and 5 would all be assigned the average rank $(2 + 3 + 4 + 5)/4 = 3.5$.

Harkins' index is computed by carrying out the following steps:[37]

1. For each pollutant variable, select a control value (usually a maximum or minimum) as a starting point.
2. Rank the observations for each pollutant variable, including the control value.
3. Compute the rank standard deviation (or the variance) for each pollutant variable using Equation 30 or 31.
4. Compute the transforms for the individual pollutant variables using Equation 28 and combine them within each set of observations using Equation 29.

The technique can be illustrated by an example (Table XXIV) based on observations from an actual stream, whose DO, ammonia and phenol values were measured by EPA's Region V in Illinois. In this example, the control values (R_c) are 10 mg/l for DO, 0 mg/l for ammonia, and 0 µg/l for phenol.

Table XXIV. Water Quality Data for Example of Harkins' Index

Day j	Dissolved Oxygen (mg/l)	Ammonia (mg/l)	Phenol (µg/l)
R_c	10.0	0.0	0.0
1	4.0	1.50	9.0
2	4.1	0.65	2.0
3	3.3	0.73	2.0
4	3.3	0.91	2.0
5	3.3	0.75	18.0
6	3.4	1.09	13.0
7	2.2	1.01	9.0

If the DO values are arranged in increasing order, the ranks for DO, R_{1j}, are as follows:

Day j	DO (mg/l)	Rank R_{1j}	
7	2.2	1	
3	3.3	2	
4	3.3	3	3.0 average
5	3.3	4	
6	3.4	5	
1	4.0	6	
2	4.1	7	
R_c	10.0	8	

258 ENVIRONMENTAL INDICES

For the three repeated values (DO = 3.3 mg/l), ranks 2, 3 and 4 are averaged to give 3.0, and this rank then is assigned to all three. The variance of the ranks, s_1^2, is calculated using Equation 31:

$$s_1^2 = \frac{1}{(12)(8)} [(8^3 - 8) - (3^3 - 3)] = \frac{480}{96} = 5$$

By repeating the ranking procedure for the other two columns in Table XXIV, one can construct a table of the ranks for each pollutant variable, listed by date (Table XXV). Unlike DO, ammonia and phenol are increasing

Table XXV. Ranks and Calculated Values for Example of Harkins' Index

Day j	R_{1j} (DO)	R_{2j} (Ammonia)	R_{3j} (Phenol)	Index I_j
R_c	8	1	1	0.0
1	6	8	5.5	14.23
2	7	2	3	1.20
3	3	3	3	6.57
4	3	5	3	8.86
5	3	4	8	16.64
6	5	7	7	15.95
7	1	6	5.5	18.66
Variance	5.000	5.250	4.938	

scale pollutant variables. Therefore, they are ranked in increasing order.

For ammonia, there are no repeated values, and the variance calculation is based on Equation 30:

$$s_2^2 = \frac{1}{12} [8^2 - 1] = \frac{63}{12} = 5.25$$

For phenol, there are two separate occurrences of repeated values—two values of 9.0 $\mu g/m^3$ and three values of 2.0 $\mu g/m^3$—and the variance calculation is based on Equation 31:

$$s_3^2 = \frac{1}{(12)(8)} [8^3 - 8 - (2^3 - 2 + 3^3 - 3)] = \frac{474}{96} = 4.938$$

To calculate the index, the ranks of the variables are aggregated according to Equation 29 by adding the square of the transforms given by Equation 28. For example, the index for day 1 is calculated as follows:

$$I_1 = \frac{(6-8)^2}{5} + \frac{(8-1)^2}{5.25} + \frac{(5.5-1)^2}{4.938} = 14.23$$

Table XXV lists the resulting values of Harkins' index calculated for all the dates in this example. The best water quality (lowest index value) was observed on day 2, with intermediate values occurring on days 3 and 4, and poorest water quality (highest index value) on day 7.

Because Harkins' index is a relative rather than an absolute index, values generated with one data set cannot be compared directly with those generated with a different data set. Thus, although one can conclude that day 1 (I_1 = 14.23) exhibits poorer water quality than day 2 (I_2 = 1.20), these values cannot be universally interpreted as "poor" or "good" water quality and cannot be compared with index values from another example. Furthermore, if observations for additional days were added to the list, index values for days 1 through 7 would change. Landwehr et al.[39] criticized this property:

> ...this limitation of Harkins' method means that it is never possible to make comparisons, which is really the primary aim of a water quality index. For example, if one wants to know how a particular river's quality compares with that of other rivers, Harkins requires that one must have a complete set of observations for all the other rivers. It is unlikely that such a full data bank could be constructed.

This property is characteristic of relative indices.

Landwehr and Deininger[40] compared the performance of Harkins' index with four variations of the NSF WQI. In this comparison, a panel of over 100 water quality experts first was asked to rate 20 water quality samples on a numerical scale from 0 (worst) to 100 (best) by examining the raw data. Then the five indices were computed from these data, and their values were compared with the panel ratings. All five water quality indices showed high correlations with each other and with the experts' ratings. A variation of the NSF WQI using a weighted product aggregation function with equal weights (w_i = 1/9 for all i) gave the best agreement with expert opinion. Correlations between Harkins' index and the experts' ratings were somewhat lower than those for the other four indices, but the differences were small and their statistical significance was not discussed.

Beta Function Index

Although relative indices usually are more difficult to apply than absolute indices, Harkins' index possesses useful statistical properties. Schaeffer and Janardan[41] of the Illinois Environmental Protection Agency exploited these properties by incorporating Harkins' approach into a statistical index which has a fixed range, the Beta Function Index.

These investigators first developed five statistically based water quality indices.[42] Two of the five used a ranking approach similar to Harkins' index, and three were based on a chi-square-like statistic. They evaluated the performance of the five indices using both real data and computer simulation, and they selected one, the Beta Function Index, for routine use.

The Beta Function Index uses the same ranking procedure employed in Harkins' index. Two additional values are computed from the ranks—the sum of the square of the z transforms given by Equation 28 and the sum of all the ranks excluding the control value:

$$S = \sum_{i=1}^{n} \sum_{j=1}^{m_i} z_{ij}^2 \qquad (32)$$

$$T = \sum_{i=1}^{n} \sum_{j=1}^{m_i - 1} R_{ij} \qquad (33)$$

where m_i = number of values for pollutant variable i
       n = number of pollutant variables

Janardan and Schaeffer[42] argue that the sum of the z values given by Equation 28 approaches, in the limit, a normal distribution according to the Central Limit Theorem. Therefore, S in Equation 32, the sum of the squares of a normally distributed random variable, approaches a chi-square distribution. They support these assumptions using the results from a computer simulation of 2,000 samples of sets of 12 observations of 4 pollutant variables.

The Beta Function Index is calculated using the following transform of S and T:

$$I = \frac{1}{b} \left[\frac{S}{T + S} \right]^{1/2} \qquad (34)$$

where

$$b = \left[\frac{2 \sum_{i=1}^{n} m_i^2}{3 \sum_{i=1}^{n} m_i^2 + \sum_{i=1}^{n} m_i - 2n} \right]^{1/2} \qquad (34a)$$

If the number of observations for each variable is the same (that is, $m_i = m$ for all i), then Equation 34a can be simplified:

$$b = \left[\frac{2m^2}{3m^2 + m - 2}\right]^{1/2} \quad (35)$$

Because S is assumed to have a chi-square distribution and T is approximately constant, the investigators conclude that the index follows a beta probability distribution. Thus, the index is nonparametric; its distribution is the same regardless of the underlying distribution of the data. The full range of the Beta Function Index, like the beta distribution, is from 0 to 1. For additional discussion of the beta distribution and probability models in general, the reader should refer to a text on engineering statistics, such as the one by Hahn and Shapiro.[43]

If the Beta Function Index is applied to the example given in Table XXIV, S is calculated by summing the values for Harkins' index in Table XXV (because each index value already is the sum of the square of the z transforms) and by adding the ranks (excluding the rank of the control value):

$$S = \sum_{j=1}^{8} I_j = 82.11$$

$$T = \sum_{j=1}^{7} R_{1j} + \sum_{j=1}^{7} R_{2j} + \sum_{j=1}^{7} R_{3j} = 28 + 35 + 35 = 98$$

Because the number of observations is the same for each variable, we use Equation 35 to calculate b:

$$b = \left[\frac{2(8)^2}{3(8)^2 + 8 - 2}\right]^{1/2} = 0.804$$

Finally, the Beta Function Index is calculated using Equation 34:

$$I = \frac{1}{0.804}\left[\frac{82.1}{98 + 82.1}\right]^{1/2} = \frac{0.675}{0.804} = 0.84$$

Notice that the entire group of observations in Table XXIV results in just one index value. Thus, the Beta Function Index can be used to compare groups of observations and different data sets.

In this example, the control value for each variable is lower (or higher) than any of the observations. If the control value lies at an intermediate level, the ranking process is altered as follows. When a particular observation X_{ij} is equal to the control value or shows "less" pollution than the control value ($X_{ij} \leq X_{ic}$ for increasing scale pollutant variables; $X_{ij} \geq X_{ic}$ for decreasing scale pollutant variables), it is ranked as though it were equal to the control value X_{ic}. In this way, only those observations which are indicative of pollution are taken into consideration.

Suppose that the control value for ammonia in Table XXIV is 5.0 mg/l instead of 0.0 mg/l. Then all seven observations are below the control value. We calculate the ranks as though all the observations are equal to the control value, giving eight repeated values. The average rank $\overline{R}$ is calculated as follows:

$$\overline{R} = \frac{1 + 2 + 3 + 4 + 5 + 6 + 7 + 8}{8} = \frac{36}{8} = 4.5$$

Then all values for ammonia are assigned the average rank of 4.5 ($R_{2j} = 4.5$ for all j).

This index has been applied routinely by the State of Illinois to examine water quality trends. In 1976, it was applied to data from over 500 stations throughout the state, and the results were reported in the Illinois annual water quality report.[44] Schaeffer and Janardan[41] have applied a biological classification system to Illinois waters and have found good correspondence between values predicted by this index and biological ratings of water quality.

Most of the general and specific-use water quality indices discussed above were designed for application to data from free-flowing surface waters. However, special problems of eutrophication (excessive plant and algal growth) occur in nonflowing water bodies, particularly in lakes. The statistical approaches are particularly useful in these situations. In 1974, EPA collected extensive data on over 800 lakes as part of the National Eutrophication Survey. Various statistical approaches were used to examine the relationships among physical, chemical and biological variables as indicators of eutrophication, and these results were compared with expert judgment of the condition of the surveyed lakes. As part of this effort, EPA also developed a Trophic Index and used it to rank the trophic condition of the lakes using six variables: total phosphorus, dissolved phosphorus, inorganic nitrogen, Secchi disk transparency, minimum DO, and chlorophyll-a.[45] It was possible to relate index ranges to three trophic states: oligotrophic, mesotrophic and eutrophic. Detailed treatment of this topic is beyond the scope of this book.

The statistical approaches, because they are general, can be applied to topic areas other than water pollution. For example, Harkins' index readily could be used to compare air quality levels at different locations or at different time periods. Similarly, the Beta Function Index could be used for analyzing air quality data, pesticide data, noise data and other types of environmental data. These statistical approaches are especially well suited for use with data from measurements of airborne particulates, in which a small number of samples are collected, and the samples are analyzed for a variety of constituents.

Biological Indices

Biological water quality indices generally evaluate water quality in terms of its impact on aquatic life in some form. There are three basic approaches.

The first approach focuses on the types and quantities of certain indicator organisms. An example is the saprobic classification systems employed in environments rich in degradable organic matter. The saprobic classification systems divide a stream into various zones of pollution depending principally on the type of organisms present. The saprobic systems are summarized by Orlando and Wrightington.[3] Another example is enteric bacteria, such as fecal coliforms, which are the normal inhabitants of the digestive tract of man and other warm-blooded animals. The presence of these indicator organisms is taken as evidence of contamination with fecal material.

The second approach concentrates on the mathematical properties of populations of organisms. For example, some techniques use information theory to describe the diversity of species within biological communities. Other species-diversity techniques employ various probabilistic models in their formulation. Pielou[46] discusses some of the statistical population techniques.

The third approach examines the physiological or behavioral responses of certain organisms to pollution. For example, pesticides are known to inhibit acetylcholinesterase activity in the brains of fish; therefore, fish-brain cholinesterase activity has been used as a monitor for pesticide pollution. Behavioral changes of certain species, such as increased activity and agitation of fish in response to toxic substances, also have been studied as an indicator of environmental pollution.[3,47]

Biological measures of pollution have the advantage that they have a "pollution integrating" tendency. Fish and other organisms tend to respond to the entire historical record of water quality. Thus, if some toxic substances are present on rare occasions and go undetected in routine water quality monitoring activities, the presence of these pollutants would still be measured in terms of their effects on aquatic organisms. Orlando and Wrightington[3] observe that this integrating feature enables biological organisms to cover more variables and conditions than conventional measurements:

> Aquatic organisms respond to a number of chemical, physical, and biological water quality variables in an integrated manner. Thus, one or a few organisms can do the work of several analysts in terms of demonstrating that a large number of variables either are or are not within the range of tolerance of aquatic life, although they do not give specific numerical values for each of the variables. A number of different industries routinely pass their effluents through a fish pond before discharge to demonstrate that the effluent is not toxic to fish.

Because of the great variety and number of biological approaches that have been developed, this book has concentrated on indices based primarily on physical and chemical variables. The reader wishing to explore biological indices and indicators in greater detail is referred to the research literature, such as the summary by Orlando and Wrightington[3] or the bibliography by Thomas, Goldstein and Wilcox.[4,7]

Comparison of Indices

The physical and chemical indices published in the literature show considerable variation in terms of the number of variables, scales and ranges (Tables XXVI and XXVII).

Indices in the general and specific-use categories (Table XXVI) share a common characteristic: they are absolute indices designed to depict the quality of free-flowing surface waters. In the 11 indices in these two groups, the number of variables included varies from 8 to 31. Most of these indices (8 of 11) have decreasing scales, and the majority (6 of 11) have fixed ranges of 0 to 100. One of these indices can be negative, and the others have ranges of 0 to 1 or above, 0 to more than 15, or 0 to more than 1,000.

By contrast, most of the planning indices (6 of 7) have increasing scales, and none has a fixed range of 0 to 100 (Table XXVII). Part of the variation among planning indices probably reflects the fact that they usually are designed for special-purpose applications. More than half of the statistical approaches also have increasing scales. The ranges generally differ from each other and from those of the general and specific-use indices. Since they are often based on published statistical techniques, the statistical approaches are relatively flexible, permitting the user to include any number of variables and define the range as he pleases. Because of their flexibility and special-purpose nature, the planning indices and statistical approaches do not lend themselves to detailed comparison.

Variables

If the variables in the 11 general and specific-use water quality indices are compared, it is seen that there is great variety (Table XXVIII). Although

Table XXVI. Summary of the General and Specific-Use Water Quality Indices Published in the Literature

Index Name[a]	Reference	No. of Variables	Scale	Range
General Water Quality Indices				
Quality Index (QI)	Horton[5]	10	Decreasing	0 to 100
Water Quality Index (NSF WQI)	Brown et al.[6]	9	Decreasing	0 to 100
Implicit Index of Pollution	Prati et al.[14]	13	Increasing	0 to 15+
River Pollution Index (RPI)	McDuffie and Haney[15]	8	Increasing	0 to 1,000+
Social Accounting System	Dinius[16]	11	Decreasing	0 to 100
Specific-Use Water Quality Indices				
Fish and Wildlife (FAWL) Index	O'Connor[18]	9	Decreasing	0 to 100
Public Water Supply (PWS) Index	O'Connor[18]	13	Decreasing	0 to 100
Index for Public Water Supply	Deininger and Landwehr[19]	11/13	Decreasing	0 to 100
Index for Recreation	Walski and Parker[17]	12	Decreasing	0 to 1
Index for Dual Water Uses	Stoner[20]	31	Decreasing	-100 to 100[b]
Index for Three Water Uses	Nemerow and Sumitomo[21]	14	Increasing	0 to 1+

[a]When the proper name for an index is unavailable, the index characteristic is listed.
[b]Index can be less than -100 and can become a large negative number.

Table XXVII. Summary of Planning Indices and Statistical Approaches Published in the Literature

Index Name[a]	Reference	No. of Variables	Scale	Range
Planning Indices				
Prevalence Duration Intensity (PDI) Index	Truett et al.[23]	b	Increasing	0 to 1
National Planning Priorities Index (NPPI)	Truett et al.[23]	b	Increasing	0 to 1
Priority Action Index (PAI)	Truett et al.[23]	b	Increasing	0 to 1
Environmental Evaluation System (EES)	Dee et al.[26]	78[c]	Decreasing	0 to 1,000
Canadian National Index	Inhaber[29]	b	Increasing	0 to 1
Potential Pollution Index (PPI)	Zoeteman[31]	3	Increasing	0 to 1,000+
Pollution Index (PI)	Johanson and Johnson[33]	b	Increasing	0 to 100+
Statistical Approaches				
Composite Pollution Index (CPI)	Shoji et al.[34]	18	Increasing	-2 to 2
Index of Partial Nutrients	Joung et al.[35]	5	Decreasing	0 to 100
Index of Total Nutrients	Joung et al.[35]	5	Decreasing	0 to 100
Principal Component Analysis	Coughlin et al.[36]	b	N.A.[d]	N.A.
Harkins' Index (Kendall ranking)	Harkins[37]	b	Increasing	0 to 100+
Beta Function Index	Schaeffer and Janardan[41]	b	Increasing	0 to 1

[a]When the proper name for an index is unavailable, the index characteristic is listed.
[b]Any number of variables can be included.
[c]Water quality variables account for 14 of the 78 variables used in this system.
[d]N.A. = not applicable.

Table XXVIII. Variables Used in General and Specific-Use Water Quality Indices

	General						Specific Use						
Variables \\ Indices	Horton[5]	NSF WQI[6]	Prati et al.[14]	McDuffie et al.[15]	Dinius[16]	Dee et al.[26]	O'Connor (FAWL)[18]	O'Connor (PWS)[18]	Deininger et al. (PWS)[19]	Walski et al.[17]	Stoner (PWS)[20]	Stoner (Irrigation)[20]	Nemerow et al.[21]
Chemical													
DO	•	•	•	•	•	•	•	•	•	•			•
BOD$_5$		•	•	•	•	•		•					
COD			•	•									
Alkalinity	•			•				•	•				•
Hardness				•					•		•		•
Iron			•						•		•	•	•
Manganese			•										•
Nitrogen:													
Ammonia			•				•				•		
Nitrites											•		
Nitrates		•	•				•		•	•			
Other				•		•							
Phosphorus:													
Phosphates		•		•			•				•		
Other						•							
Chlorides	•		•		•			•			•		•
Fluorides								•	•		•	•	
Sulfates								•			•		•
Oil and Grease										•			
Phenol							•	•	•				
ABS				•									
CCE	•			•									
Other				•	•	•					•	•	
Physical													
pH	•	•	•		•	•	•	•	•	•	•		•
Temperature	•	•		•	•	•	•		•	•			•
Specific Cond.	•			•	•							•	
Turbidity		•			•	•	•	•	•	•			•
Dissolved Solids						•	•	•	•				
Suspended Solids				•							•		•
Total Solids		•											•
Color					•			•			•	•	•
Other						•							•
Biological													
Fecal Coliforms		•			•	•		•	•		•	•	•
Total Coliforms	•			•	•					•			
Other	•												

Dee's water quality index is part of a larger planning system called the Environmental Evaluation System, the index shares many characteristics with general water quality indices, and it is included in this table.

For convenience, the variables in Table XXVIII are grouped into three categories: (1) chemical, (2) physical and (3) biological. By examining the number of dots in each row of this table, it is seen that DO is the most common chemical variable; pH is the most common physical variable; and fecal coliforms is the most common biological variable. All the indices but two, Prati's index and O'Connor's FAWL, include either fecal or total coliforms. The diversity of the variables included in these indices apparently results from different assumptions by the index developers regarding the importance of each variable.

One would expect to find different variables included in each specific-use water quality index, because the importance of a variable depends on the intended water use. Sayers and Ott[48] have independently classified the key variables according to five water uses (Table XXIX). These are the same five use categories discussed in Chapter I (page 40):

- public water supply
- industrial water supply (*e.g.*, cooling)
- agricultural water supply (*e.g.*, crop irrigation)
- maintenance of aquatic life and wildlife
- recreation and aesthetics (*e.g.*, boating and swimming)

Examination of Tables XXVIII and XXIX suggests that each of the published water quality indices may be missing important variables in a certain use category. For example, the NSF WQI of Brown *et al.*[6] does not include color or oil and grease, variables that are important for the recreational and aesthetic uses of water:

> First, water should be pleasant to the senses of man and capable of supporting life forms of aesthetic value. Visible pollution such as unnatural color, sludge banks, floating solids, foam, oil films, algal blooms, and excessive aquatic weed growths should be absent. There should also be an absence of constituents in amounts sufficient to produce undesirable tastes or odors, alone or in combination with other constituents present or added in the treatment process.[48]

Thus, it is possible for waters to have an unappealing color or floating oil and still be regarded as "good" (I = 100) using this index. Because of this tendency for certain variables to be excluded, it is important to use judgment in applying today's water quality indices.

Table XXIX. Key Water Quality Variables Classified by Sayers and Ott[48] According to Various Water Uses

Public Water Supply	Industrial Water Supply	Agricultural Water Supply	Aquatic Life and Wildlife Maintenance	Recreation and Aesthetics
Coliform Bacteria	Processing (except foods)	Farmstead	Temperature	Recreation
Turbidity	pH	(same as for public supply)	DO	Coliforms
Color	Turbidity	Livestock	pH	Turbidity
Taste-Odor	Color	(similar to that for public supply)	Alkalinity/Acidity	Color
Trace Metals	Hardness	Irrigation	Dissolved Solids	pH
Dissolved Solids	Alkalinity/Acidity	Dissolved Solids	Salinity	Odor
Trace Organics	Dissolved Solids	Specific Conductance	Carbon Dioxide	Floating Materials
Chlorides	Suspended Solids	Sodium	Turbidity	Settleable Materials
Fluorides	Trace Metals	Calcium	Color	Nutrients
Sulfates	Trace Organics	Magnesium	Settleable Materials	Temperature
Nitrates	Cooling	Potassium	Floating Materials	Aesthetics
Cyanides	pH	Boron	Tainting Substances	Turbidity
Radioactivity	Temperature	Chlorides	Toxic Substances	Color
	Silica	Trace Metals	Nutrients	Odor
	Aluminum			Floating Materials
	Iron			Settleable Materials
	Manganese			Nutrients
	Hardness			Temperature
	Alkalinity/Acidity			Substances Adversely
	Sulfates			Affecting Wildlife
	Dissolved Solids			
	Suspended Solids			
	Sanitary			
	(same as for public supply)			

Mathematical Structures

The mathematical structures of the published water pollution indices also are quite varied. However, all of the general and specific-use water quality indices (including Dee's index) can be analyzed by the mathematical system introduced in Chapter II.

As shown in Table XXX, water quality indices frequently use nonlinear (implicit and explicit) subindex functions. Nonlinear subindices are much more common in water indices than in air pollution indices (see page 121). Horton's index[5] uses staircase step functions (segmented linear); the NSF WQI[6] uses implicit nonlinear functions based on expert judgment; Dinius' index[16] uses a mixture of linear and power (explicit nonlinear) functions; and Walski and Parker's index[17] uses exponential and parabolic (explicit nonlinear) functions. One index, Prati's Implicit Index of Pollution,[14] uses segmented nonlinear subindex functions.

Although the subindex functions usually are more complex than those used in air pollution indices, the aggregation functions are simpler in form (Table XXX). Most of the general and specific-use water quality indices (9 of 12) use the weighted linear sum aggregation function. As discussed in Chapter II (pages 80 to 82), the weighted linear sum has serious eclipsing problems when it is used in decreasing scale indices. If a single subindex exhibits poor water quality ($I_i = 0$ for some i), the weighted linear sum unfortuntely does not exhibit poor water quality. The weighted product aggregation function, which was evaluated by Landwehr[1] for use in the multiplicative NSF WQI_m, is designed to circumvent this problem. Although it reduces eclipsing, it becomes a nonlinear transform when the weights are small (see pages 87 to 88 and Problem 10, page 94). Nevertheless, indices using the weighted product have given good correlations with independent expert opinion.[40] Nemerow and Sumitomo[21] offer a more complex aggregation function, the root-mean-square of the maximum subindex and mean of subindices. This aggregation function reduces, but does not eliminate, the eclipsing problem.

To make detailed comparisons of individual subindex functions, it is first necessary to compensate for differences in scales and ranges. To convert an increasing scale subindex which ranges from 0 to k into a decreasing scale subindex which ranges from 0 to 100, the following transform can be used:

$$I = -\frac{100}{k} I' + 100 \tag{36}$$

where: I = transformed decreasing scale index ($0 \leqslant I \leqslant 100$)
I' = original increasing scale index ($0 \leqslant I' \leqslant k$)
k = constant equal to the maximum value of the original index

Table XXX. Mathematical Characteristics of General and Specific-Use Water Quality Indices Published in the Literature

Index	Subindices[a]	Aggregation Function	Comments
General Water Quality Indices			
Horton[5]	Segmented Linear (Step Functions)	Weighted Sum Multiplied by Two Dichotomous Terms	Eclipsing Region
Brown et al.[6] (NSF WQI$_a$)	Implicit Nonlinear[b]	Weighted Sum	Eclipsing Region
Landwehr[1] (NSF WQI$_m$)	Implicit Nonlinear[b]	Weighted Product	Nonlinear
Prati et al.[14]	Segmented Nonlinear	Weighted Sum (Arithmetic Mean)	Eclipsing Region
McDuffie and Haney[15]	Linear	Weighted Sum	Eclipsing Region
Dinius[16]	Nonlinear (Linear, Power)	Weighted Sum	Eclipsing Region
Dee et al.[27]	Implicit Nonlinear	Weighted Sum	Eclipsing Region
Specific-Use Water Quality Indices			
O'Connor[18] (FAWL, PWS)	Implicit Nonlinear[b]	Weighted Sum	Eclipsing Region
Deininger and Landwehr[19] (PWS)	Implicit Nonlinear[b]	Weighted Sum/Weighted Product	Eclipsing/Nonlinear
Walski and Parker[17]	Nonlinear	Weighted Product (Geometric Mean)	Nonlinear
Stoner[20]	Nonlinear	Weighted Sum	Negative Values
Nemerow and Sumitomo[21]	Segmented Linear	Root-Mean-Square of Maximum and Arithmetic Mean	Minimizes Eclipsing

[a]Unless listed as *implicit*, subindices are *explicit* mathematical functions.
[b]Based on a panel of water quality experts; NSF WQI$_a$ is the additive form, and NSF WQI$_m$ is the multiplicative form.

Noitce that Equation 36 gives I = 0 for I' = k and I = 100 for I' = 0. Equation 36 also can be used to transform a decreasing scale subindex which ranges from 0 to k into an increasing scale subindex which ranges from 0 to 100.

To compare different subindices for DO, Equation 36 was used to convert Prati's DO subindex function (k = 12.25) and McDuffie's DO subindex function (k = 100) into decreasing scale forms. When plotted alongside the DO subindices from McDuffie's index, Dinius' index and the NSF WQI, the five DO subindices appear reasonably similar to each other (Figure 25).

Subindex functions for pH, when plotted on the same graph, also are similar to each other (Figure 26). Of the four pH subindices shown in this figure, Walski and Parker's gives the widest range of pH values. This subindex is from a specific-use index intended for recreational water use. The other three pH subindex curves, which come from general water quality indices, lie within the 80% confidence intervals shown for the NSF WQI in Figure 4.

Figure 27 shows selected subindex functions for coliform organisms. Although the curves show a greater variation than the curves for DO or pH, one must distinguish between fecal coliforms (dotted lines) and total coliforms (solid lines). The fecal coliform subindex curves lie approximately within the 80% confidence intervals shown for the NSF WQI in Figure 3.

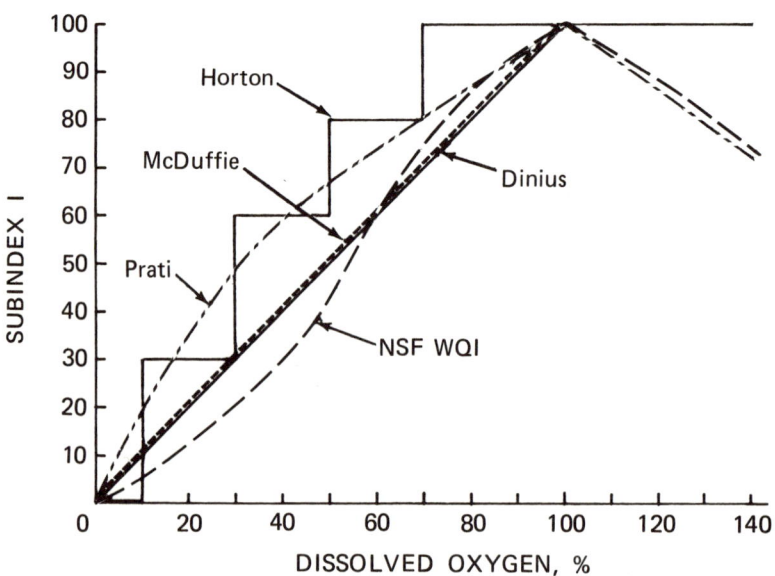

Figure 25. Subindex functions for DO from five water quality indices.

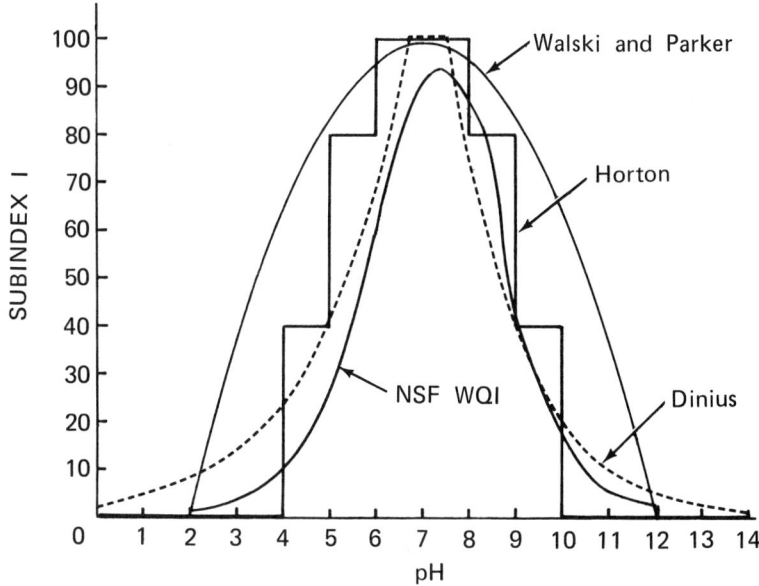

Figure 26. Subindex functions for pH from four water quality indices.

If the BOD_5 subindices from four different water quality indices are plotted on one graph (Figure 28), they all have sharp downward slopes at values less than about 10 mg/l. Although two of these subindices are represented by straight lines and two are represented by curves, all four lie approximately within the 80% confidence intervals shown for the NSF WQI in Figure 5.

INDICES IN USE

Although the literature reveals that considerable effort has gone into the development of water quality indices, it does not indicate which indices are being used in practice. To determine more about the actual applications of water quality indices, it is necessary to survey the index users.

In 1977, Ott[12] conducted a national survey of U.S. water pollution control agencies to determine: (1) which agencies were using indices, (2) the type of indices being used, (3) the purposes for which they were being used, and (4) the attitudes of agency personnel toward indices. The approach was similar to that used to survey air pollution control agencies (Chapter III, page 123). Telephone calls were made to all 60 U.S. state and interstate water pollution

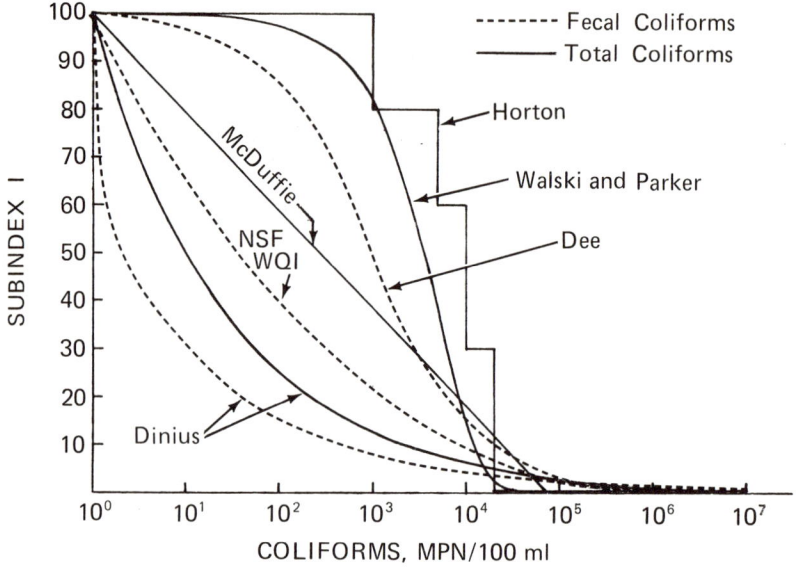

Figure 27. Subindex functions for coliform organisms from six water quality indices.

control agencies from October 1976 to February 1977 to ask about their index applications and to solicit written materials. The 10 EPA Regional Offices also were surveyed. In general, all respondents were asked whether they used an index, and, if so, what type, how it was calculated, its purpose and its usefulness.

On the basis of the survey results, two general categories of respondents could be identified: (1) index user agencies and (2) index nonuser agencies. To be classified as an index user, the agency must have used a water quality index in an official capacity over an extended time period. The index must have been applied to a large-scale water quality study extending over a year or more, or it must have appeared in an official publication of the agency. Nonuser agencies were subdivided into four categories:

- agencies unfamiliar with indices
- agencies that had considered or evaluated indices
- agencies planning to evaluate or develop indices
- agencies currently evaluating or developing indices

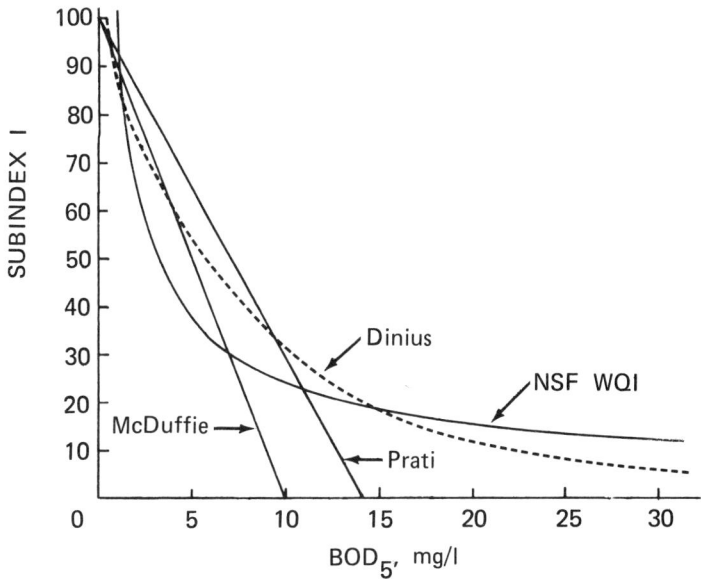

Figure 28. Subindex functions for BOD$_5$ from four water quality indices.

These categories were intended to form a continuum, beginning with agencies least familiar with current water quality index approaches and ending with those most familiar; namely, agencies currently evaluating or developing indices.

Table XXXI shows the classification of 51 U.S. state agencies (including the District of Columbia) according to these categories, and Table XXXII shows a similar classification of the 9 interstate commissions. Ten of the 51 states and 2 of the 9 interstate commissions were classified as index users (Table XXXIII). Thus, 12 out of 60 agencies (20%) were classified as index users. However, one agency, the Interstate Commission on the Potomac River Basin, had used Harkins' index in just one report and was no longer using it. Thus, 11 state and interstate agencies (18%) were actually using water quality indices at the time of the survey. In addition, 2 of the 10 EPA Regional Offices were classified as index users.

Of the nonuser agencies, approximately one-fifth (11 states and 2 interstate commissions) indicated that they were, to a varying degree, unfamiliar with water quality indices. A slightly larger group (14 states and 4 interstate commissions) had previously considered or evaluated indices but had

276 ENVIRONMENTAL INDICES

Table XXXI. Status of Water Quality Index Utilization by States[12]

State Agency	Index Nonuser				Index User
	Unfamiliar with Indices	Considered or Evaluated	Planning to Evaluate or Develop	Currently Evaluating or Developing	
Alabama	●				
Alaska		●			
Arizona	●				
Arkansas	●				
California		●			
Colorado					●
Connecticut	●				
Delaware		●			
District of Columbia	●				
Florida		●			
Georgia					●
Hawaii		●			
Idaho			●		
Illinois					●
Indiana					●
Iowa			●		
Kansas				●	
Kentucky			●		
Louisiana			●		
Maine		●			
Maryland				●	
Massachusetts		●			
Michigan					●
Minnesota		●			
Mississippi	●				
Missouri	●				
Montana					●
Nebraska			●		
Nevada				●	
New Hampshire		●			
New Jersey				●	
New Mexico		●			
New York					●
North Carolina	●				
North Dakota			●		
Ohio			●		
Oklahoma					●
Oregon					●
Pennsylvania		●			
Rhode Island			●		
South Carolina				●	
South Dakota			●		
Tennessee	●				
Texas		●			
Utah		●			
Vermont			●		
Virginia	●				
Washington		●			
West Virginia	●				
Wisconsin				●	
Wyoming					●

Table XXXII. Status of Water Quality Index Utilization by Interstate Commissions[1,2]

Commission	Index Nonuser				Index User
	Unfamiliar with Indices	Considered or Evaluated	Planning to Evaluate or Develop	Currently Evaluating or Developing	
Delaware River Basin		•			
Great Lakes	•				
International Joint		•			
Interstate (CT, NJ, NY)		•			
Klamath River	•				
New England Interstate					•
Ohio River				•	
Potomac River Basin					•
Susquehanna River Basin		•			

Table XXXIII. Summary of Water Quality Index Utilization by States and Interstate Commissions[1,2]

	Index Nonusers				Index Users	Total
	Unfamiliar with Indices	Considered or Evaluated	Planning to Evaluate or Develop	Currently Evaluating or Developing		
States	11	14	10	6	10	51
Commissions	2	4	0	1	2	9
Total	13	18	10	7	12	60

decided not to become index users. However, 10 agencies indicated that they were planning to evaluate or develop water quality indices in the near future, and 7 additional agencies were actually evaluating or developing indices at the time of the survey. Thus, 17 state and interstate agencies could be classified as "potential users" of water quality indices, because they may become users in the near future. Most of the agencies in this group were considering the NSF WQI, but one, Nevada, was developing its own index.

Overall, a total of 14 U.S. agencies could be classified as index users at the time of the survey: 10 states, 2 interstate commissions, and 2 EPA Regional Offices (Table XXXIV). Of the 14, 6 states and 1 interstate commission were using the NSF WQI. Two states had selected Harkins' index, and three states—Georgia, Illinois and Oregon—had developed their own indices. Two

Table XXXIV. Agencies Using Water Quality Indices, June 1977[12]

Agency	Index	No. of Variables	No. of Years in Use
Colorado	Modified NSF WQI	8	3
Georgia	Developed own index	8	1
Illinois	Developed own index	4	2
Indiana	Modified NSF WQI	7	1
Michigan	Modified NSF WQI	8	3
Montana	Standard NSF WQI	9	1½
New York	Standard NSF WQI	9	1
Oklahoma	Harkins' index	7	1½
Oregon	Developed own index	6	½
Wyoming	Modified NSF WQI	8	2
New England Interstate	Standard NSF WQI	9	1
Potomac River Interstate	Harkins' index	4	a
EPA Region VIII (Denver)	Developed own index	8	2
EPA Region X (Seattle)	Developed own index	≥10	2½

aIndex used for one report only.

EPA Regional Offices—Region VIII in Denver and Region X in Seattle—also had developed their own indices. The user-developed indices are discussed in greater detail in the next section. Most agencies had implemented their indices within a 3-year period prior to the survey, and five agencies had implemented their indices during the previous year. Thus, routine application of water quality indices was a very recent phenomenon. The state agencies usually applied their indices on a state-wide basis, generally selecting those monitoring stations with the most data available.

Although the NSF WQI was the most widely used index, the users often made minor modifications to it, usually deleting one or more variables because of data limitations (Table XXXV). The most frequently deleted NSF WQI variable was (departure from equilibrium) temperature (3 agencies), although total solids and turbidity were deleted in Indiana. In Wyoming, BOD_5 was deleted whenever it was missing, and the weights were altered. With most deletions of variables, the user altered the weights so that they retained the same ratios as before, but still added to 1.0. One agency, the New England Interstate Water Pollution Control Commission, deleted temperature from the NSF WQI and conducted a special study to examine the effect of this deletion on calculated values.[49] This study generally found a less than one-half point difference between the modified and standard NSF WQI. The agency concluded that the differences were negligible, and, therefore, to achieve greater uniformity with other agencies, it decided to use the

Table XXXV. Variables Included by Users of the NSF WQI[1,2]

Agency	Variables[a]								
	DO	FC	pH	NO_3	PO_4	BOD_5	Temp	TS	Turb
Colorado	•	•	•	•	•	•		•	•
Indiana	•	•	•	•	•	•	•		
Michigan	•	•	•	•	•	•		•	•
Montana	•	•	•	•	•	•	•	•	•
New York	•	•	•	•	•	•	•	•	•
Wyoming[b]	•	•	•	•	•	•		•	•
New England Interstate	•	•	•	•	•	•	•	•	•

[a]FC, fecal coliforms; TS, total solids.
[b]In Wyoming, BOD_5 was deleted and weights were altered when missing values were encountered. Total dissolved solids was substituted for total solids when suspended solids was missing.

standard version of the NSF WQI rather than the modified version. Three agencies used the additive form of the NSF WQI (NSF WQI_a), three others used the multiplicative form (NSF WQI_m), and one agency, Wyoming, used both.[12]

The primary purposes expressed by the index users for applying their indices were quite varied (Table XXXVI). Three purposes were mentioned most frequently: analysis of water quality trends (8 agencies), preparation of the Section 305(b) annual water quality reports (9 agencies), and public information (8 agencies). Section 305(b) of the Federal Water Pollution Control Act Amendments[32] requires each state to submit a report describing its water quality to EPA on an annual basis. Each report includes a "...description of the water quality of all navigable waters in such State during the preceding year, with appropriate supplemental descriptions as shall be required to take into account seasonal, tidal, and other variations...."[32] In turn, EPA analyzes this information and prepares a summary report which it submits, along with the states' reports, to Congress. EPA permits a state to use an index in its report, but it neither encourages nor discourages the practice.

Although preparation of the Section 305(b) reports was the most frequently cited purpose for applying a water quality index, examination of these reports reveals that the index results usually constituted only a part of the overall report. For example, Michigan's Section 305(b) report[50] applies the NSF WQI on a state-wide basis to give the reader an overview of the status of Michigan's waters (see map in Figure 29):

Table XXXVI. Purposes for Which Water Quality Indices Are Being Used[12]

Agency	Trend Analysis	Intensive Surveys	305(b) Report	Other Reports	Public Hearings	Public Information
Colorado	•		•			•
Georgia	•		•			•
Illinois	•		•			•
Indiana			•			
Michigan	•	•	•			•
Montana	•			•		•
New York	•		•	•	•	•
Oklahoma			•			
Oregon			•	•		
Wyoming	•	•	•			•
New England[a]	•					•
Potomac River[a]			•			

[a]Interstate commissions; the Interstate Commission on the Potomac River Basin used the index in one report only.

> Michigan's abundant natural resources include over 36,000 miles of rivers and streams, more than 11,000 inland lakes, and 38,500 square miles of Great Lakes waters. Michigan has selected the Water Quality Index developed by the National Sanitation Foundation to present a summary of stream quality. As shown [in Figure 29], most of Michigan's river basins rate good to excellent on the water quality index scale for water year 1975....

To deal with the problem of geographical differences between waters in Michigan and those in other states, the report explains that the maximum attainable upper limit of the NSF WQI will be slightly lower for Michigan than for other areas:

> The water quality index scale is designed to accommodate a wide range of water quality nationwide. Because of weather and natural geological conditions, it is unlikely that Michigan waters in even the most remote and natural settings will reach 100 units on the water quality index scale. By the same token, values near 90 show that Michigan water quality is within 10 units of a national ideal of the maximum attainable limit.[50]

Thus, the NSF WQI range is simply redefined for waters in the State of Michigan. This approach would allow the same index to be used in other areas of the U.S. without changing its basic structure.

The Michigan Section 305(b) report also uses the NSF WQI to display high and low values observed for a number of streams throughout the state (Figure 30). Some streams show great variability; index values for the Carp

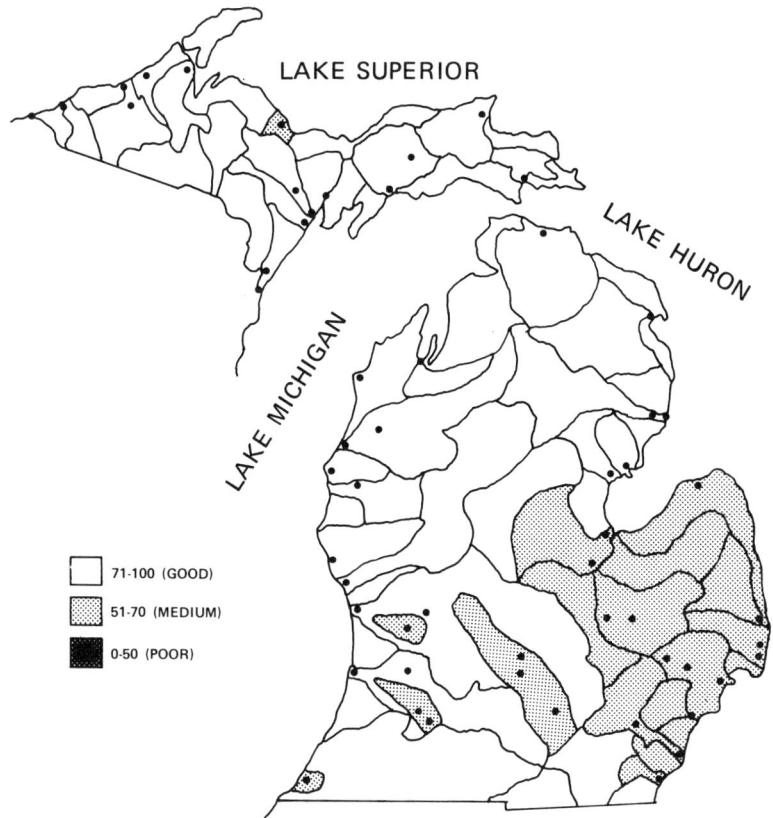

Figure 29. Map showing average NSF WQI values for the State of Michigan.[50]

River, for example, vary from "poor" to "good." The Michigan report also uses the NSF WQI to show water quality trends over time (Figure 31). The gradual upward trend from 1969 to 1976 shows an improvement in water quality over the period. New York's Section 305(b) report to Congress[51] also uses the NSF WQI to show water quality trends for rivers throughout the state (Figure 32).

One important purpose for using a water quality index was to examine changes in water quality in response to water pollution control efforts. The New England Interstate Water Pollution Control Commission, for example, applied the NSF WQI to help assess the impact of an expenditure of $30 million for wastewater treatment facilities.[12] By applying the index both

282 ENVIRONMENTAL INDICES

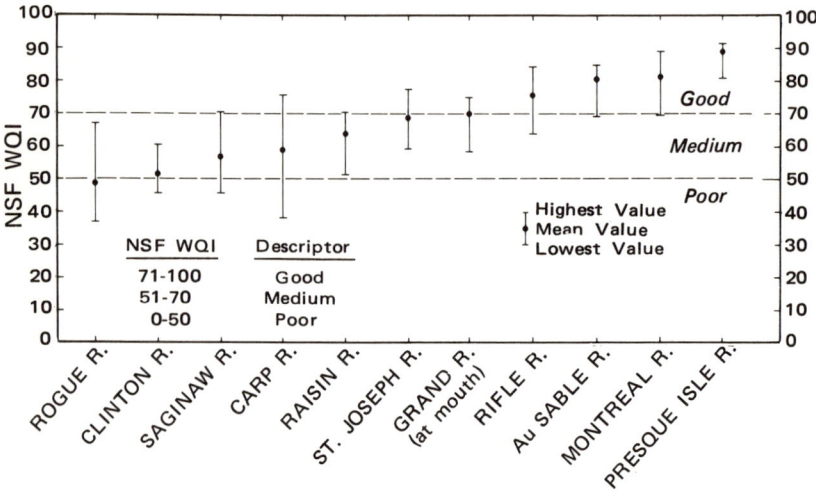

Figure 30. Ranges of the NSF WQI for selected rivers in Michigan.[50]

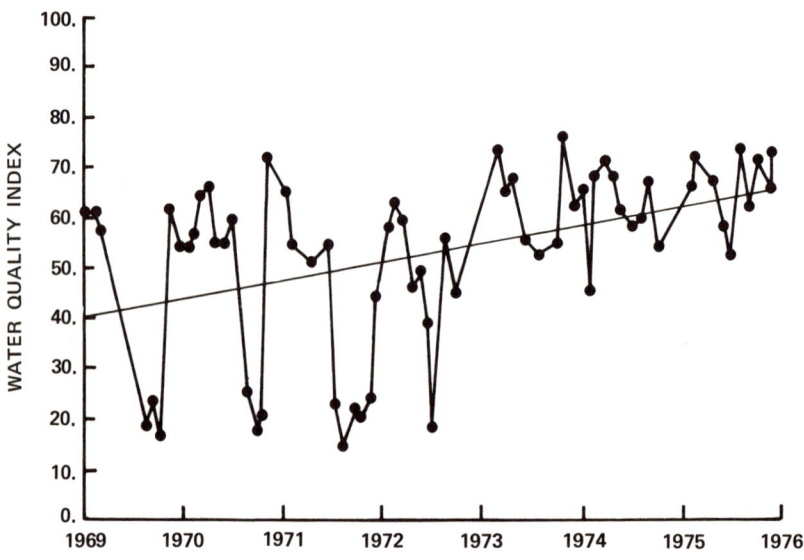

Figure 31. Example of water quality trend report using the NSF WQI on Michigan's Raisin River at Monroe (Station No. 580046).[50]

WATER POLLUTION INDICES 283

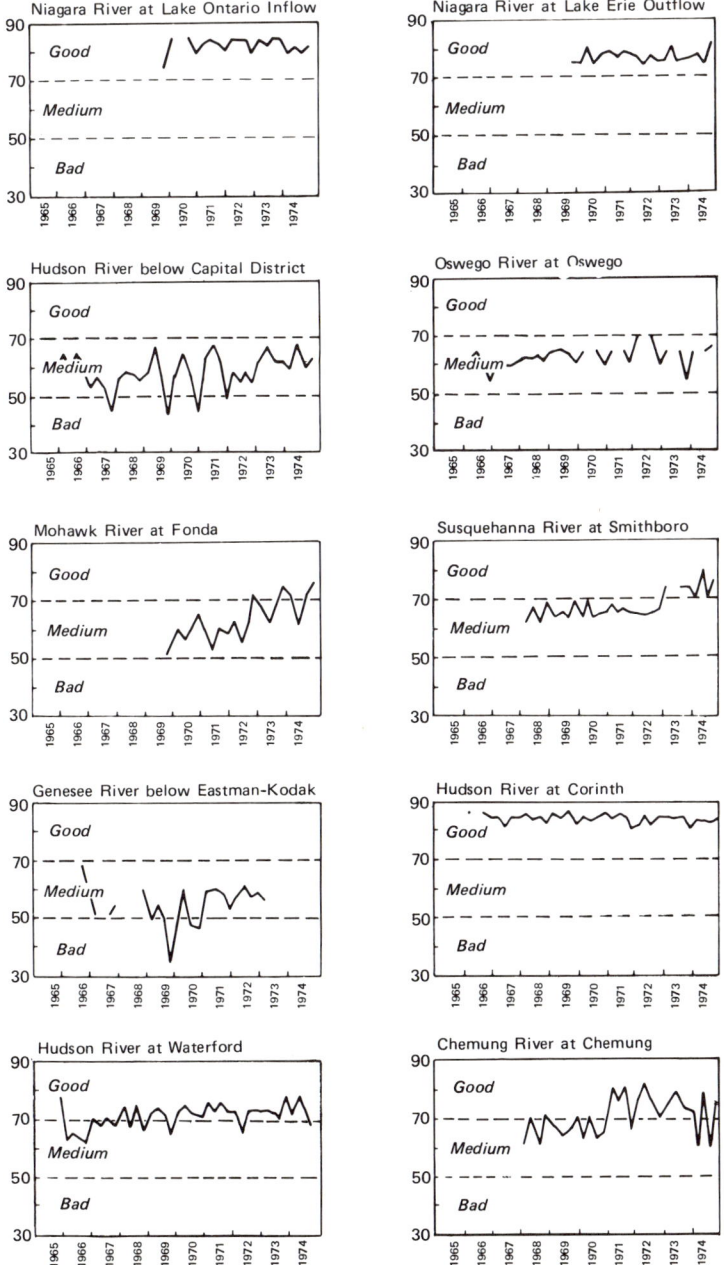

Figure 32. Example of water quality trends using the NSF WQI on rivers in New York State, from the annual Section 305(b) water quality report.[51]

before and after construction of these facilities, the agency planned to assess the impact of this project on water quality.

The most commonly mentioned public information use of a water quality index was to communicate water quality data to the "layman." A typical comment of the survey respondents was, "We feel that the general public needed something more in keeping with the layman's knowledge than the raw data would provide." However, the survey came across few documented examples in which an index had been used to report water quality data directly to members of the general public through radio, television or newspapers. Judging by the content of the Section 305(b) reports, the audience more often appeared to be the informed water quality professional or pollution control official. Unlike the air pollution indices, water quality indices were never used to report data to the general public on a daily basis. Apparently, indices serve a more limited audience in the water pollution field than they do in the air pollution field.

Indices Developed by Agencies

The national survey of water quality indices[1,2] found that six new indices had been developed by water pollution control agencies. Four of these were developed by state agencies—Georgia, Illinois, Oregon and Nevada—and two were developed by EPA's Regional Offices. Some of these were similar to, or had evolved from, the NSF WQI, but others were quite different from the NSF WQI.

Georgia

Georgia's Department of Natural Resources (Environmental Protection Division) developed two water quality indices and selected one, the Trend Monitoring Index (TMI), for routine use.[52]

The TMI uses eight implicit nonlinear subindex functions similar to those in the NSF WQI, but the curves are based on the expert judgment of the Department of Natural Resources staff. Unlike the NSF WQI, the TMI uses a novel aggregation function called "variable weighting." This method is designed to give greatest emphasis to the "worst" of the eight water quality variables by weighting subindices with the lowest values (worst water quality) more heavily than those with the highest values (best water quality). This is accomplished as follows. The highest subindex is written down once, the second highest twice, the third highest three times, and so on, until the eighth highest subindex (which is the lowest subindex) is written down eight times. This procedure gives 36 numbers that are multiplied together, and the 36th root then is taken of the product. The purpose of these calculations is to substitute a new set of weights for the subindices. The weight of the highest subindex becomes 1/36 = 0.0278, and the weight of the lowest subindex becomes 8/36 = 0.2222. Thus, individual weights are no longer associated with a particular variable but are applied in a way that always gives greatest weight to the variable with the lowest subindex and least weight to the variable with the highest subindex.

Mathematically, this aggregation function is expressed as follows:

$$TMI = \prod_{i=1}^{8} I_i^{[i/36]} \tag{37}$$

where $I_1 > I_2 > ... > I_8$

Like other forms of the weighted product (pages 82-89), the sum of the weights is 1.0, and the index ranges between 0 and 100.

The TMI was applied in Georgia's 1975 Section 305(b) report[53] and was used to "compare water quality trends throughout the State on a common basis." It was computed for data covering a 2-year period at most of the monitoring stations in Georgia and a 5-year period at selected stations. At the latter stations, index values were compared with expert evaluations of water quality based both on chemical and biological data. To cope with the variation in natural water quality from one Georgia stream to another, different "natural TMI ranges" were defined for different streams.

When applied to Georgia's streams, most readings in the "poor" (0-40) or "fair" (41-60) ranges could be attributed to municipal or industrial wastewater problems. However, nonpoint sources of pollution, such as urban runoff, also could reduce the index to the "fair" or even to the "poor" range for limited periods of time. TMI values calculated for sampling sites upstream and downstream of Georgia's major urban and industrial areas showed dramatic differences—sometimes as much as 50 points.

Illinois

Historically, governmental agencies within the State of Illinois have shown considerable interest in water quality indices. In 1972, Barker and Kramer,[54] of the Illinois Department of Transportation (Division of Water Resource Management), developed a simple water quality index, the Pollution Index (PI). In the process, they identified six criteria that a meaningful index should possess:

- include water quality variables that are widely and regularly measured
- include variables that have clear effects on aquatic life, recreational use, or both
- include variables that have manmade sources as opposed to natural sources
- include variables that are amenable to control through pollution abatement programs

- use realistic ranges for each variable—from no pollution to gross pollution
- be sensitive to reasonably small changes in water quality

The PI included five variables: DO, COD, fecal coliforms, ammonia nitrogen, and nutrients (either phosphates or nitrates). Its subindices consisted of staircase step functions, and they were aggregated using a linear sum. The index had an increasing scale ranging from 0 to 30, and five descriptor words were used: "light" (0-7), "moderate" (8-9), "heavy" (10-11), "severe" (12-15), and "gross" (16-30). This index was applied on a pilot basis to data from monitoring stations throughout the State of Illinois. Lake Michigan had PI values in the range 0-3, but most of the waterways in Chicago fell into the range 20-25. Index values also were compared with the number and kinds of fish found in Illinois stream surveys, and the number of fish species tended to decrease as the index increased. Barker and Kramer[54] also developed a Treatment Index (TI) intended as a simplified first step toward computing the cost of treating water for public water supplies. Although these indices were applied by their developers to water quality data in Illinois, no subsequent applications are known.

In 1976, Schaeffer and Janardan,[41,42] members of the Illinois Environmental Protection Agency, developed five statistically based water quality indices and selected one, the Beta Function Index, for routine use. The structure of this index, including an example of how to apply it, was discussed above as part of the survey of the literature (pages 259-263). The Illinois Environmental Protection Agency applied the Beta Function Index on a state-wide basis using four variables (DO, fecal coliforms, dissolved solids and ammonia nitrogen) and presented the results in its 1976 Section 305(b) report to Congress.[44] Numerical values of the index were compared for 522 stations over the 5-year period from 1971 to 1975. Of the total, 93 stations had improved over the period, 50 had deteriorated, and 379 had remained the same. Based upon the index values, each station was assigned one of four ratings in each year: "good," "average," "semipolluted," and "polluted." Agency personnel appeared satisfied with the performance of the index and planned to use it in future annual water quality reports.[12] The index also has been used to determine whether a statistically significant difference in violations of water quality standards took place as a result of a mild drought which occurred during 1976.[55]

Nevada

At the time of the national survey of water quality indices by Ott,[12] Nevada was developing and evaluating a simple water quality index. The index used an increasing scale in which 1 or less represented "good" water quality, and 10 or more represented "poor" water quality. It included 15 variables which were represented by explicit linear and nonlinear subindex functions (Table XXXVII). As noted by Sheen,[56] the subindex equations were based largely on the subjective judgment of the index developers and on water quality conditions in Nevada:

> ...I point out that all these relationships are purely empirical and not very objective. In other words, they are subject to change depending on a particular user's feel of the actual water quality at a particular sampling site. I feel that even in the seemingly infant stages in developing these index programs, they hold a certain degree of promise for making a complex list of data a little easier to handle and understand relative to points up or downstream in the same analysis network.

The subindices were aggregated using a linear sum:

$$I = \frac{2}{3} \sum_{i=1}^{15} I_i \qquad (38)$$

Although the index was being tested by officials from the Nevada Department of Human Resources (Environmental Protection Services) in 1977, it had not yet been applied routinely.

Oregon

Oregon's approach to index development is of particular interest, because it offers an example of a "case study" procedure in which an index is adapted systematically to the geographical area under study. In this approach, candidate indices first are carefully reviewed. Then an index structure is selected, variables are chosen, and the resulting index is tailored, or "custom fit," to the area in which it is to be used. Different investigators following this procedure would develop indices with different weights, variables and subindex functions. Presumably, this approach provides the best means possible for incorporating geographical differences of water quality into the resulting index.

Using this approach, Oregon's Department of Environmental Quality (DEQ) developed a water quality index that was uniquely tailored to the Willamette River Basin. The overall approach was similar to the technique used by Brown *et al.*[6] to develop the NSF WQI, except that it was carried out on a state-wide basis. As reported in a draft manuscript by Dunnette,[57] the evolution of the index consisted of four basic steps:

1. evaluation of water quality indices previously proposed in the literature
2. development of criteria for rejecting unsuitable variables
3. survey of the DEQ staff using a modified Delphi opinion assessment technique

Table XXXVII. Subindex Equations Used in Nevada's Water Quality Index[12]

Variable	Equation
Temperature (°C)	$I = 0.0016 \, X^2$
Dissolved Oxygen (mg/l)	$I = 0.0125 \, (X - 10)^2$
pH (standard units)	$I = 0.25 \, (X - 7)^2, \, 5 \leq X \leq 9$
5-Day Biochemical Oxygen Demand (mg/l)	$I = 0.125 \, X$
Chlorides (mg/l)	$I = 0.140 \, X^{0.5}$
Total Phosphates (mg/l)	$I = 0.33 \, X$
Orthophosphates (mg/l)	$I = 0.40 \, X$
Nitrates (mg/l)	$I = 0.20 \, X$
Total Dissolved Solids (mg/l)	$I = 8 \times 10^{-6} \, X^2$
Alkalinity (mg/l CaCO$_3$)	$I = 0.005 \, X$
Bicarbonates (mg/l)	$I = 0.0067 \, X$
Carbonates (mg/l)	$I = 0.20 \, X$
Total Coliforms (no./100 ml)	$I = 0.037 \, X^{0.333}$
Fecal Coliforms (no./100 ml)	$I = 0.058 \, X^{0.333}$
Turbidity (JTU)	$I = 0.08 \, X$

4. classification of variables into four general "impairment" categories to eliminate redundancy

Oregon's evaluation of candidate water quality indices began by considering five water quality index criteria that previously had been proposed by the Council on Environmental Quality[58]:

- facilitate communication of environmental quality information to the public
- be readily derived from available monitoring data
- strike a balance between oversimplification and complex technical conceptualizations
- impart an understanding of the significance of the data they represent
- be objectively designed but amenable to comparison with expert judgment so that their validity can be assessed

A total of 12 published indices were evaluated using these criteria (Table XXXVIII). According to Dunnette,[57] only 2 of the 12 indices—Harkins' index and the NSF WQI—met all five of the CEQ criteria, as interpreted for Oregon. Because Harkins' index uses a ranking procedure in which all previously calculated values must be recalculated to compare new data (that is, it is a relative index), it was eliminated from further consideration. The remaining index, the NSF WQI, was used as the basis for developing the Oregon Water Quality Index (OWQI) using the case study procedure.

The process of selecting variables began with a list of 90 candidate variables. Sixty were rejected because they were (1) not present in significant or

Table XXXVIII. Evaluation by Dunnette[57] of 12 Water Quality Indices According to CEQ Criteria

	Proposed Indices											
Criterion[a]	Region X	Truett	NSF WQI	Nemerow	Harkins	Dee	McDuffie	Inhaber	Dinius	Prati	Walski	Horton
1	X	X	X	X	X	X	X	X	X	X	X	X
2	X	X	X	O	X	O	X	O	X	O	O	O
3	O	X	X	X	X	O	O	O	X	X	O	O
4	X	O	X	X	X	X	O	X	X	X	O	O
5	O	O	X	O	X	O	O	O	O	O	O	O

[a]Key to criteria: X = met; O = not met. The criteria, from the 1974 CEQ Annual Report,[58] are as follows: (1) facilitate improved communication of environmental quality information to the public; (2) be readily derived from available monitoring data; (3) strike a balance between oversimplification and complex technical conceptualizations; (4) impart an understanding of the significance of the data they represent; and (5) be objectively designed but amenable to comparison with expert judgment in order that their validity can be assessed.

harmful amounts, (2) limited by insufficient data, or (3) of questionable significance. The remaining 30 variables were sent to 22 professional DEQ staff members involved in water quality work, and each member was asked to select 10 variables from the list and rate each one numerically (on a scale of 1 to 5) in terms of its reflection of general water quality in Oregon's Willamette River Basin. This survey yielded 15 rating forms; the results were tabulated, and a summary of the findings was distributed to the participants. After reviewing the group's ratings, each respondent was allowed to change his original ratings. This reduced the list from 30 to 14 variables.[5,7]

Finally, to further reduce the number of redundant variables, five general "impairment categories" were identified: (1) oxygen status; (2) eutrophication, or potential for excess algae and plant growth; (3) physical characteristics; (4) dissolved substances; and (5) health hazards. To the extent possible, variables were selected that would give suitable representation to each impairment category (Table XXXIX). In some cases, a given variable was represented in more than one impairment category. In summary, the choice of variables was based on the importance ratings given to them by the DEQ staff, the desire to reduce redundancy in each of the impairment categories, and the judgment of the investigator regarding the availability of data and significance of the variable for Oregon streams.

The resulting index consisted of six exponential (explicit nonlinear) subindex functions. The functions were determined by examining historical records for six monitoring stations on the Willamette River for the period 1973-75. For DO, BOD_5, total solids, and nitrate-plus-ammonia, a subindex value of $I = 80$ was chosen to coincide with the mean value of each variable over the historical period. For fecal coliforms, a subindex value of $I = 70$ was chosen to correspond to the mean coliform level over the 3-year period. For pH, $I = 60$ was assigned to both pH = 6 and pH = 9. Graphs of the subindex curves are included in the report by Ott.[12] The overall index is computed using the weighted linear sum aggregation function:

$$OWQI = \sum_{i=1}^{6} w_i I_i \qquad (39)$$

The study also examined correlations among several variations of the OWQI and several other indices. This comparison included a multiplicative version of the OWQI, Prati's index, McDuffie's index, and six- and eight-variable versions of Harkins' index. Correlation coefficients ranged from 0.56 to 0.99, with a mean of 0.87. Dunnette attributes these high correlations, in part, to the similar finding by O'Connor (page 228) that selection of the variables which go into an index tends to be more important than the particular shapes of the subindex curves.

Table XXXIX. Weighting Factors Used in Oregon's Water Quality Index.[57]

Variable	Impairment Category[a]	Normalized NSF WQI Weight[b]	Final Variable Weight
Dissolved Oxygen	1	0.24	0.40
Fecal Coliforms	5	0.21	0.20
pH	4	0.17	0.10
5-Day Biochemical Oxygen Demand	1	0.14	0.10
Total Solids	3,4	0.10	0.10
Nitrate + Ammonia	2	0.14	0.10

[a] 1 = oxygen depletion; 2 = eutrophication; 3 = physical characteristics; 4 = dissolved substances; 5 = health hazards.
[b] A normalized NSF WQI weight was obtained by deleting three of the original NSF WQI variables and calculating new weights that add to 1 and preserve the ratios among the remaining NSF WQI weights.

The OWQI was implemented in Oregon in 1978 as a monthly reporting system to provide information on water quality conditions at each station on the Willamette River Basin. The DEQ ultimately plans to apply the index to waters throughout the state.

Region VIII

The water quality indices developed by EPA's Regions VIII and X both have increasing scales, but they are different in design than the indices discussed above. The subindices are based not on mathematical functions but on the frequency of violation of water quality standards.

In 1975, EPA's Region VIII in Denver, Colorado, developed a water quality index that included many of the variables in the NSF WQI. However, temperature and pH were not considered significant problems in the Region VIII geographical area (Colorado, Montana, North Dakota, South Dakota, Utah and Wyoming), and they were not included. As reported by Cogger, Payne and Sprenger,[59] five subindices were used, each of which was calculated as the percentage of the observations of a pollutant variable (or group of variables) exceeding state water quality standards.

Subindex weights were chosen by reevaluating the weights used in the NSF WQI in light of water quality conditions in the Region VIII geographical area. DO and BOD_5 were treated as a group and were represented by one subindex. The sum of the NSF WQI weights for DO and BOD_5 (0.17 + 0.10 = 0.27) was considered too high; because few DO and BOD_5 violations are measured by fixed stations on Region VIII mainstreams, the weight for this group was reduced to 0.25. Fecal coliforms has an NSF WQI weight of 0.15, and total coliforms ordinarily is not included in the NSF WQI. This emphasis was considered too low, because total coliforms (as a group) includes organisms that are somewhat more resistant to chlorination and may indicate, at least for waters influenced by sewage treatment plant wastewater, the presence of more tolerant and long-lived pathogens. Thus, the weight for fecal and total coliforms, which also were grouped, was raised to 0.25. Similarly, the NSF WQI weights of 0.10 for nitrates (nitrogen) and phosphates (phosphorus) were considered too low, and they were raised to 0.125 for each. Physical and aesthetic factors, represented in the NSF WQI by turbidity (w = 0.08) and total solids (w = 0.08), were treated as a group, and the weight for this group was increased to 0.25.[59]

The five subindices were aggregated using the weighted product aggregation function:

$$I = I_1^{0.25} I_2^{0.25} I_3^{0.125} I_4^{0.125} I_5^{0.25} \tag{40}$$

where: I_1 = percent violation of DO and BOD standards
I_2 = percent violation of fecal and total coliform standards
I_3 = percent violation of nitrogen standards
I_4 = percent violation of phosphorus standards
I_5 = percent violation of criteria for physical and aesthetic factors

Note that the numerical values of the weights, which add to 1.0, are given in Equation 40. Because the subindices are percentages which range from 0 to 100, the overall index also ranges from 0 to 100.

Occurrences of pesticides and toxic substances were too rare to be included directly in the index. Thus, they were listed separately in a "severe events" list, which gave qualitative information about the date of the occurrence, the composition of the chemical observed, and the location of the measurement.[59]

This index was applied to data from the six states comprising the Region VIII area, and changes in water quality between two 3-year time periods (July 1969 to June 1972 and July 1971 to June 1974) were examined.[59] A total of 87 stations were identified as having sufficient data for both periods. Index values were presented on maps of the rivers in these states using a color coding system consisting of four different colors (Table XL). The index probably will be used in future reports and trend studies in Region VIII.

Region X

An index developed by Beebe[60-62] incorporates both "time" (proportion of observations in which recommended limits were exceeded within a

Table XL. Descriptive Categories Used for Reporting the Region VIII Water Quality Index[59]

Index	Color	Significance
>15	Red	Waters or areas that have significant water quality problems
5-15	Yellow	Waters or areas that have intermittent water quality problems
0-5	Blue	Waters or areas that have infrequent water quality problems
Insufficient Data	Green	Waters or areas requiring data for determination of status

certain time interval and "space" (length of stream affected by a given water quality). Designed for use on a computer, the index is somewhat complex and includes both water quality standards and implicit subindex functions similar to those in the NSF WQI.

The variables used in the index are grouped into 10 different pollutant categories. For each category i and station j, the subindex $I_{i,j}$ is calculated as the weighted product of a quality term $Q_{i,j}$ (from implicit subindex functions) and a frequency term $F_{i,j}$:

$$I_{i,j} = w_i Q_{i,j} F_{i,j} \tag{41}$$

where: w_i = weight for pollutant category i
 $Q_{i,j}$ = quality term for category i and station j
 $F_{i,j}$ = proportion of observations which exceed recommended limits for category i and station j.

Next, subindex I_i is formed for category i by weighting the values calculated in Equation 41 by the distance in river miles between adjacent stations and adding the results for all stations along the river:

$$I_i = \frac{1}{2} \sum_{i=1}^{m-1} (I_{i,j+1} + I_{i,j})(d_{j+1} - d_j) \tag{42}$$

where: I_i = water quality subindex for category i
 $I_{i,j}$ = water quality subindex for category i and station j
 d_j = river mile distance of jth station $(d_{j+1} > d_j)$
 m = number of stations

Finally, the index is formed by adding the subindices for all 10 categories:

$$I = \sum_{i=1}^{10} I_i \qquad (43)$$

In recent years, the index has been modified for use in water quality profile reports for the various states within Region X (Alaska, Idaho, Oregon and Washington).[63] A quality term (Q_{ij}) now is generated for each individual pollutant variable, and the time-weighted average of these terms is computed over the reporting period. The frequency term (F_{ij}) no longer is included in the index calculation. The distance estimates now are based on the professional judgment of the user regarding the representativeness of each station. These distance estimates of "affected river miles" also are used to weight the index computation. This modified version of the Region X index currently is being used to prepare environmental quality profiles on an annual basis.

PROSPECTS FOR A NATIONALLY UNIFORM WATER QUALITY INDEX

A question which arises in the study of water quality indices is whether a nationally uniform water quality index is desirable, assuming that a suitable index could be found. Although much effort has been directed toward development and adoption of a uniform air pollution index in the U.S. (Chapter III), similar activity has not occurred in the water pollution field. However, adoption of a uniform water quality index faces a variety of problems which are more serious than those in the air pollution field.

Respondents in the national survey of water pollution control agencies by Ott[12] expressed a great variety of opinions about prospects for a uniform water quality index. A majority of the respondents that were using indices felt that the federal government should recommend a uniform index for agencies wishing to use indices. Of the 11 state and interstate agencies using indices, for example, 8 expressed views favoring a recommended uniform index. Some felt that considerable uniformity already was occurring because of the widespread use of the NSF WQI, but they were willing to consider a slightly different index if it were recommended for national use. The following comments from three different index users were typical:

> A uniform index that would be recommended by the federal government would be a good idea. One of the best features of the NSF WQI is that it already has received widespread application. Thus, a certain amount of uniformity already has occurred. If EPA were to recommend a uniform index, EPA should probably first poll the users. It would not upset me greatly to use a slightly different index, unless some new variables were included. If the index software were available, it would be very easy to recalculate the values with the new index.[12]

An index like the NSF WQI would be better for national use than an index based on water quality standards. Violation of water quality standards is not a good basis for an index for national use because each state has different water quality standards.[12]

I feel that a uniform index is a good concept. I would support the idea that EPA should support a uniform index. It would provide a basis to "educate the public" that a certain number means a certain quality. The standard NSF WQI appears to be a good choice for a uniform index and would allow comparisons to be made of water quality in different states.[12]

One reason given for favoring a uniform index was to avoid index "customizing," in which the user deletes variables, alters the subindices, or modifies the aggregation function:

It probably would be a good idea for EPA to recommend a standardized index to avoid user "customizing." The chief problem with customizing is that people will say, "You customized just to make yourself look good."[12]

Although most of the agencies classified as index users held favorable views about water quality indices and supported a uniform approach, one user felt that a uniform index would not work because of a variety of problems:

The index really is too simplistic. It's useful only when you understand its limitations. I would not favor adoption by the federal government of a uniform index, because the waters vary considerably in different areas. Also, stream flow rates differ, and different streams react differently to different waste loads. Possibly, application of indices on a regional basis would be useful, however.[12]

When considering the problems that must be overcome in developing a uniform water quality index, a number of differences between the air and water pollution fields must be noted. These differences make the attainment of index uniformity more difficult in the water pollution field than it was in the air pollution field.

Geographical variation in water quality, for example, means that levels of a certain variable normally may be insignificant in one geographical area, while concentrations generally exceed recommended limits in another area simply because of natural conditions. This is one reason that different water quality standards have been adopted in different states, or sometimes in different areas within a state. One survey respondent, who had developed his own index, felt that geographical variation makes it necessary for each index to evolve from unique conditions in the area using a case study approach:

We don't feel that an absolute type of index can work; there are too many geographical differences, even within a state. An index must be tailored to the specific location and it must be empirically based. We based our index on the data we have been collecting in the state over a period of years. We were able to assess the index performance in light of our actual knowledge of the geographical variation in water quality and, thus, to come up with a suitable means for detecting changes in water quality over time.[12]

In the air pollution field, by contrast, most pollutants are measured in urban areas where natural sources are negligible. Furthermore, the NAAQS, because they apply nationally, greatly facilitate development of a uniform index based on NAAQS values.

Another problem stems from different water uses. What is considered "good" water quality for one use may not be "good" for another use. In the air pollution field, this problem does not arise, because air has one primary use: it is consumed by living systems to support life. The air comes to its users in raw form, is breathed directly, and has direct effects on public health. This use is analogous to only one water use: supply of public drinking water. However, drinking water usually is treated before it is consumed, and pollutants known to affect public health are removed. Because air pollution affects the public directly, often on a daily basis, the public's interest in air quality reports probably is more pressing and urgent than its interest in water quality reports. Of course, the public's interest in water quality often becomes more intense in situations that affect it immediately, such as a spill of toxic substances upstream.

Although public information was a commonly cited purpose for using a water quality index, few examples are known in which a water quality index has been used to report water quality directly to the public through radio, television and newspapers. Usually, the audience consists of water pollution professionals or government officials, rather than television viewers or newspaper readers. By contrast, one of the most important purposes for using an air pollution index is to report air quality data directly to the public through the mass media. Of course, lack of public dissemination of water quality information (by means of indices or any other approaches) may be evidence of a lack of effort on the part of water pollution control agencies to interest the media in presenting water quality information to the public. Possibly, water quality index reports could be presented in newspapers on a seasonal or monthly basis along with summary information about climate, tides and air quality trends.

A final problem is the large number of constituents that affect water quality. An index that includes only eight or nine variables may miss an important pollutant that could kill fish or harm humans. As noted by Ficke[64] of the U.S. Geological Survey, the large number of variables that are important makes it difficult to represent water quality by a simple index:

... In rivers, the concentrations of the metals vary as a function of the geology of the terrain and individual pollutants. They do not correlate well with each other or with other water quality characteristics, and in order to judge the suitability of a water through an index, that index must consider every one of these metals. In order to use the index it is necessary to have data for every water quality characteristic for which there are standards or criteria. It is just plain foolish to narrow down to an index that considers only eight variables and claim that people generally agree that these are the important ones. If they were the only important ones, we wouldn't have the long lists of standards and criteria that we now have.

Water Pollution Index

One way to incorporate a large number of variables is to develop a water quality index that uses a more flexible aggregation function than the weighted sum or weighted product. For example, an index using an aggregation function similar to that in the PSI air pollution index would be able to accommodate any number of variables without changes in its basic structure.

For illustrative purposes, we now introduce the Water Pollution Index (WPI), a decreasing scale index that uses the minimum operator as its aggregation function:

$$\text{WPI} = \min \{I_1, I_2, \ldots, I_n\} \tag{44}$$

The WPI is intended as a possible approach rather than a proposed index that is presented in detail. Its subindices can be selected from the other indices presented in this chapter. A companion form, the WPI′, is an increasing scale index that is similar to PSI; it employs the maximum operator as its aggregation function:

$$\text{WPI}' = \max \{I_1, I_2, \ldots, I_n\} \tag{45}$$

Because of the flexibility of these aggregation functions, any variable which is considered important (and for which data are available) can be included. For example, all nine of the NSF WQI subindices can be included in the WPI, if desired. Stoner's subindices for toxic substances also can be included in the index, although the behavior of the overall index then will be similar to Stoner's index, becoming negative when a subindex for any toxic substance becomes negative. Various combinations of subindices from different indices also can be included. In fact, it would be possible to construct a version of the WPI which includes all of the variables listed in Table XXVIII as well as any other variables of importance. Of course, the scales and ranges of some subindices would have to be modified using Equation 36 or a similar transformation before they are included in the WPI.

The WPI concept can be illustrated by plotting the minimum of the nine NSF WQI subindices calculated from McClelland's[9] data for the Wakarusa and Delaware Rivers in Kansas (Figures 33 and 34). In these examples, the WPI exhibited greater variation over time than either the NSF WQI_a or the NSF WQI_m, with values that were occasionally quite low. The smaller variation of the NSF indices was attributable to the tendency for the additive and multiplicative forms to dampen out changes in the individual variables. The pollutant variable responsible for each WPI value (*i.e.*, the pollutant with the minimum subindex), which is analogous to PSI's "critical pollutant," also varied over time. Of the 10 observations on the Wakarusa River (Figure 33), the most frequently occurring critical pollutant was fecal coliforms (6 instances), followed by total solids (2 instances), DO (1 instance), and turbidity (1 instance). For these data, the correlation coefficient between the WPI and the NSF WQI_a was r = 0.60, and the correlation coefficient between the WPI and the NSF WQI_m was r = 0.80. For the 19 values in Figure 34, the correlation coefficient between the WPI and the NSF WQI_a was r = 0.95; the correlation coefficient between the WPI and the NSF WQI_m was r = 0.92.

The WPI gives considerable insight into the reasons for changes in water quality. In Figure 33, for example, the NSF WQI_a declined from 74.9 on November 5 to 60.3 on November 13, a change of about 15 points. Over the same period, the WPI decreased from 41.0 to 5.0, a change of 36 points. Examination of the WPI values shows that this decrease in water quality was clearly attributable to turbidity. In Figure 34, the NSF indices showed relatively low values on November 1 and November 13 (NSF WQI_m = 66.5 and

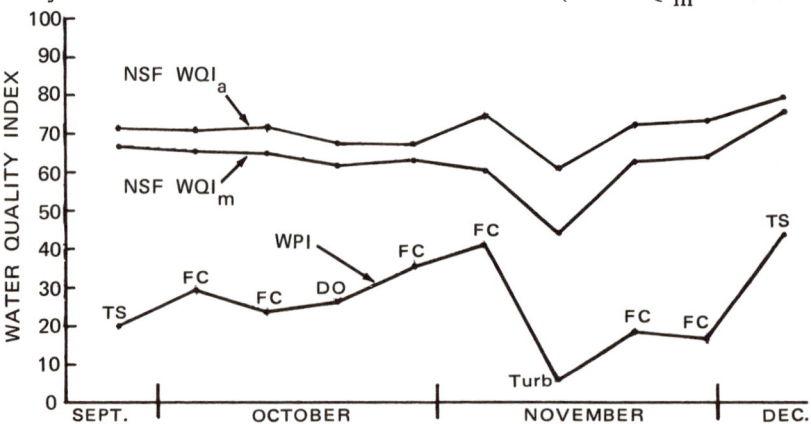

Figure 33. Comparison by the author of the WPI with the NSF WQI_a and the NSF WQI_m for data from the Wakarusa River, Kansas. (TS = total solids; DO = dissolved oxygen; FC = fecal coliforms; Turb = turbidity)

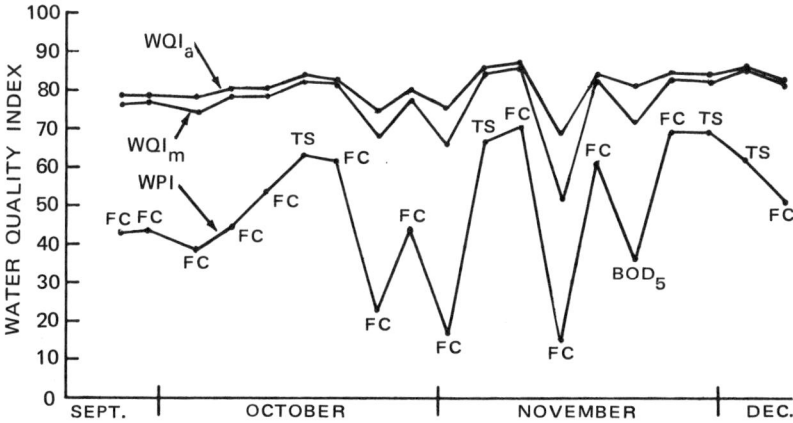

Figure 34. Comparison by the author of the WPI with the NSF WQI_a and the NSF WQI_m for data from the Delaware River, Kansas. (FC = fecal coliforms; TS = total solids.)

51.4, respectively), but the values did not differ strikingly from those observed on the other dates. By contrast, the WPI gave values for these two dates that were extremely low (WPI = 16.8 and 14.6, respectively). These low WPI values can be attributed to fecal coliforms, for which the actual observations were 2,000 organisms/100 ml and 2,900 organisms/100 ml, respectively. Thus, the potential problem from high fecal coliform levels is partially eclipsed in the NSF WQI_m (and even more so in the NSF WQI_a), but it shows up clearly in the WPI.

In general, a water quality index that is based on either Equation 44 or Equation 45 will be free of both eclipsing and ambiguity problems (see Chapter II) and will have a linear (one-to-one) relationship between subindex values and overall index values. In addition, when the index is applied to a given location, any variable which is not suspected of being a problem can be deleted quite easily without affecting the overall index computation. Whenever the index indicates "poor" water quality, the pollutant variable responsible also can be identified along with the index report, giving an indication of the nature of the problem. Because the WPI gives considerable insight into the behavior of the individual pollutant variables, it probably should be computed whenever the NSF WQI or similar indices are computed, with the resulting WPI values plotted alongside the index being studied. It is hoped that readers will consider the WPI concept as a starting point and will be encouraged to explore other possible uses of the WPI and the WPI .

Criteria for an Ideal Water Quality Index

Many index developers had proposed criteria to assist them in the development and evaluation of water quality indices. By examining the criteria proposed by Schaeffer and Janardan,[41] Barker and Kramer,[54] Dunnette,[57] and the 1974 CEQ Annual Report[58] and by taking into account the mathematical analysis presented in Chapter II, it was possible to assemble a list of 21 characteristics that an ideal water quality index should possess:

1. be developed from a logical scientific rationale or procedure
2. strike a reasonable balance between oversimplification and technical complexity
3. be sensitive to small changes in water quality
4. avoid eclipsing
5. avoid ambiguity
6. avoid nonlinearity in the aggregation process
7. be dimensionless
8. employ a clearly defined range
9. impart an understanding of the significance of the data
10. be relatively easy to apply
11. easily accommodate new variables
12. permit probabilistic interpretations to be made
13. include variables that are widely and routinely measured
14. include toxic substances
15. include variables that have clear effects on aquatic life, recreational use, or both
16. be tested in a number of geographical areas
17. show reasonable agreement with expert opinion
18. show reasonable agreement with biological measures of water quality
19. be compatible with water quality standards
20. include guidance on how to handle missing values
21. clearly document the limitations

No single index is expected to meet all 21 criteria completely. However, different candidate indices will vary in their proximity to the ideal, and the criteria are intended to provide a uniform system by which to judge indices. Each criterion varies, of course, in terms of its relative importance and the significance that it should receive when indices are being evaluated.

CONCLUSION

A number of ingenious water quality indices have been developed. About one-fifth of the U.S. water pollution control agencies are actually using water quality indices, and many additional agencies are evaluating indices or including them in their future plans. The most commonly used index is the NSF WQI, followed by Harkins' index. Most of the other indices in the literature are not being used in practice, but several water pollution control agencies have developed indices of their own. The purposes cited for using an index include public information, analysis of water quality trends, and preparation of the annual Section 305(b) reports to Congress. Most index users seem satisfied with the performance of their indices, but there is considerable controversy about the practicability of a nationally uniform water quality index similar to the uniform air pollution index. Although water quality index usage is a relatively recent phenomenon, it appears to be gaining in popularity.

PROBLEMS FOR STUDY

1. Horton's index, while very simple, represents an early example of a water quality index. Assume that the following data are from a river in which temperatures less than 93°F (34°C) are considered acceptable. (This example appears in the original article by Horton.[5])

Pollutant Variable	Stream 1	Stream 2
Sewage Treatment (%)	96%	75%
Dissolved Oxygen (%)	60%	60%
pH	7.3	3.6
Total Coliforms (MPN/100 ml)	4,000	<1,000
Specific Conductance (μmho/cm)	700	650
Carbon Chloroform Extract (10^{-3} mg/l)	150	90
Alkalinity (mg/l)	70	0
Chlorides (mg/l)	50	20
Temperature	<93°F (34°C)	<93°F (34°C)
Obvious Pollution	OK	OK

(a) From an examination of the raw data, which stream has poorer quality? (b) Calculate the sum of the subindices using Horton's index. (c) Calculate Horton's Quality Index. [Answers: Stream 2 is of poorer quality than Stream 1, largely because of acid conditions and poor sewage treatment; 1,660, 1,100; QI = 92, QI = 61]

2. The observations in the following table are based on data from the Wakarusa River in Kansas which appear in a report by McClelland.[9] Assume that the temperature subindex $I_7 = 93$ for all sets of observations.

Date in 1972	DO (%)	Fecal Coliforms (no./100ml)	pH	BOD$_5$ (mg/l)	Nitrates (mg/l)	Phosphate (mg/l)	Turbidity (JTU)	Total Solids (mg/l)
10/04	80	300	7.9	3.9	0.34	0.30	17	400
10/12	97	620	7.9	5.3	0.02	0.24	12	460
10/28	50	180	7.8	2.6	0.36	0.22	28	430
11/13	86	2,300	7.9	5.5	0.50	0.38	155	570
11/21	96	1,700	8.0	2.5	1.33	0.05	32	410

(a) Using the NSF WQI subindex curves given in this chapter, calculate the additive NSF WQI$_a$ from these data. [Answers: 71, 73, 66, 61, 72]

(b) Calculate the multiplicative NSF WQI$_m$ from these data. [Answers: 66, 66, 60, 45, 63] Note: Your answers may differ slightly from these answers due to differences in estimating values from the graphs.

3. Discuss the basic differences between Stoner's index and the NSF WQI with regard to overall philosophy, variables included, structure and subindex functions.

4. Plot the following subindices for the public water supply version of Stoner's index: (a) ammonia nitrogen, (b) nitrate nitrogen, (c) fecal coliforms and (d) pH.

5. Consider the water quality index proposed by Nemerow and Sumitomo.[21] (a) Show that Equation 20 is just another form of Equation 19. (b) By similar triangles, show that Equations 19 and 20 describe the straight lines plotted in Figure 23. (c) Assume that a pH subindex is developed in which $X_a = 6.5$, $X_o = 7.5$, and $X_s = 8.5$; plot this subindex and calculate I for $X = 5.5$ and $X = 10$. [Answers: 2.0, 2.5]

6. The Beta Function Index, like Harkins' index from which it evolved, has important statistical properties. Assume that the control values for four variables are as follows: DO, 5 mg/l; fecal coliforms, 200/100 ml; ammonia, 1.5 mg/l; dissolved solids, 1,000 mg/l. Assume that the following observations are available. (This example appears in papers by Schaeffer and Janardan.[41,42])

DO (mg/l)	Fecal Coliforms (no./100 ml)	Ammonia (mg/l)	Dissolved Solids (mg/l)
9.1	140	0.1	240
9.6	470	1.15	390
12.8	30	0.0	330
7.4	10	0.15	360
7.6	1,100	0.05	280
6.9	100	0.35	280
7.4	430	0.19	250
10.4	10	0.20	240
9.1	100	0.50	140
11.5	100	0.26	280
10.9	800	0.18	250

(a) Determine the ranks for each variable. [Answers: All ranks for DO, ammonia and dissolved solids are 6.5, because these observations show better water quality than the control values; for fecal coliforms, the ranks are 4.5, 10, 4.5, 4.5, 12, 4.5, 9, 4.5, 4.5, 4.5, 11] (b) Calculate the standard deviation of the ranks for each variable. [Answers: $s_1 = 0$, $s_2 = 2.9$, $s_3 = 0$, $s_4 = 0$] (c) Calculate Harkins' index for the fecal coliform observations. [Answers: 0, 3.6, 0, 0, 6.7, 0, 2.4, 0, 0, 0, 5.0] (d) Calculate the Beta Function Index. [Answers: S = 17.7, T = 288, b = 0.807, I = 0.298]

7. In the ranking process for the Beta Function Index, suppose that all observations of a particular increasing scale pollutant variable i are less than the control value (that is, $X_{ij} \leq X_{ic}$ for all j). Show that the average rank $\bar{R}_i$ is computed as follows:

$$\bar{R}_i = \frac{m_i + 1}{2}$$

8. Show that the minimum variance for pollutant i in Harkins' index, which occurs when all observations are tied with each other but differ from the control value, is given by:

$$Var(R_i) = s_i^2 = \frac{m_i - 1}{4}$$

[Hint: set $t = m_i - 1$ in the variance equation.]

9. For the Beta Function Index, assume that the same number of observations are available for every pollutant variable. In this situation, show that Equation 34a reduces to Equation 35.

304 ENVIRONMENTAL INDICES

10. Drawing upon O'Connor's comparison of various indices, Dunnette's correlations in Oregon, and the mathematical treatment given in Chapter II, discuss the relative importance for a water quality index of the following characteristics: aggregation function, weights and shapes of subindex curves.

11. The national survey of water pollution control agencies revealed that water quality indices serve a more limited audience than air pollution indices. Discuss the possible reasons for this.

12. Most water quality indices have been applied to free-flowing streams and not to the quality of drinking water at the tap, which, like air, is consumed directly by humans. What problems would arise in developing a water quality index for tap water?

REFERENCES

1. Landwehr, Jurate Maciunas. "Water Quality Indices—Construction and Analysis," Ph.D. Dissertation, University of Michigan, University Microfilms No. 75-10,212 (1974).
2. Rosen, Richard H., Ron Beck, Valerie P. Bennett, Joseph A. Orlando and Robert B. Wrightington. "A Review and Evaluation of Water Quality Indices and Similar Indicators, Volume I: Summary and Users Guide," prepared for the Council on Environmental Quality by Energy Resources Co., Inc., and Mathematica, Inc. (September 1976).
3. Orlando, Joseph A., and Robert B. Wrightington. "A Review and Evaluation of Water Quality Indices and Similar Indicators, Volume II: A Review of Available Indices," prepared for the Council on Environmental Quality by Mathtech, Inc., a subsidiary of Mathematica, Inc. (September 1976).
4. Orlando, J. A., R. B. Wrightington and L. D. Maxim. "Water Quality Indicators—A Review of Available Indicators," presented at the 171st National Meeting of the American Chemical Society, New York, NY, April 8, 1976.
5. Horton, Robert K. "An Index-Number System for Rating Water Quality," *J. Water Poll. Control Fed.* 37(3):300-306 (March 1965).
6. Brown, Robert M., Nina I. McClelland, Rolf A. Deininger and Ronald G. Tozer. "A Water Quality Index—Do We Dare?" *Water Sewage Works*, pp. 339-343 (October 1970).
7. Dalkey, Norman C. "Delphi," Report No. P-3704, The Rand Corporation, Santa Monica, CA (October 1967).
8. Linstone, Harold A., and Murray Turoff. *The Delphi Method: Techniques and Applications*, (Reading, MA: Addison-Wesley Publishing Co., 1975).
9. McClelland, Nina I. "Water Quality Index Application in the Kansas River Basin," U.S. Environmental Protection Agency, Kansas City, MO, EPA-907/9-74-001 (February 1974).

10. Brown, Robert M., Nina I. McClelland, Rolf A. Deininger and Jurate M. Landwehr. "Validating the WQI," presented at the National Meeting on Water Resources Engineering of the American Society for Civil Engineers, Washington, DC, January 29, 1973.
11. McClelland, Nina I., Robert M. Brown and Rolf A. Deininger. "WQI Enhancing Appreciation of Quality Improvement," presented at the 171st National Meeting of the American Chemical Society, New York, NY, April 8, 1976.
12. Ott, Wayne R. "Water Quality Indices: A Survey of Indices Used in the United States," U.S. Environmental Protection Agency, Washington, DC, EPA-600/4-78-005 (January 1978).
13. Brown, Robert M., Nina I. McClelland, Rolf A. Deininger and Michael F. O'Connor. "A Water Quality Index—Crashing the Psychological Barrier," in *Indicators of Environmental Quality*, William A. Thomas, Ed. (New York: Plenum Press, 1972).
14. Prati, L., R. Pavanello and F. Pesarin. "Assessment of Surface Water Quality by a Single Index of Pollution," *Water Research* 5:741-751 (1971).
15. McDuffie, Bruce, and Jonathan T. Haney. "A Proposed River Pollution Index," presented at the spring 1973 meeting of the American Chemical Society, Division of Water, Air, and Waste Chemistry, New York, NY, April 13, 1973.
16. Dinius, S. H. "Social Accounting System for Evaluating Water Resources," *Water Resources Research* 8(5):1159-1177 (October 1972).
17. Walski, Thomas M., and Frank L. Parker. "Consumers Water Quality Index," *J. Environ. Eng. Div.*, Am. Soc. Civil Eng., pp. 593-611 (June 1974).
18. O'Connor, Michael Fredrick. "The Application of Multi-Attribute Scaling Procedures to the Development of Indices of Water Quality," Ph.D. Dissertation, University of Michigan, University Microfilms No. 72-29,161 (1972).
19. Deininger, Rolf A., and Jurate Maciunas Landwehr. "A Water Quality Index for Public Water Supplies," unpublished report, Department of Environmental and Industrial Health, School of Public Health, University of Michigan, Ann Arbor, MI (July 1971).
20. Stoner, Jerry D. "Water Quality Indices for Specific Water Uses," U.S. Geological Survey, Reston, VA, Circular No. 770 (1978).
21. Nemerow, Nelson L., and Hisashi Sumitomo. "Benefits of Water Quality Enhancement," Syracuse University, Syracuse, NY, Report No. 16110 DAJ, prepared for the U.S. Environmental Protection Agency (December 1970).
22. "Water Quality Criteria 1972," a report of the Committee on Water Quality Criteria, National Academy of Sciences, Washington, DC (1972).
23. Truett, J. B., A. C. Johnson, W. D. Rowe, K. D. Feigner and L. J. Manning. "Development of Water Quality Management Indices," *Water Resources Bull.* 11(3):436-448 (June 1975).
24. Greeley, R. S., A. Johnson, W. D. Rowe and J. Bruce Truett. "Water Quality Indices," Report No. M72-54, MITRE Corporation, McLean, VA (April 1972).

25. "Environmental Quality: The Third Annual Report of the Council on Environmental Quality," U.S. Government Printing Office, Washington, DC (August 1972), p.12.
26. Dee, Norbert, Janet K. Baker, Neil L. Drobny, Kenneth M. Duke and David C. Fahringer. "Environmental Evaluation System for Water Resource Planning," Report No. PB 208 822, Battelle Columbus Laboratories, Columbus, OH (January 1972).
27. Dee, Norbert, Janet Baker, Neil Drobny, Kenneth M. Duke, Ira Whitman and Dave Fahringer. "An Environmental Evaluation System for Water Resource Planning," *Water Resources Research* 9(3):523-535 (June 1973).
28. Andrews, Richard N. L. "Comments on 'An Environmental Evaluation System for Water Resource Planning' by Norbert Dee *et al.*," *Water Resources Research* 10(2):376-378 (April 1974).
29. Inhaber, H. "Environmental Quality: Outline for a National Index for Canada," *Science* 186(29):798-805 (November 1974).
30. Inhaber, H. "A National Environmental Quality Index for Canada," Technical Edition, Internal Report to the Canadian Department of the Interior, Ottawa (1974).
31. Zoeteman, B. C. J. "The Potential Pollution Index as a Tool for River Water Quality Management," Technical Paper No. 6, Government Institute for Water Supply, World Health Organization, International Reference Centre for Community Water Supply, The Hague, The Netherlands (August 1973).
32. "Federal Water Pollution Control Act Amendments of 1972," *Public Law 92-500.*
33. Johanson, Edward E., and Jaret C. Johnson. "Identifying and Prioritizing Locations for the Removal of In-place Pollutants," Contract No. 68-01-2920, prepared for the U.S. Environmental Protection Agency, Office of Water Planning and Standards, Washington, DC (May 1976).
34. Shoji, Hikaru, Takeo Yamamoto and Takakazu Nakamura. "Factor Analysis on Stream Pollution of the Yodo River System," *Air and Water Pollution International Journal,* 10:291-299 (1966).
35. Joung, H. M., W. W. Miller, C. N. Mahannah and J. C. Guitjens. "A Water Quality Index Based on Multivariate Factor Analysis," *Experiment Station J.,* Series No. 378, Nevada Agricultural Experiment Station, University of Nevada, Reno, NV (1978).
36. Coughlin, Robert E., Thomas R. Hammer, Thomas G. Dickert and Sallie Sheldon. Regional Science Research Institute Discussion Paper No. 53, Philadelphia, PA (March 1972).
37. Harkins, Ralph D. "An Objective Water Quality Index," *J. Water Poll. Control Fed.* 46(3):588-591 (March 1974).
38. Kendall, Maurice. *Rank Correlation Methods* (London: Charles Griffin & Co., Ltd., 1975).
39. Landwehr, J. Maciunas, R. A. Deininger, N. L. McClelland and R. M. Brown. "Discussion: An Objective Water Quality Index," *J. Water Poll. Control Fed.* 46(7):1804-1809 (July 1974).
40. Landwehr, J. Maciunas and R. A. Deininger. "A Comparison of Several Water Quality Indexes," *J. Water Poll. Control Fed.* 48(5):954-958 (May 1976).

41. Schaeffer, David J., and Konanur G. Janardan. "Communicating Environmental Information to the Public: A New Water Quality Index," *J. Environ. Educ.* 8(4):18-26 (Summer 1977).
42. Janardan, Konanur G., and David J. Schaeffer. "Development and Application of Five New Water Quality Indices," Illinois Environmental Protection Agency, Springfield, IL, unpublished report (1974); preprint available (1977).
43. Hahn, Gerald J., and Samuel S. Shapiro. *Statistical Models in Engineering* (New York: John Wiley and Sons, Inc., 1967).
44. "Illinois Water Quality Inventory Report 1976," report prepared pursuant to Section 305(b) of *Public Law 92-500*, Illinois Environmental Protection Agency, Springfield, IL (1976).
45. "The Relationships of Phosphorus and Nitrogen to the Tropic State of Northeast and Central Lakes and Reservoirs," U.S. Environmental Protection Agency, Environmental Monitoring and Support Laboratory, Las Vegas, NV, Working Paper No. 23 (1974).
46. Pielou, E. C. *Mathematical Ecology* (New York: John Wiley and Sons, Inc., 1977).
47. Thomas, William A., Gerald Goldstein and William H. Wilcox. *Biological Indicators of Environmental Quality* (Ann Arbor, MI: Ann Arbor Science Publishers, Inc., 1976).
48. Sayers, William T., and Wayne R. Ott. "Use and Interpretation of Water Quality Data," Proceedings of the 7th Materials Research Symposium *Accuracy in Trace Analysis: Sampling, Sample Handling, and Analysis*, National Bureau of Standards Publication No. 422, Gaithersburg, MD (August 1976).
49. "Application of the Water Quality Index, 'Nashua River Basin,' " interim report of the New England Interstate Water Pollution Control Commission, Boston, MA (December 1976).
50. "Water Quality and Pollution Control in Michigan," report prepared pursuant to Section 305(b) of *Public Law 92-500*, Michigan Department of Natural Resources, Bureau of Water Management, Lansing, MI (April 1976).
51. "New York State Water Quality Inventory Report," report prepared pursuant to Section 305(b) of *Public Law 92-500*, New York Department of Environmental Conservation, Division of Pure Waters, Albany, NY (May 1976).
52. Neal, Larry A. "A Water Quality Index for Georgia—Update of Technical Memorandum of October 29, 1975," Technical Memorandum, Georgia Department of Natural Resources, Environmental Protection Division, Atlanta, GA (December 30, 1975).
53. "Water Quality Control in Georgia 1975," report prepared pursuant to Section 305(b) of *Public Law 92-500*, Georgia Department of Natural Resources, Environmental Protection Division, Atlanta, GA (1975).
54. Barker, Bruce, and Paul Kramer. "Water Quality Conditions in Illinois," Appendix A to Chapter III, "Water Quality Management," State-wide Water Resource Development Plan, Illinois Department of Transportation, Division of Water Resource Management, Springfield, IL (1972).
55. Hudson, LeVerne D., and David J. Schaeffer. "Effects of the 1976 Drought on Illinois Water Quality," Illinois Environmental Protection Agency, presented at the American Water Resources Association,

Illinois Section, Fifth Annual Water Resources Conference, June 10-11, Chicago, IL (February 1977).

56. Sheen, Jackson R. State of Nevada Department of Human Resources, Environmental Protection Services, Carson City, NV, letter to Wayne Ott dated January 19, 1977.

57. Dunnette, David A. "Water Quality Index Development and Application for the Willamette Basin, Oregon," unpublished report, Department of Environmental Quality, Water Quality Division, Portland, OR (July 1976).

58. "Environmental Quality: The Fifth Annual Report of the Council on Environmental Quality," U.S. Government Printing Office, Washington, DC (December 1974), pp. 333-334.

59. Cogger, William J., Marshall L. Payne and Lester D. Sprenger. "Water Quality Inventory," U.S. Environmental Protection Agency, Region VIII, Surveillance and Analysis Division, Denver, CO (October 1975).

60. Beebe, James A. "WQI, A Water Quality Data Analysis System," unpublished report, U.S. Environmental Protection Agency, Region X, Seattle, WA (May 1, 1975).

61. Beebe, James A. "WQI, A Water Quality Data Analysis System User Guide," unpublished report, U.S. Environmental Protection Agency, Region X, Seattle, WA (July 1975).

62. Beebe, James A. "WQI, A Water Quality Data Analysis System Program Documentation," unpublished report, U.S. Environmental Protection Agency, Region X, Seattle, WA (July 1975).

63. Peterson, Ray. U.S. Environmental Protection Agency, Region X, Surveillance and Analysis Division, Seattle, WA, personal communication to Wayne Ott dated August 1977.

64. Ficke, John F. Quality of Water Branch, U.S. Geological Survey, Reston, VA, letter to Wayne Ott dated June 28, 1977.

CHAPTER V

CONCEPTUAL APPROACHES

This book has focused on quantitative techniques for interpreting and presenting information on the state-of-the-environment. These techniques generally are based on data on physical, chemical and biological variables from existing environmental monitoring activities. This chapter presents techniques for representing environmental phenomena which are somewhat more abstract and therefore do not necessarily depend on monitoring data. These approaches really are conceptual models designed more for gaining insight into basic relationships than for interpreting data for decision-making purposes. Because these approaches do not depend on existing data, they often extend beyond the air and water pollution fields and sometimes seek to include information on all matters of concern to the total environment of man.

QUALITY OF LIFE

Development of the Quality of Life (QOL) concept parallels the development of "social indicators" in the U.S. in the 1960s. QOL is a much broader concept than the notion of air quality or of water quality. In general, it is supposed to reflect all aspects of a person's sense of well-being; that is, it includes all factors which contribute to human satisfaction or dissatisfaction, factors which determine a person's "happiness" or "unhappiness."

In 1973, EPA sponsored a symposium on the QOL concept, and the published proceedings[1] offered a number of definitions for QOL. Apparently, the concept has evolved from the need to provide a more complete measure of social well-being than could be obtained from traditional economic indicators:

> For many years only economic indicators, such as the Gross National Product and the Consumer Price Index, have been available to decision makers

to measure "progress" and the health of the nation. The failure of these indicators to account for noneconomic factors has led to the quest for QOL indicators.[1]

However, some critics have objected to the vague and subjective nature of this concept, suggesting that QOL means one thing to one person but something else to another person:

> Those who argue that one cannot possibly quantify QOL in a meaningful way raise a number of compelling points. The one voiced most frequently is that the parameters, or factors, which are important in determining QOL are so highly individual that any attempt to describe QOL for a group, no matter how small, will inevitably miss the mark as far as the individuals in that group are concerned. Virtually everybody acknowledges that there are distinct differences between, for example, urban slum dwellers, midwest farmers, and suburban housewives. While the broad differences between cultures, races, economic strata, age groups, and so on are widely recognized, even by the most enthusiastic advocates of QOL quantification, there are some who will admit to no common groupings of QOL parameters at all. These champions of the entirely individualistic view of QOL argue that each person's view of life is unique and that it is not possible, or acceptable, to compare that person's satisfactions, happiness, achievement, or whatever with anybody else's. In brief, this view holds that any aggregation of QOL measures, even if such measures could be obtained, would so totally distort the individual's situation as to make the aggregated measure meaningless.[1]

Liu[2] has suggested a model in which QOL "isoquality" curves similar to the indifference curves used in microeconomic theory[3,4] are plotted on a two-dimensional graph (Figure 1). In this model, each person's QOL is expressed as a function of two variables: P_1, which reflects physical factors, and P_2, which reflects psychological factors. The physical factors include material goods and public services (housing, schooling, medical care, police protection), and the psychological factors include intangible items (esteem, dignity, lack of anxiety). The model assumes that the relative quantities of P_1 and P_2 determine an individual's QOL:

> The physical inputs consist of the bundles of material goods and services which satisfy most of the basic needs of human beings, while the psychological inputs are mostly self-actualized and developed. It is possible that the former inputs can be used as substitutes to a certain extent for the latter inputs, such as lack of fear, . . . feelings of being loved and respected, and awareness of beauty. Although deprivations of one's ownership of physical goods and services below the subsistence level are most serious . . . , depreciations in psychological inputs could also impoverish considerably the affluent society. That both P_1 and P_2 play an important role in determining the quality of life is vividly manifested by the growing discontent of today's Americans.[2]

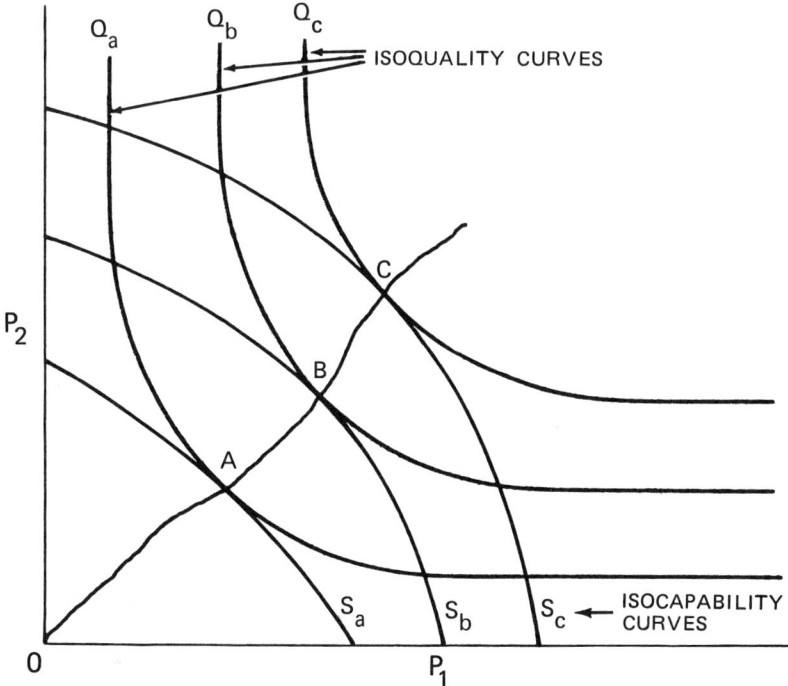

Figure 1. Diagram proposed by Liu[2] for maximization of an individual's Quality of Life (QOL), subject to his capability constraint.

Figure 1 shows three curves—Q_a, Q_b and Q_c—from the family of isoquality curves for a particular individual. Each isoquality curve denotes the combinations of P_1 and P_2 which give a particular level of the QOL. Isoquality curves that are further from the origin denote a greater QOL than those that are nearer to the origin (that is, $Q_c > Q_b > Q_a$). The isoquality curves are convex to the origin, indicating that greater and greater quantities of P_2 must be substituted for P_1 as P_1 decreases. For each QOL, there is a minimum possible quantity of P_1 and P_2. As P_1 decreases, for example, each isoquality curve becomes vertical and parallel to the P_2-axis. Here, the individual possesses such a large quantity of P_2 that P_2 no longer is of importance to him, and P_1 becomes the critical determinant of his QOL.

The isoquality curves reflect the needs of an individual and the way in which various combinations of P_1 and P_2 satisfy these needs. Liu[2] depicts each individual as further constrained by a limitation of the possible quantities of P_1 and P_2 that he can afford. This limitation is similar to the budget constraint in utility theory,[3,4] which limits the quantities of goods that a consumer can afford. The capability constraint is reflected by a family of "isocapability" curves, and three of these curves—S_a, S_b and S_c—are shown in Figure 1. Liu[2] depicts the isocapability curves either as straight lines or as lines that are concave to the origin. In general, a person seeks to maximize his QOL by reaching the highest isoquality curve, subject to his isocapability constraint. By doing so, he selects the proper combination of P_1 and P_2 that maximizes his QOL within his capability:

> A rational individual attempts to maximize his overall QOL production, subject to certain capability constraints. Perceive a situation of no constraints of any form, or of limitless capability of a human being; each individual would move to the bliss point at which all his desires are fully satisfied. Unfortunately, that is not the case in reality. Each one has only 24 hours a day to spend in securing his P_1 and P_2 inputs for production of his QOL. Observe [that] an individual's capability to exchange P_1 and P_2 inputs is limited by the social, economic, political conditions, and environments in which he lives. In addition, the ability to acquire and to share with others the total P_1 goods and services available in a society depends strategically upon the individual's own economic wealth. On the other hand, there are restrictions on each individual's effort to secure P_2 goods. For example, the amount of P_2 acquired is determined in part by one's degree of willingness to exchange resources and efforts for spiritual and psychological inputs, such as esteem, belovedness, belongingness, feeling of security, individual dignity and integrity, etc., that other people in the society are willing to render him. As expected, the esteem, security and dignity also depend, to some extent, on P_1.[2]

The individual's equilibrium at a certain time is determined by the point at which the isocapability curve is just tangent to the highest isoquality curve. Point A in Figure 1, for example, shows the combinations of P_1 and P_2 which maximize the QOL for an individual's capability constraint S_a. It lies on the isoquality curve Q_a. At some later time, a person's capability may increase, and his QOL may rise to a higher isoquality curve, such as the curve Q_b passing through point B. The many possible QOL equilibria for a particular individual give a locus of points, as illustrated by the wavering line drawn through points A, B and C.

QOL in Metropolitan Areas

Because QOL is a relatively abstract concept, it is difficult to apply it to real data. The concept appears more appropriate as a general framework for describing the interaction among factors than a methodology which can be applied to actual situations.

Although QOL is intended as a conceptual framework, Liu has used it as a basis for designing a study to rank conditions in 50 U.S. states[5] and 243 U.S. Standard Metropolitan Statistical Areas (SMSAs).[2,6] After reviewing the data from these cities, he found that the only data available were for physical factors (P_1), and data on psychological factors (P_2) were ". . . either not measurable or not existent for all SMSAs."[2] He identified five QOL components or "principal goal areas" which could be represented by the available data:

- Economic Component
- Political Component
- Environmental Component
- Health and Education Component
- Social Component

These QOL components were selected to provide as broad a concept of well-being as possible.

Working at the Midwest Research Institute (M.R.I.) under an EPA grant, Liu[2,6] divided the 243 SMSAs into three groups according to their population: "large" (above 500,000), "medium" (200,000 to 500,000), and "small" (under 200,000). There were 65 large SMSAs, accounting for a total population of 102.6 million; 83 medium SMSAs, accounting for 24.9 million people; and 95 small SMSAs, accounting for 11.9 million people. For each QOL component, he selected variables for which data were available and assigned arbitrary weights to them. The environmental component was represented by 17 variables that were grouped into 7 categories (Table I). These variables were chosen to reflect air pollution, visual pollution, noise, solid waste, water pollution, climate and recreation. Air pollution was represented by SO_2 and total suspended particulates (TSP), and water pollution was represented by the MITRE PDI index described in Chapter IV (pages 248 to 249). The sum of the absolute values of the weights was unity.

To calculate the QOL index, Liu[2] computed a normalized transform for each variable and SMSA; then the transforms were aggregated for all variables and components. To compute the transform (Z_{ij}) for variable i and SMSA j, he first computed the arithmetic mean $\bar{x}_i$ and arithmetic standard deviation s_i for all SMSAs:

$$Z_{ij} = \frac{X_{ij} - \bar{x}_i}{s_i} \qquad (1)$$

where Z_{ij} = normalized transform for the i*th* variable and j*th* SMSA
X_{ij} = value of the i*th* variable for the j*th* SMSA
$\bar{x}_i$ = arithmetic mean of the i*th* variable
s_i = standard deviation of the i*th* variable

Notice that this transform is similar to Harkins' z transform (Chapter IV, page 256); however, raw observations are used here instead of rank orders, and the arithmetic mean is used instead of a control value.

Table I. Variables Used in Liu's[2] Environmental Component

Variable	Effect and Weight	
Air Pollution		
Mean Annual TSP Concentration ($\mu g/m^3$)	(−)	0.05
Mean Annual SO_2 Concentration ($\mu g/m^3$)	(−)	0.05
Visual Pollution		
Mean Annual Inversion Frequency	(−)	0.033
Housing Units Dilapidated (%)	(−)	0.033
Parks and Recreational Areas (acres/1,000 people)	(+)	0.033
Noise		
Population Density in the Central City of the SMSA (persons/mi^2)	(−)	0.033
Motor Vehicle Registrations (no./1,000 people)	(−)	0.033
Motorcycle Registrations (no./1,000 people)	(−)	0.033
Solid Waste		
Solid Waste Generated by Manufacturing (tons/million dollars of value added)	(−)	0.10
Water Pollution		
MITRE Prevalence Duration Intensity (PDI) Index	(−)	0.10
Climate		
Mean Annual Inversion Frequency	(−)	0.05
Annual Sunshine (no. of days/year)	(+)	0.05
Thunderstorms (no. of days/year)	(−)	0.05
Warm Temperatures (no. of days $\geq 90°F$)	(−)	0.05
Cold Temperatures (no. of days $\leq 32°F$)	(−)	0.05
Recreational Areas and Facilities		
Parks and Recreational Areas (acres/1,000 people)	(+)	0.125
Trails (miles/100,000 people)	(+)	0.125

This transform was selected to eliminate the units of measurements among different variables so that they could be combined more easily. Because this transform is related linearly to the raw data, the transformed values are normally distributed about $\bar{x}_i$ if the raw data are normally distributed:

> One of the most significant characteristics of this transformation is that the Z scores are normally distributed with almost 99.8 percent of the transformed observations falling between values of ($\bar{x}_i \pm 3s_i$) or "± 3," 95.0 percent between ($\bar{x}_i \pm 2s_i$), and 68.3 percent between ($\bar{x}_i \pm s_i$) or "± 2" or "± 1," respectively, given that the original distribution is also normal.[2]

Although the assumption of normality may be valid for some of these variables, it is dubious for the air and water quality data. Ott and Mage[7] have found that air and water pollutant concentration data, because they result from diffusion processes, are better represented by skewed distributions (for example, the two- and three-parameter lognormal probability models) than by symmetrical distributions.

For each SMSA j and component k, the subindex I_{jk} was computed as the weighted sum of the Z values obtained for each variable:

$$I_{jk} = \sum_{i=1}^{n} w_i Z_{ij} \qquad (2)$$

where I_{jk} = subindex for the k*th* component and j*th* SMSA
  w_i = weight for the i*th* variable
  Z_{ij} = transform of the i*th* variable for the j*th* SMSA
  n = number of variables in the k*th* component

Finally, the overall index was computed as the weighted sum of the five subindices:

$$QOL_j = \sum_{k=1}^{5} w_k I_{jk} \qquad (3)$$

where QOL_j = Quality of Life for the j*th* SMSA
  w_k = weight for the k*th* component

When this approach was applied to U.S. SMSAs, the weights for each component in Equation 3 were set to 1/5 (w_k = 1/5 for all k), giving the arithmetic mean of component subindices.

Figure 2 shows Liu's ranking of the 65 large SMSAs based on calculations of the environmental component using 1970 census data. The SMSAs were grouped according to five descriptors: (A) "outstanding," (B) "excellent," (C) "good," (D) "adequate," and (E) "substandard." Using this system, the best environmental QOL was observed in Sacramento, California, and the worst environmental QOL was in Pittsburgh, Pennsylvania. Apparently, Sacramento's high rating can be attributed to the fact that this SMSA generated the least amount of solid waste from manufacturing (350 tons per million dollars of value added) and had the longest trail mileage (about 2 miles per 1,000 people). Nevertheless, Sacramento did not rate high for all variables:

> Although Sacramento ranked first in the environmental component, this does not mean that it has all the best [attributes] in every category. For instance, it had nearly the worst noise problem in that year because of its high motorcycle and vehicle registrations per 1,000 population and high population density in the central city. Admittedly, these are only crude indicators of noise pollution, which in reality depends on the number of motorcycles and vehicles used per day, and their capacity of noise generation such as the age, size, etc. In comparison, Miami SMSA had the best

316 ENVIRONMENTAL INDICES

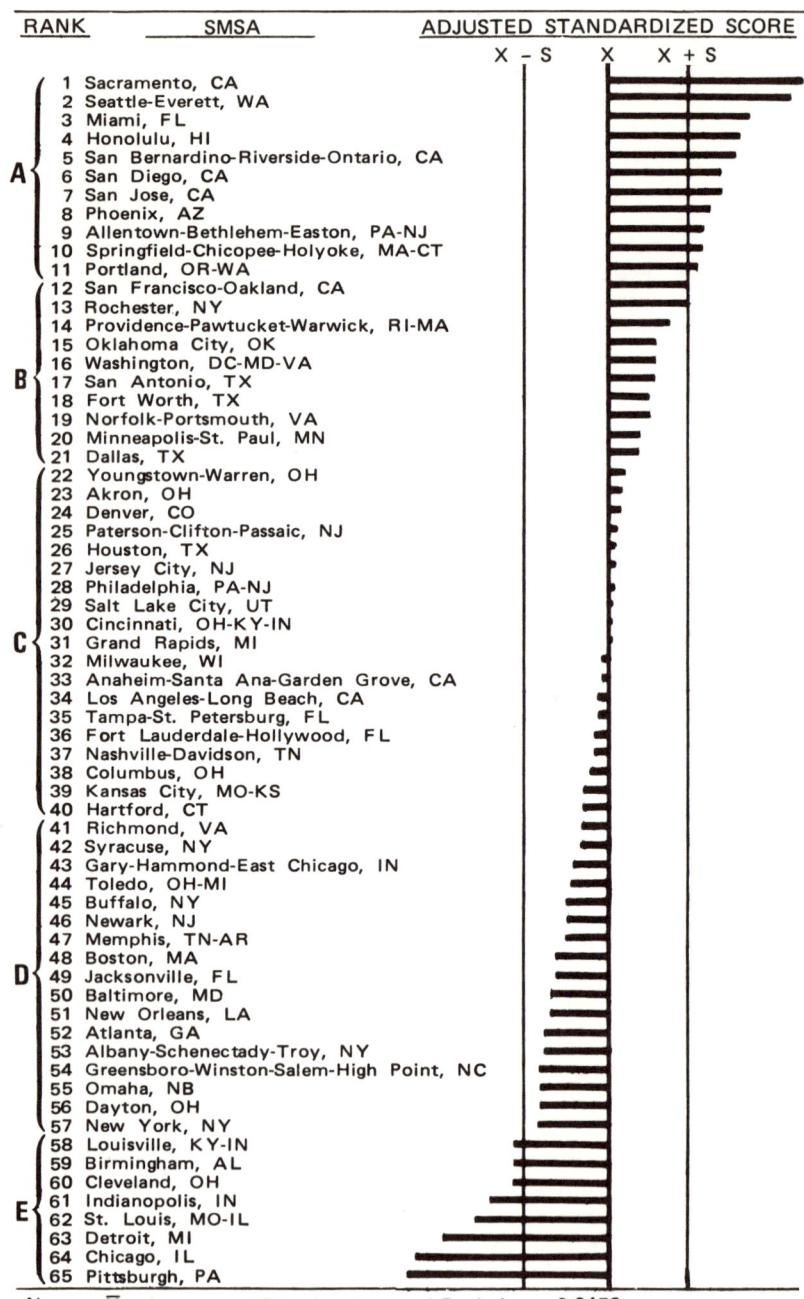

Figure 2. Ranking of environmental component for 65 large U.S. metropolitan areas, as reported by Liu.[2]

natural environment and had virtually no visual pollution, but its water pollution and solid waste problems were considerably worse than most SMSAs under discussion. Seattle/Everett SMSA had very little air, visual, and water pollution, but its noise pollution was worse than average.[2]

This failure of the index to reflect the noise variable demonstrates, once again, the eclipsing problem that is characteristic of weighted linear sum aggregation functions such as Equation 2. As discussed in Chapter II (page 80), eclipsing is particularly serious if this aggregation function is used in decreasing scale indices, such as the QOL system.

In Liu's ranking, the most serious environmental problems appeared to be in SMSAs in the northeast central regions of the U.S.:

While noise pollution did not seem to be a problem in Pittsburgh, the worst water pollution, plus very serious air and visual pollution, push the rating for Pittsburgh down to the bottom. For instance, the mean level for sulfur dioxide in Pittsburgh was 63.0 $\mu g/m^3$, lower only than Cleveland (113.0 $\mu g/m^3$) and Providence/Pawtucket/Warwick (64.0 $\mu g/m^3$); the water pollution index was 48.0 for Pittsburgh, substantially higher than the second and third worst SMSAs of Detroit (31.06) and Boston (24.00), and much higher than the majority of the SMSAs with indexes ranging from 0.68 (Anaheim/Santa Ana/Garden Grove) to 9.78 (Columbus). People in both Chicago and Detroit suffered seriously from the air and water pollution; however, people in Detroit enjoyed a relatively better natural environment and saw fewer dilapidated housing units than citizens in Chicago. St. Louis was observed to have little solid waste problem, but its very small park and recreational area (2.3 acres per 1,000 people) and bad climatological data forced its rating down.[2]

When Liu's evaluation of the QOL in U.S. SMSAs was completed in 1975,[2,6] the press seemed very interested in the results, and the rankings were published widely by newspapers throughout the country. Some newspapers, such as the *Dallas Times Herald*,[8] listed the numerical values of the overall QOL rating for all three groups of SMSAs. Other newspapers gave a general discussion of the way in which the study was done, indicating why their particular area ranked high or low on the list. Typical headlines were as follows: "Our Quality of Life Rated Low: Area Flunks Tests for Social Life, Environment,"[9] "Charlotte 59*th* in Life Quality Study,"[10] "New Orleans' Quality of Life Ranked 63 of 65 Cities,"[11] "Dallas Gets 'Good' Rating,"[12] "Quality of Life in Topeka Rated 'Outstanding'."[13]

Newspapers tended to interpret the rankings as overall measures of the livability of each area, rather than merely the demonstration of a scientific methodology:

Each urban area was given a rating by M.R.I. in five performance categories —economic, political, environmental, health and education, and social— as well as an overall rating. Overall, Gary-Hammond-East Chicago ranked

52nd in quality of life among the urban areas. Chicago and its suburbs didn't do too much better, coming in 37th out of 65, with an overall evaluation of "good."[14]

The newspapers sometimes discussed each component of the ranking on an individual basis as it applied to their areas:

> In a word, the quality of life in the Gary-Hammond-East Chicago area is "adequate." It's not an especially good place to earn a living. Performance by governmental agencies in the area is just so-so. Water, air, noise, and visual pollution is just bearable, and job and educational opportunities, especially for minority group members, aren't bad but aren't good either. These conclusions can be drawn from a recently completed study on the quality of life in 243 urban areas in the United States with a population of more than 50,000.[14]

Typically, the articles compared the rankings of cities in other parts of the country with the rankings of cities in their immediate geographical area, as evident from the following quotation from the *Chicago Tribune:*

> One of the most fascinating conclusions of the study is that while West Coast major urban areas seem to have a far superior quality of life compared to large cities in other parts of the country, the reverse is true with respect to medium-sized and smaller metropolitan areas. In fact the majority of the "outstanding" medium-sized and smaller urban areas, according to the M.R.I. report, are right here in the Midwest.[14]

There was evidence of serious concern among those cities which received low rankings. For example, newspapers in San Antonio, Texas (environmental component was rated "excellent," but overall rating was "substandard") disagreed with the rating, noting that many persons had chosen to live there:

> That livability study by Midwestern (sic) Research Institute of Kansas City doesn't sit well in San Antonio. That is because it rates our city 62 in a list of 65 major American cities in terms of livability. . . . We think this is a good place to live. So do nearly a million other people and for many years after World War II, San Antonio was the fastest growing major city in America. It provided enough livability for a lot of people from elsewhere to settle here.[15]

An official from the San Antonio Chamber of Commerce strongly criticized both EPA and the report:

> One of the sad features of such a ridiculous report is that it was funded by federal tax money. . . . If government agencies like EPA are using 1970 data to make 1975 policy, then we are in greater danger than I thought.[16]

In Detroit, which was rated "substandard" in the environmental and social components, the ranking provoked resentment, as indicated by the comments of a local theater owner:

> I think it's outrageous. . . . Outside of New York and Washington, we have more legitimate theater than any other city. We've got a top symphony and a great art gallery. I really resent people who don't even come into the town and make these observations.[9]

The rankings also provoked negative comments in the medium and small SMSAs. For example, the study ranked Charlotte, North Carolina, as 59 on the list of 83 medium SMSAs, and its environmental component was rated "substandard." The president of the Charlotte Chamber of Commerce questioned the assumptions used in the ranking process:

> I think we rank a whole lot higher than that. . . . I suppose I disagree with the assumptions they used to base the ratings on. Now of course, there might be 58 cities that are unbelievably good in front of us. . . . We've got the purest water in the South. . . . The EPA has given us an excellent rating for our quality. Our sewer system is pretty doggone adequate.[10]

In an interview with members of the press, Liu explained that the methodology was not really intended to show which city was best and which was worst:

> This study is like the general physical exam of the individual by the doctor. . . . He can tell you what you have but he can't tell you what you feel. We are just trying to say here are the measurable things. Our purpose is not to say who is the best and who is the worst. Our purpose is to try to give you (the cities) a very clear picture of your weaknesses and your strengths, so you can see areas for improvements.[12]
>
> Quality of life is a notion of many dimensions and each individual has his own views as to what is important to him.[14]

Officials at Midwest Research Institute generally defended the approach, saying, "The methodology, given the state-of-the-art, is as sound as it can be."[17] However, they indicated that the QOL concept has limitations: "We recognized that there are many subjective factors to quality of life that are not quantifiable and are certainly beyond the scope of the study."[17] Liu indicated that the data used in the rankings, although 5 years old at the time of the study, were the most complete data available: "Nobody would have better information than 1970."[9]

Probably the most critical reaction to the report came from Congress. Congressman Buchanan of Alabama described the report as ". . . a libel and a slander against a number of areas."[18] Among other things, he objected to the use of 5-year-old data and was concerned about the low ratings given to cities within his district:

> It is ironic that while the data in this study were taken primarily from the 1970 census and, hence, are 5 years-and-more old, the study itself sought to stand in judgment on the quality of life in cities in 1975, 5 years later. The persons who made the study never set foot in Birmingham, Alabama, which is one they rated low; they never set foot in my lovely port city of Mobile, Alabama, with all of its old South charm and all of its new South industry as a great and growing port city in America. They never saw the charm and greatness of New Orleans, which they graded down.[18]

Congressman Buchanan also regarded the variables included in the ranking process as "most peculiar" and felt that many important factors had been left out:

> I note among the factors that were not considered was whether or not the area was a place where there were religious institutions, such as churches. My seven colleges and universities also were not included as a factor in the quality of life. In Birmingham we do have a city of churches, a city where the Community Chest every year goes over the top in its fund drive, and one filled with eleemosynary institutions offering help to those in need and hope to those who are hopeless. This is not considered relevant by the Midwest Research Institute to the quality of life. We disagree.[18]

Congressman Dominick V. Daniels of New Jersey, whose Congressional District included Jersey City (overall rating of "substandard"), also criticized the diverse variables that were considered in the ranking process:

> In the development of the environmental classification, the research institute used such criteria as the number of trail-miles per capita, the number of campsites per capita, the number of public swimming pools per capita, and the number of tennis courts open to the public. Obviously, Jersey City lags behind considerably in the number of campsites and trails available for its citizens. However, I think that it is unfortunate that this sort of criteria was averaged in with such compelling environmental data as photochemical oxidants and the level of water pollution. There is no way you can compare a city like Denver, Colorado, with Jersey City, New Jersey, and come up with anything worthwhile, statistically speaking. I think we should question this type of comparison, and discard it as being devoid of meaning.[18]

Congressman Daniels' comments suggest that any study which ranks the characteristics of urban areas enters the realm of politics whether the author intends it or not:

> Unfortunately, the Midwest Research Institute did not consider it necessary to notify the mayors of the cities and towns that received less than favorable ratings so that they might have an opportunity to study the

report before they were besieged by local press and angry constituents. The extension of this basic courtesy could have saved the institute and the EPA a lot of trouble, because they will now have to deal with upset people calling them from all across the Nation, including Members of Congress.[18]

Congressman Daniels also felt that the low ranking given to Jersey City was a suitable rationale for receiving additional pollution control funds from EPA:

> ... if this report is to have any positive impact on improving the quality of life in American cities, it should be used as a guideline by the Environmental Protection Agency and other Government agencies in developing a priority system for funding of vital programs. Those cities that have greater needs should receive more funds. Currently, this is not the case. Jersey City has to fight for every Federal dollar it receives, despite overwhelming evidence of compelling need. ... I shall be watching the EPA to see how responsive they are to the environmental needs of Jersey City and other cities classified in the report as being "substandard."[18]

The rankings for the medium and small groups of SMSAs also elicited severe criticism from Congress, as indicated by Congressman Kazen's comments about Laredo, Texas (ranked 95 on the list of 95 small SMSAs):

> Madam Speaker, I accuse the Environmental Protection Agency of fouling the air and discouraging important citizen effort to improve the quality of life through flagrant carelessness. My attitude stems from a report by the Midwest Research Institute of Kansas City, operating a project funded by the Environmental Protection Agency, that examined our Nation's urban areas. Ranked worst among small cities was Laredo, Texas, which I am proud to call my home.[18]

The reactions of the public and the Congress to this report were embarrassing to EPA's research staff. One official was quoted in the press as saying, in a telephone conversation to a congressional critic, "We're embarrassed by the study ourselves. Quite frankly, it was a failure."[17] EPA responded to the uproar created by Liu's study by indicating its own disappointment with the rankings and the way in which they were interpreted. The Agency indicated that the report was intended only as the demonstration of a methodology and should not have been used to draw conclusions about the livability of different cities. EPA's Assistant Administrator for Research and Development, Wilson K. Talley, said that the study ". . . was useful in terms of its limited scope and it's unfortunate that its limitations were not more readily apparent."[19] He felt that ". . . broad generalizations such as the rankings of cities should not have been made from this sort of study."[20] Talley also indicated that the methodology had serious limitations: "I regret it if the report appears, or is used to deprecate one city or to praise another. . . . My conclusion is that the methodology is not well developed enough to be used in assessing qualitative factors."[21] Finally, he criticized the report's failure to give concrete recommendations, stating that even if the rankings based on the 123 variables could be viewed as accurate, "... they don't give you much guidance on how to improve the quality of life."[22]

Several methodological problems are apparent in Liu's study. One of these is the limited number of variables that are included. Liu[2] based his rankings on those variables for which data from U.S. cities were readily available. The result was that the variables were few in number and not necessarily the right ones. For example, noise was represented in the environmental component by relatively crude measures (population density, number of vehicles and motorcycles per 1,000 people) instead of the number of noisy vehicles in traffic or actual ambient noise measurements. In the case of air pollution, Liu chose only two variables—ambient SO_2 and TSP measurements—and decided not to include other air pollution data, such as source emissions of various air pollutants. Because source data on the major air pollutants are available for most urban areas, future ranking efforts should include source data and should cover more pollutants. Another problem has to do with the aggregation function. In decreasing scale systems, the weighted linear sum can have serious eclipsing problems, and it appears that more effort should be devoted to the design of the aggregation function and the choice of the weights, which apparently were arbitrary. Despite these methodological problems, Liu's study is an interesting example of a uniform approach applied on a national scale. With better data, refined assumptions, and improvements in mathematical procedures, the approach could be useful for characterizing geographical differences and trends. However, future investigators should always be aware of the possible political ramifications of such rankings and should emphasize the limitations of their approaches. In general, ranking the environmental characteristics of metropolitan areas is a serious technical challenge that must be approached with both caution and thoroughness.

Other QOL Approaches

There have been several other attempts to define the Quality of Life in America. Wingo and Evans[23] have published a collection of papers discussing experimental methods, social surveys and other approaches for evaluating the QOL. Campbell, Converse and Rodgers[24] used social research methods to assess the sense of satisfaction or dissatisfaction that persons draw from several critical "domains" of their lives, such as their marriages, jobs and housing. They conducted interviews of a national probability sample of 2,720 households, asking respondents about the quality of streets and roads in their neighborhood, police protection, comfort, climate, public transportation, security, housing, education, job satisfaction, pay, spare-time activities, membership in clubs and organizations, health, and personal friendships.

CONCEPTUAL APPROACHES   323

They developed an Index of Well-Being which accounted for most of the variance in 17 specific domains of life, which were "areas of experience which have significance for all or most people and which may be assumed to contribute in some degree to the general quality of life experience."[24]

OTHER ENVIRONMENTAL INDICES

Few environmental indices have been developed for areas other than air or water pollution. Therefore, this book has devoted two full chapters to air and water pollution indices (Chapters III and IV). For completeness, examples of indices from other topic areas are discussed briefly here.

Solid Waste

Collins[25] has developed an index called the Solid Waste Environmental Quality Index (SWEQI), which is based on standard decision analysis techniques. SWEQI evaluates the outcome of each solid waste disposal alternative in terms of its impact on aesthetics, air pollution, water pollution, land use, noise, odors, physical hazards, and resource conservation. Utility functions are developed for each variable in the index, and they are multiplied by probabilities to calculate the expected utilities, which then are aggregated.

Noise

In the area of noise pollution, it is not necessary to aggregate different variables, because measurements involve one variable, environmental sound. As indicated by Manns,[26] the effects of environmental noise on people involve three factors: (1) the frequency spectrum, (2) the overall sound level, and (3) the temporal variation of the frequency and sound level. To simplify the approach, environmental noise measurements use a frequency spectrum that is weighted to reflect human hearing sensitivity, or "A-weighted," and levels are expressed in terms of their equivalent energy over a specified time period. Over time period $t_2 - t_1$, the observed sound pressure $P(t)$ relative to a reference level P_0 is averaged to give the equivalent sound energy, L_{eq}, expressed in decibels (dB):

$$L_{eq} = 10 \log_{10} \left[\frac{1}{t_2 - t_1} \int_{t_1}^{t_2} \frac{P^2(t)}{P_0^2} \, dt \right] \tag{4}$$

Because the population is more adversely affected by sound in the evening hours than in the daytime hours, EPA uses an indicator which weighs observed sound levels more heavily at night than in the daytime. In a 24-hour period, 10 dB are added to measurements in the 9 nighttime hours (10 PM to 7 AM) to give the day-night energy equivalent noise level, L_{dn}:

$$L_{dn} = 10 \log_{10} \left\{ \frac{15}{24} \left[10^{L_d/10} \right] + \frac{9}{24} \left[10^{(L_n+10)/10} \right] \right\} \quad (5)$$

where L_{dn} = day-night energy equivalent noise level (dB)
 L_d = energy equivalent noise level for the 15 daytime hours (dB, from Equation 4)
 L_n = energy equivalent noise level for the 9 nighttime hours (dB, from Equation 4)

Environmental control officials also are interested in the number of people exposed to different sound levels. It has been estimated that 100 million people reside in areas where the L_{dn} exceeds 55 dB, a level identified with marked annoyance.[26-28]

Total Environment

There have been few efforts to develop indices of the total environment, and one reason is the complexity of the problem. Pikul[29] has suggested that there are over 100 factors, which he ranked and grouped into 14 categories, for which indices of environmental quality could be constructed. His 14 categories consist of water pollution, hazardous substances (radioactivity levels, pesticide residues, etc.), land management (wetlands, erosion potential, dam siltage, urban green, etc.), solid waste disposal, recycling, resources (timber, agriculture, fish catch, minerals, etc.), natural phenomena (climate, solar radiation, floods, runoff, etc.), social-aesthetic conditions (rat infestation, outdoor recreation, housing, noise, urban sprawl, traffic congestion, street maintenance, odor, roadside litter, etc.), population, human health, biological health and ecological balance (endangered species, fish kills, algal blooms, plant growth, etc.), economic loss, and pollution control measures. It is clear that development of indices to reflect all possible factors that affect the environment of man would be a formidable undertaking.

Each year the National Wildlife Federation publishes the Environmental Quality (EQ) index, which reports the environmental status of seven topic areas: wildlife, air, water, minerals, soil, forests, and living space in America.[30] The EQ is not really an index which aggregates variables but rather a narrative report which discusses the current status of environmental activities in each of these topic areas. Although graphs are presented to illustrate trends in each topic area, the curves are based on subjective estimates rather than actual observations, and the Federation does not claim the curves to be accurate measures of environmental conditions. Although not analytically rigorous, the EQ publication presents a number of striking, and often alarming, facts and illustrations, and the text provides an instructive discussion of the problems and progress in each environmental topic area.

Species Diversity

Species diversity is a concept that is based on the richness of a habitat in terms of the number of species present and the number of individuals within each species. The best known species diversity indices are based on concepts of information theory initially introduced by Shannon and Weaver,[31] which express the information content of a system mathematically. Information content usually is computed by taking logarithms of the ratio of the number of individuals of each species present to the total population and adding the result.

Hurlbert[32] has reviewed the literature on species diversity indices and concludes that it contains ". . . many semantic, conceptual, and technical problems." One problem stems from the varied definitions of "species diversity" used by different authors:

> The term "species diversity" has been defined in such various and disparate ways that it now conveys no information other than "something to do with community structure"; species diversity has become a nonconcept.[32]

Hurlbert also suggests that indices based on information theory, although widely used, have not been shown to have any particular relevance to biological communities:

> Although these information theoretic indices have been examined and applied to ecological problems by many ecologists, no one has yet specified exactly what significance the "number of bits per individual" has to the individuals and populations in a community. It has not been shown that information theoretic indices have any greater biological relevance than do the infinite number of other potential indices which have a minimum value when $S = 1$ and a maximum value when $S = N$.[32]

Considerable literature exists on the species diversity concept, and the reader wishing to explore this topic further is referred to authors such as Hurlbert,[32] Hamilton[33] and Pielou.[34]

ENVIRONMENTAL DAMAGE FUNCTIONS

Quality of Life is essentially a decreasing scale system. By contrast, the concept of environmental damage is an increasing scale system. As discussed in Chapter I (pages 27 to 36), development of meaningful damage functions is severely handicapped by lack of knowledge of the effects of pollutant variables on health and welfare as a function of concentration. If sufficient knowledge were available to construct meaningful damage functions, how would they be used? How would individual damage functions be aggregated? How would an environmental index based on damage functions be constructed? The answers to these questions will give insight into the properties of existing index structures and help identify the data that are needed for meaningful damage functions to be constructed. Although some of the following concepts are similar to those used in microeconomic theory,[3,4] the general conceptual framework is original and is presented in this book for the first time.

A general damage function of n pollutant variables is written as follows:

$$D = D(X_1, X_2, \ldots, X_i, \ldots, X_n) \qquad (6)$$

where X_i = pollutant variable i ($X_i \geq 0$ for all i)
D = overall damage ($D \geq 0$)

Equation 6 assumes that some *common measure* of damage can be found. For example, one air pollutant may affect the function of the lung, while another, such as carbon monoxide, may affect the blood and thereby the central nervous system. Unless a common measure of damage or injury can be identified, it will be difficult to develop a single relationship that includes both these pollutant variables. One possible common measure of damage, for example, is the number of individuals per unit time that die as a result of environmental pollution, or the mortality rate. Sometimes, this measure is described as the number of "excess" deaths per unit time to indicate that it refers only to those persons who die as a result of environmental pollution.

Univariate Damage Functions

In Chapter I, various authors offered arguments for the shapes that univariate (single-variable) damage functions should possess. One characteristic shape is that of the hockey stick function (page 30). Some authors extend the range of the damage function to include a saturation level at which the effects of the pollutant are assumed to taper off (page 31). A typical logistic, or sigmoid, damage function with this characteristic shape has the following equation:

$$D(X_1) = \frac{b}{1 + e^{-a(X_1-c)}} \qquad (7)$$

This function is plotted in the top part of Figure 3 for the case in which $a = 1$, $b = 100$, and $c = 5$.

Although similar in shape, this curve should not be confused with the curves that appear in population growth models, such as those proposed by Pielou[34] and others. In the population models, the vertical axis usually represents total population, and the horizontal axis represents time. By contrast, damage functions are similar to the curves used routinely in laboratory experiments in which organisms are exposed to increasing levels of toxins while death rates are observed. Results from such experiments show that the sigmoid curve is a suitable model for studies in which different doses of an insecticide, for example, are applied under standardized conditions to members of an insect species. Finney[35] has proposed formal methods for analyzing the data obtained from such experiments. In this approach, *probit analysis*, the logarithm of the concentration at which members of the population die (*i.e.*, the "tolerance level") is assumed to have a Gaussian, or normal, distribution. Because the cumulative percentage of the population killed is calculated as the integral of this distribution, the approach produces an s-shaped curve similar to that in the top portion of Figure 3. Finney's approach also treats the random nature of such deaths in considerable detail.

The slope of the damage function plotted in Figure 3 is a bell-shaped curve that resembles the normal distribution (bottom portion of Figure 3). It is obtained by taking the first derivative of Equation 7. The term "marginal damage (MD)," which is patterned after "marginal utility" in economics, will be defined as the first partial derivative of the damage function with respect to pollutant variable i:

$$MD_i = \frac{\partial D}{\partial X_i} \qquad (8)$$

where MD_i = marginal damage of the ith pollutant variable

For the univariate example in Equation 7, the marginal damage is expressed as follows:

$$MD_1 = \frac{\partial D}{\partial X_1} = \frac{abe^{-a(X_1-c)}}{1 + 2e^{-(X_1-c)} + e^{-2a(X_1-c)}} \qquad (9)$$

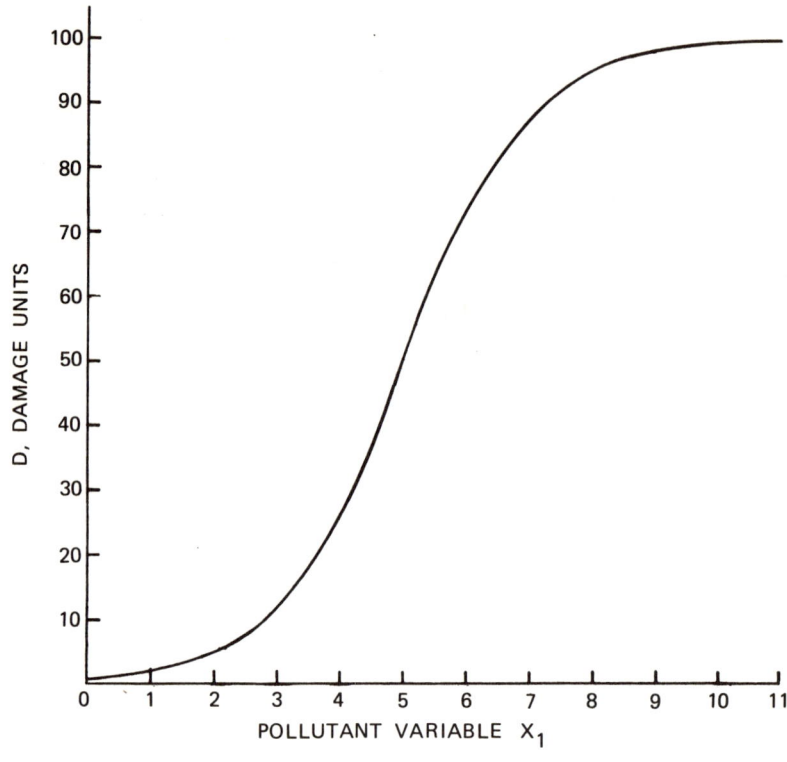
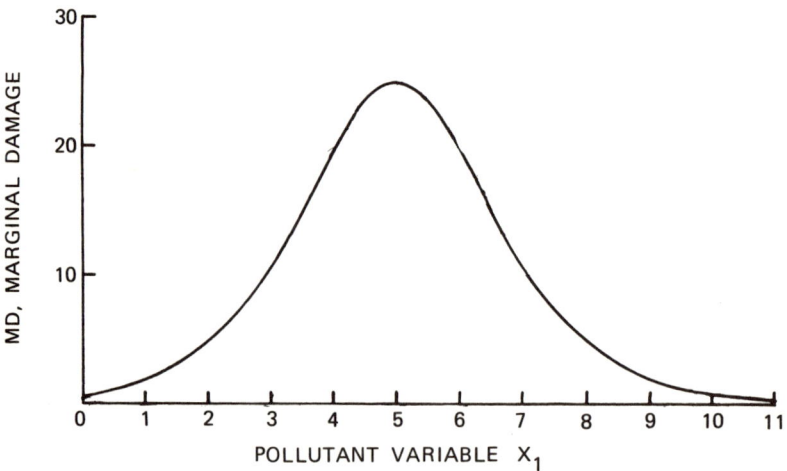

Figure 3. Logistic damage function $D = b/[1 + e^{-a(X_1-c)}]$, for $a = 1$, $b = 100$, and $c = 5$, showing damage function (top) and marginal damage function (bottom).

At low values of the pollutant variable, the marginal damage in Figure 3 is near zero. It reaches a maximum when the s-shaped damage function is at its point of inflection (X = c), and it then again approaches zero as the damage function approaches saturation. The bell-shaped marginal damage curve shows that, in this example, incremental increases in the pollutant variable have the greatest relative impact on damage in the middle range of the pollutant variable.

Multivariate Damage Functions

Damage functions involving just one pollutant variable are relatively easy to handle analytically. When two or more pollutants are present, however, the analytical treatment becomes more complex.

If, when each pollutant variable is multiplied by a constant factor λ, the resulting damage function changes by a factor λ^k, the damage function is said to be homogeneous of degree k. That is, a damage function containing n variables is homogeneous of degree k if:

$$D(\lambda X_1, \lambda X_2, \ldots, \lambda X_n) = \lambda^k D(X_1, X_2, \ldots, X_n) \tag{10}$$

If a damage function is homogeneous of degree 1, then multiplication of each pollutant variable by 3 will increase damage by a factor of 3. If a damage function is homogeneous of degree 2, then multiplication of each pollutant variable by 3 will cause damage to increase by a factor of $3^2 = 9$.

Consider a bivariate (two-variable) damage function $D = D(X_1, X_2)$. How does inclusion of both variables affect overall damage, and how does overall damage relate to the individual damage functions when one or the other variables are set to zero? Three possible situations can occur: (1) superposition, (2) synergism and (3) antagonism.

Superposition

Suppose that an experiment is conducted in which the two pollutant variables are set such that $X_1 = a$ and $X_2 = 0$, and damage $D(a,0)$ is observed. Next, the pollutant variables are set such that $X_1 = 0$ and $X_2 = b$, and damage $D(0,b)$ is observed. Finally, both pollutant variables are set such that $X_1 = a$ and $X_2 = b$. If the observed damage with both pollutants present equals the sum of the individual damages of X_1 and X_2 acting alone, we say that superposition applies. Mathematically, superposition is represented as follows:

$$D(a,b) = D(a,0) + D(0,b) \tag{11}$$

In such an experiment, superposition has been shown to apply only for the specific values $X_1 = a$ and $X_2 = b$. It would be necessary to conduct many more experiments to establish that superposition applies over the entire range of X_1 and X_2.

Consider the following example of a bivariate damage function consisting of the weighted sum of two pollutant variables, in which ϕ_1 and ϕ_2 are weights:

$$D(X_1,X_2) = \phi_1 X_1 + \phi_2 X_2 \tag{12}$$

If $X_1 = a$ and $X_2 = b$, then $D(a,b) = \phi_1 a + \phi_2 b$. However, if $X_1 = a$ and $X_2 = 0$, then $D(a,0) = \phi_1 a$, and if $X_1 = 0$ and $X_2 = b$, then $D(0,b) = \phi_2 b$. If the two individual damage functions, $D(a,0)$ and $D(0,b)$, are added together, the result is the same as the overall damage function, $D(a,b)$:

$$D(a,0) + D(0,b) = \phi_1 a + \phi_2 b = D(a,b) \tag{13}$$

Thus, superposition applies in this example.

Synergism

Suppose that the two pollutants interact in some fashion, thereby imparting more deleterious effects upon the organism than the sum of their individual effects. An example is the interaction between SO_2 and particulate matter, in which small particles transport SO_2, and sometimes sulfuric acid, deep into the lungs where damage is increased (page 18). Another example is the catalytic effect of certain pollutants, which can increase chemical reaction rates and thereby make damage more severe.

If, in the above experiment, the observed damage with both pollutants present exceeds the sum of the individual damages of X_1 and X_2 acting alone, we say that synergism applies. Mathematically, synergism is represented as follows:

$$D(a,b) > D(a,0) + D(0,b) \tag{14}$$

Antagonism

If two pollutants interact such that overall damage is lessened, antagonism applies. For example, two pollutants may react chemically with each other, creating a compound that is less injurious to health than the sum of the effects of each pollutant acting alone. If the observed damage with both pollutants present is less than the sum of the individual damages of X_1 and X_2 acting alone, we say that antagonism applies. Mathematically, antagonism is represented as follows:

$$D(a,b) < D(a,0) + D(0,b) \qquad (15)$$

Note that Equation 15 does not mean that introduction of a second pollutant, once the first one is present, necessarily reduces overall damage. When a second pollutant is introduced, overall damage usually increases; however, if overall damage does not increase sufficiently to reach the sum of the two individual damage functions, $D(a,0) + D(0,b)$, then antagonism applies.

This gives rise to a special case of antagonism. Suppose that the first pollutant is present, giving damage $D(a,0)$, and then a second pollutant is introduced at level $X_2 = b$. If the overall damage is less than the damage with the first pollutant alone, we say that the antagonism is "diminutive" with respect to X_1:

$$D(a,b) < D(a,0) \qquad (16)$$

In an extreme case of diminutive antagonism, the antagonism is diminutive with respect to both X_1 and X_2. That is, damage with both pollutants present is less than the damage associated with either pollutant acting alone. This gives rise to a second possibility, in which we say that the antagonism is diminutive with respect to X_2:

$$D(a,b) < D(0,b) \qquad (17)$$

Antagonism can be diminutive with respect to X_1 alone, X_2 alone, or both X_1 and X_2. Of course, it is possible that introduction of a second pollutant, once the first one is present, has no effect whatever—overall damage neither increases nor decreases. This neutral situation is represented as a borderline case, where the antagonism just becomes diminutive. The diminutive cases are included for completeness and may or may not represent conditions that commonly occur.

General Criteria

The above analysis considers only the damage associated with only two points, $X_1 = a$ and $X_2 = b$. More general criteria covering the entire ranges of the pollutant variables can be specified:

Synergism: $\qquad D(X_1,X_2) > D(X_1,0) + D(0,X_2) \qquad (18)$

Superposition: $\qquad D(X_1,X_2) = D(X_1,0) + D(0,X_2) \qquad (19)$

Antagonism: $\qquad D(X_1,X_2) < D(X_1,0) + D(0,X_2) \qquad (20)$

$\qquad$ Diminutive with Respect to X_1: $D(X_1,X_2) < D(X_1,0) \qquad$ (20a)

$\qquad$ Diminutive with Respect to X_2: $D(X_1,X_2) < D(0,X_2) \qquad$ (20b)

$\qquad$ for $X_1 > 0, X_2 > 0$

Note that X_1 and X_2 are defined over the range of all positive real values but that zero is specifically excluded from this range. This is because $X_1 = 0$ and $X_2 = 0$ are specific conditions contained within the criteria themselves.

If either Equation 20a or Equation 20b is satisfied, then Equation 20 is satisfied. This is verified by noting that $D(X_1,0) \geqslant 0$ and $D(0,X_2) \geqslant 0$ from the definition of a damage function given in Equation 6. By this condition, the right-hand side of Equation 20 must equal or exceed the right-hand sides of Equation 20a or 20b.

These criteria, although useful for describing the theoretical properties of damage functions, specify rigorous conditions. It is doubtful that real damage functions, if they were available, would meet just one criterion over the entire range of the pollutant variable. Rather, different criteria would apply to different ranges of the pollutant variable. Although the conditions specified by the criteria are exacting, the criteria are useful for describing the properties of idealized damage functions.

This analysis deals with damage functions involving just two pollutant variables for illustrative purposes only. These concepts would apply equally well to cases involving n pollutant variables. However, the analysis and notation become increasingly complex as more pollutant variables are included.

Graphical Representation

How can damage functions be represented graphically? For the bivariate case, one possible approach is to plot the damage function in three dimensions, with the vertical axis representing overall damage $D = D(X_1,X_2)$ and the two horizontal axes representing pollutant variables X_1 and X_2. For the linear damage function given in Equation 12, the three-dimensional plot yields a single inclined plane (plane A in Figure 4). This plane is oriented such that its projection on the (X_1-D)-plane is a straight line with slope ϕ_1 and the equation $D = D(X_1,0) = \phi_1 X_1$. Similarly, the projection on the (X_2-D)-plane is a straight line with slope ϕ_2 and the equation $D = D(0,X_2) = \phi_2 X_2$. Any plane that is parallel to the (X_1-D)-plane will intersect plane A along a line that is parallel to the line $D = \phi_1 X_1$. Likewise, any plane that is parallel to the (X_2-D)-plane will intersect plane A along a line that is parallel to the line $D = \phi_2 X_2$.

Equation 12 is linear with respect to the individual pollutant variables X_1 and X_2 in the univariate case. A damage function that is linear with respect to X_1 and X_2 but is synergistic would appear as a curved surface for which all points lie above plane A, except at the edges where the surface joins the lines $D = \phi_1 X_1$ and $D = \phi_2 X_2$. Conversely, an antagonistic damage function would be represented by a surface that joins the lines $D = \phi_1 X_1$ and $D = \phi_2 X_2$, but all other points would lie below plane A. Any points

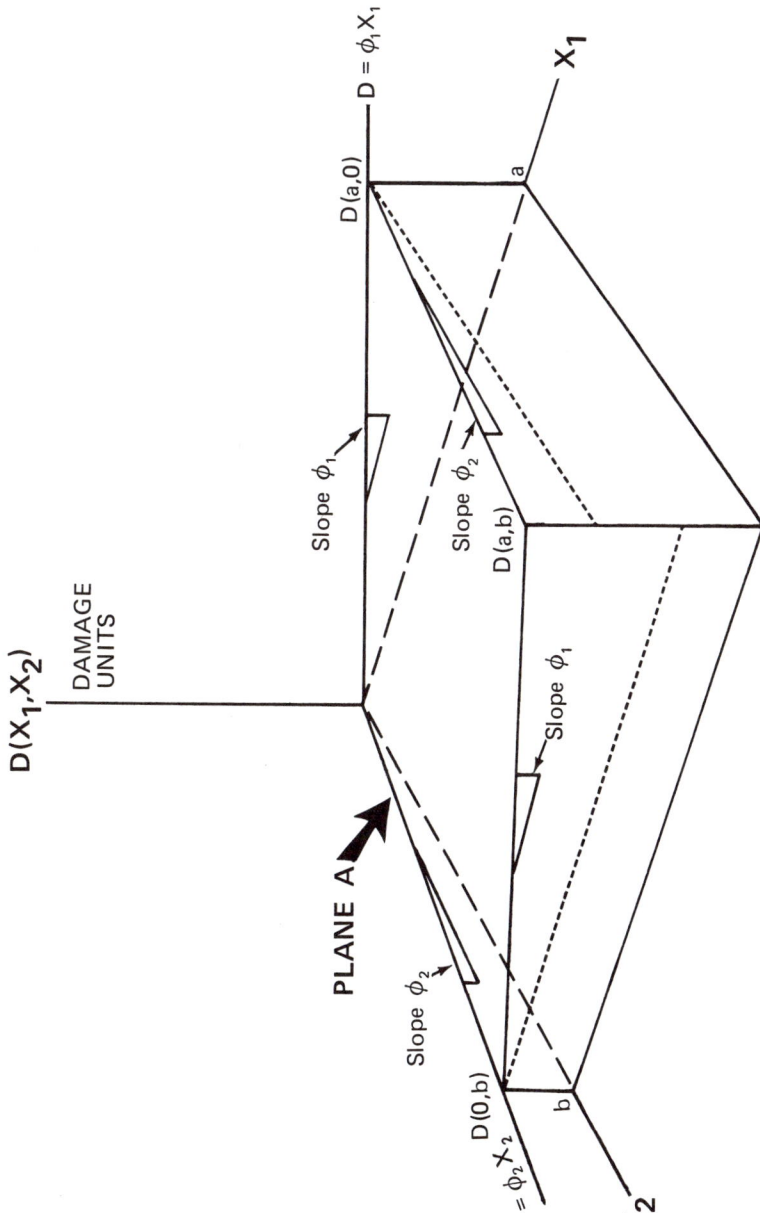

Figure 4. Damage function $D(X_1, X_2) = \phi_1 X_1 + \phi_2 X_2$ illustrating superposition case plotted in three dimensions.

below a horizontal plane of height $D(a,0)$ would be diminutive with respect to X_1, while any points below a horizontal plane of height $D(0,b)$ would be diminutive with respect to X_2.

Because of the difficulty of drawing and visualizing three-dimensional surfaces, a simpler graphical technique is desired which can describe environmental damage functions when more than one pollutant is present. We therefore introduce the concept of the "damage space," which is a two-dimensional system similar to the (I_1, I_2)-plane introduced previously in Chapter II. In the damage space, the two pollutant variables X_1 and X_2 are plotted on the two axes instead of the subindices I_1 and I_2 plotted in the (I_1, I_2)-plane. Therefore, the damage space reflects the overall behavior of the damage function with respect to the raw observations. Like the (I_1, I_2)-plane, a two-dimensional system is used for illustrative purposes, but the analysis applies equally well to the n-dimensional case.

The damage space is assumed to be a continuum. That is, every possible combination of X_1 and X_2 can occur, and each quantity is assumed to be infinitely divisible. The following formal assumptions apply to the damage space:

> 1. **Continuity** Any continuous curve in the damage space connecting point A, which has more damage than point B, to point C, which has less damage than point B, must pass through point D, which has damage equivalent to point B.

From this assumption, if D is equivalent to B, then there exists a continuous curve passing through D and B, every point of which is equivalent in damage to B (Figure 5). Thus, the locus of points which produce exactly the same damage is a continuous curve in the damage space. We shall refer to this locus of points as an "equivalence curve," because it identifies all combinations of X_1 and X_2 for which pollutant variables produce equivalent effects.

> 2. **Transitivity** If B is equivalent to D and D is equivalent to E, then B is equivalent to E.

This assumption emphasizes the unrestricted nature of equivalence. In Assumption 1, for example, B and D both lie on the same equivalence curve. Thus, the damage at point B is indistinguishable from the damage at point D, because B and D are equivalent in every respect.

> 3. **Monotonicity** Given any point A, such that A lies on an equivalence curve further from the origin than the curve passing through point B, then A represents greater damage than B.

By "further from the origin," we mean that each successive equivalence curve intersects each axis at successively increasing distances from the origin.

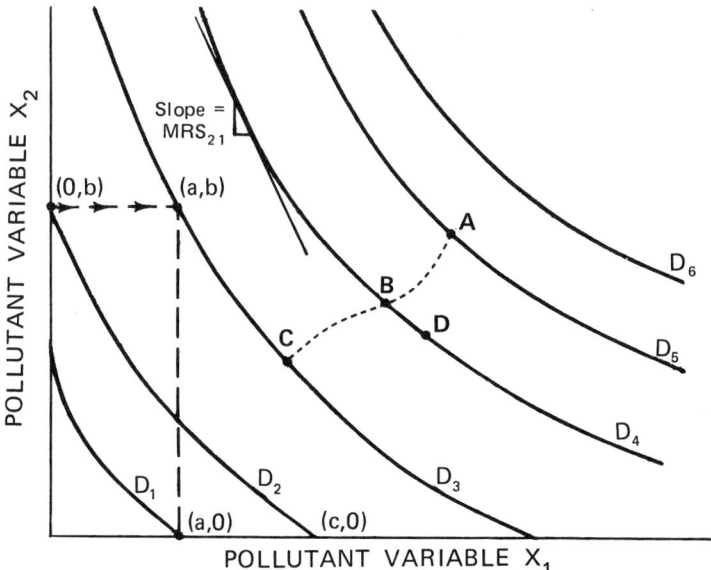

Figure 5. Equivalence curves plotted in the damage space.

That is, $D(c,0) > D(a,0)$ if $c > a$, which is a normal property of univariate increasing scale damage functions.

To visualize the damage space, imagine an infinite number of equivalence curves forming a surface in the (X_1, X_2)-plane. Point A in Figure 5 lies on one of these curves (D_5), and point B lies on another curve (D_4). Because point A in Figure 5 lies on an equivalence curve that is further from the origin, it represents greater damage than point B; that is, $D_A > D_B$ because $D_5 > D_4$. Through every point in the damage space (except the origin) there passes an equivalence curve which divides the damage space into three parts: (1) the equivalence curve itself, (2) the region of less damage than the equivalence curve, and (3) the region of greater damage than the equivalence curve.

To examine the properties of equivalence curves in the damage space, it is useful to employ the concept of marginal rate of substitution (MRS). This concept, borrowed from economics, allows the relative contribution of each pollutant variable to overall damage to be examined in detail. The MRS has been applied by Meisel and Horowitz[36] to air pollution indices in the same manner that we apply it here to damage functions.

Assuming that X_1 changes very slightly, how much would X_2 have to change to keep $D(X_1, X_2)$ exactly the same? The marginal rate of substitution of X_2 for X_1, MRS_{21}, is the incremental change in pollutant variable X_2 necessary to compensate for a change in pollutant variable X_1 such that damage remains constant, $D = D_0$. Mathematically, the MRS_{21} is the partial derivative of X_2 with respect to X_1:

$$MRS_{21} = \left.\frac{\partial X_2}{\partial X_1}\right|_{D=D_0} \qquad (21)$$

The MRS_{21} is the slope of the equivalence curve in the damage space.

The MRS_{21} is a measure of the relative importance of the two pollutant variables in terms of their contribution to overall damage. If, for example, the MRS_{21} is a large negative number, then a small increase in X_1 would have to be accompanied by a large decrease in X_2 if damage $D(X_1,X_2)$ is to remain the same. In this case, X_1 would be the more important pollutant variable, because a small change in X_1 is "worth" a large change in X_2 in terms of its impact on damage. Conversely, if MRS_{21} is near zero (and still negative), a large increase in X_1 would have to be accompanied by a small decrease in X_2 for damage to remain the same, and X_2 would be the more important variable. That is, a small amount of X_2 would be "worth" a large amount of X_1 in terms of its impact on damage.

The slope of the MRS_{21}, which is the second partial derivative of X_2 with respect to X_1, gives an indication of the way in which the relative contributions of X_1 and X_2 to damage change with changes in X_1:

$$\frac{\partial(MRS_{21})}{\partial X_1} = \frac{\partial^2 X_2}{\partial X_1^2} \qquad (22)$$

If the MRS_{21} is negative and its derivative, calculated from Equation 22, is positive, the equivalence curve will be convex to the origin like the curves shown in Figure 5. This happens because the slope of each equivalence curve begins as a large negative value, and it increases (becomes less negative) as X_1 increases.

Consider a simple damage function that is linear with respect to the univariate cases of X_1 and X_2—that is, $D(X_1,0)$ and $D(0,X_2)$ are linear functions of X_1 and X_2 individually. If the equivalence curves are convex to the origin in this case, the damage function is synergistic (Figure 6). Introduction of a second pollutant once the first is present has considerable impact, but the relative impact decreases as the second pollutant continues to increase. If the equivalence curves are straight lines, Equation 22 is zero, and superposition applies. If the equivalence curves are concave to the origin, Equation 22 is negative, and antagonism applies. That is, introduction of a second pollutant once the first is present has relatively little impact, but the relative impact increases as the second pollutant continues to increase. In the

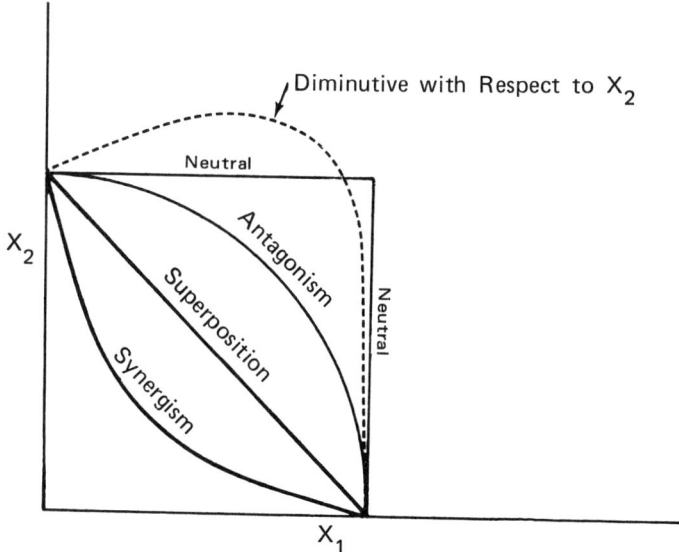

Figure 6. Characteristic shape of equivalence curves in the damage space for a damage function that is linear with respect to X_1 and X_2 acting individually.

antagonistic case, both X_1 and X_2 must become increasingly large for damage to remain the same because of the tendency for each pollutant to reduce the other's deleterious effects. If antagonism becomes diminutive, the slope (MRS_{21}) actually becomes positive. Figure 6 shows an equivalence curve in which antagonism is diminutive with respect to pollutant variable X_2 (dotted line). The borderline between antagonism and diminutive antagonism is represented by horizontal and vertical straight lines marked "neutral." In this neutral case, introduction of a second pollutant has no effect whatever —until a critical point is reached at which the second pollutant variable exceeds the first one. These neutral equivalence curves correspond to the maximum operator damage function $D = \max\{X_1, X_2\}$.

In the examples given in Figure 6, the assumption that damage is linear with respect to X_1 and X_2 acting alone is restrictive, and it will not apply to all damage functions. More general criteria for synergism, superposition and antagonism can be developed which will apply to all damage functions, and these criteria also can be illustrated in the damage space. Consider a situation in which $X_1 = a$ and $X_2 = 0$. This situation is represented in Figure 5 as the point (a,0) on the horizontal axis, and it lies on equivalence curve D_1. Next consider a situation in which $X_1 = 0$ and $X_2 = b$. This situation is represented as the point (0,b) on the vertical axis, and it lies on equivalence curve D_2. Now suppose that X_1 increases from zero to $X_1 = a$ while X_2

remains fixed at $X_2 = b$. In Figure 5, this change is represented by the horizontal movement from the point $(0,b)$ to the point (a,b). Because the point (a,b) lies on equivalence curve D_3, the change $\Delta X_1 = a$ corresponds to a change in damage of $\Delta D = D_3 - D_2$. By definition, superposition applies if and only if $D_3 = D_1 + D_2$. This condition is satisfied only if $D_3 - D_2 = \Delta D = D_1$. That is, the horizontal movement $\Delta X_1 = a$ gives the same change in damage $\Delta D = D_1$ for $X_2 = 0$ as it does for $X_2 = b$. In general, for the superposition case, a given change in pollutant variable X_1 produces the same change in damage regardless of the value of X_2. Similarly, a given change in pollutant variable X_2 produces the same change in damage regardless of the value of X_1. Stated another way, the marginal damage with respect to X_1 will be independent of X_2, and the marginal damage with respect to X_2 will be independent of X_1:

$$MD_1 = \frac{\partial D}{\partial X_1} = g(X_1) \tag{23}$$

$$MD_2 = \frac{\partial D}{\partial X_2} = g(X_2) \tag{24}$$

Here, $g(X_1)$ is any function, including a constant, which does not contain X_2, and $g(X_2)$ is any function, including a constant, which does not contain X_1.

If we now take partial derivatives of either Equation 23 or Equation 24 with respect to the other pollutant variable, the following result is obtained:

$$\frac{\partial (MD_1)}{\partial X_2} = \frac{\partial^2 D}{\partial X_1 \partial X_2} = \frac{\partial (MD_2)}{\partial X_1} = 0 \tag{25}$$

Equation 25 is an important condition for superposition. By differentiating both sides of Equations 18, 19 and 20, we obtain the following general criteria for synergism, superposition and antagonism:

Synergism: $\qquad\qquad\qquad \dfrac{\partial^2 D}{\partial X_1 \partial X_2} > 0 \qquad\qquad (26)$

Superposition: $\qquad\qquad\quad \dfrac{\partial^2 D}{\partial X_1 \partial X_2} = 0 \qquad\qquad (27)$

Antagonism: $\qquad\qquad\qquad \dfrac{\partial^2 D}{\partial X_1 \partial X_2} < 0 \qquad\qquad (28)$

Because these criteria allow any damage function to be tested for the interactive properties of its pollutant variables, we refer to Equations 26, 27 and 28 as the "criteria for interaction." The criteria for interaction are applied by first differentiating the damage function with respect to X_1 and then with respect to X_2. Then we test to see if the result is either positive, zero, or negative.

Examples

Several examples of theoretical damage functions will illustrate these concepts. The simple linear damage function given in Equation 12 is plotted in the damage space by solving for X_2 as a function of X_1 and $D = D(X_1, X_2)$, in which D is treated as a constant which takes on selected values, one for each equivalence curve:

$$X_2 = -\frac{\phi_1}{\phi_2} X_1 + \frac{D}{\phi_2} \tag{29}$$

The marginal rate of substitution, obtained by differentiating Equation 29 with respect to X_1, is a constant: $MRS_{21} = -\phi_1/\phi_2$. This function plots in the damage space as a series of straight lines of slope $-\phi_1/\phi_2$ and X_2-intercept D/ϕ_2. Figure 7 shows this damage function plotted in the damage space for the case $\phi_1 = \phi_2 = 1$. The marginal damages are obtained by differentiating Equation 12 with respect to X_1 and X_2:

$$MD_1 = \frac{\partial D}{\partial X_1} = \phi_1 \tag{30}$$

$$MD_2 = \frac{\partial D}{\partial X_2} = \phi_2 \tag{31}$$

The criteria for interaction, which are applied by differentiating either Equation 30 or 31 with respect to the other pollutant variable, give zero in this example, showing that superposition applies:

$$\frac{\partial (MD_1)}{\partial X_2} = \frac{\partial^2 D}{\partial X_1 \partial X_2} = 0 \tag{32}$$

Consider the following multiplicative damage function, in which θ_1 and θ_2 are constants:

$$D = X_1^{\theta_1} X_2^{\theta_2} \text{ for } \theta_1 > 0, \theta_2 > 0 \tag{33}$$

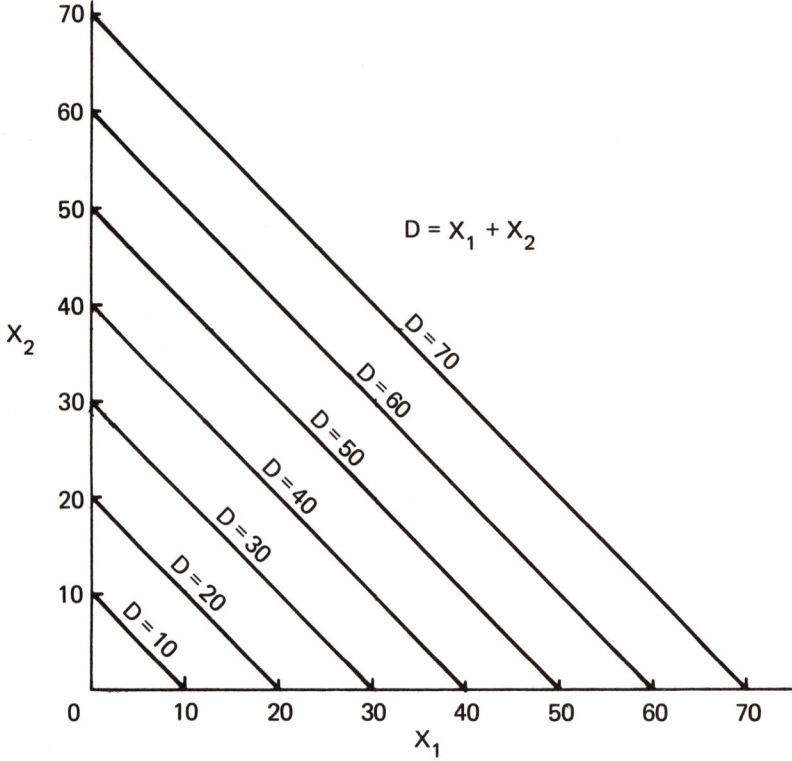

Figure 7. Example of the linear damage function $D = X_1 + X_2$ plotted in the damage space.

Like the example in Equation 29, this function is plotted in the damage space by solving Equation 33 for X_2 as a function of X_1 and D and then specifying selected values for D:

$$X_2 = D^{1/\theta_2} X_1^{-\theta_1/\theta_2} \qquad (34)$$

The marginal rate of substitution and its derivative are obtained directly by taking partial derivatives of X_2 with respect to X_1:

$$\text{MRS}_{21} = \frac{\partial X_2}{\partial X_1} = -\frac{\theta_1}{\theta_2} D^{1/\theta_2} X_1^{-(\theta_1/\theta_2 + 1)} \qquad (35)$$

$$\frac{\partial (\text{MRS}_{21})}{\partial X_1} = \frac{\partial^2 X_2}{\partial X_1^2} = D^{1/\theta_2} \left[\left(\frac{\theta_1}{\theta_2}\right)^2 + \frac{\theta_1}{\theta_2} \right] X_1^{-(\theta_1/\theta_2 + 2)} \qquad (36)$$

Because θ_1 and θ_2 are defined as positive constants (and $D \geq 0$, $X_1 \geq 0$, and $X_2 \geq 0$), Equation 35 is negative, indicating that the equivalence curves will be negatively sloped when plotted in the damage space. Because Equation 36 is positive, the curves will exhibit increasing rates of marginal substitution and will be convex to the origin.

The marginal damage functions are obtained by differentiating Equation 33 with respect to X_1 and X_2:

$$MD_1 = \frac{\partial D}{\partial X_1} = \theta_1 X_1^{\theta_1-1} X_2^{\theta_2} \tag{37}$$

$$MD_2 = \frac{\partial D}{\partial X_2} = \theta_2 X_1^{\theta_1} X_2^{\theta_2-1} \tag{38}$$

The criteria for interaction are applied by taking the partial derivative of Equation 37 with respect to X_2:

$$\frac{\partial (MD_1)}{\partial X_2} = \frac{\partial^2 D}{\partial X_1 \partial X_2} = \theta_1 \theta_2 X_1^{\theta_1-1} X_2^{\theta_2-1} = \frac{\theta_1 \theta_2}{X_1 X_2} D. \tag{39}$$

Because the result given by Equation 39 is positive, the criterion for interaction given by Equation 26 is satisfied, and the damage function is synergistic. This synergistic property can be verified by showing that Equation 33 also satisfies the general synergistic criterion given by Equation 18.

Figure 8 shows a plot of Equation 33 in the damage space for the case in which $\theta_1 = \theta_2 = 1$. The equivalence curves are hyperbolas. The synergistic property of this function has been used in the Pollutant Standards Index for the subindex TSP x SO_2, which is based on the product of the two observed concentrations (see Chapter III).

As a next example, consider the root-sum-power damage function, in which ρ is a constant:

$$D = (X_1^\rho + X_2^\rho)^{1/\rho} \quad \text{for } \rho > 1 \tag{40}$$

As in the above examples, this function can be plotted in the damage space by first solving for X_2:

$$X_2 = (D^\rho - X_1^\rho)^{1/\rho} \tag{41}$$

The MRS and its derivative are obtained as follows:

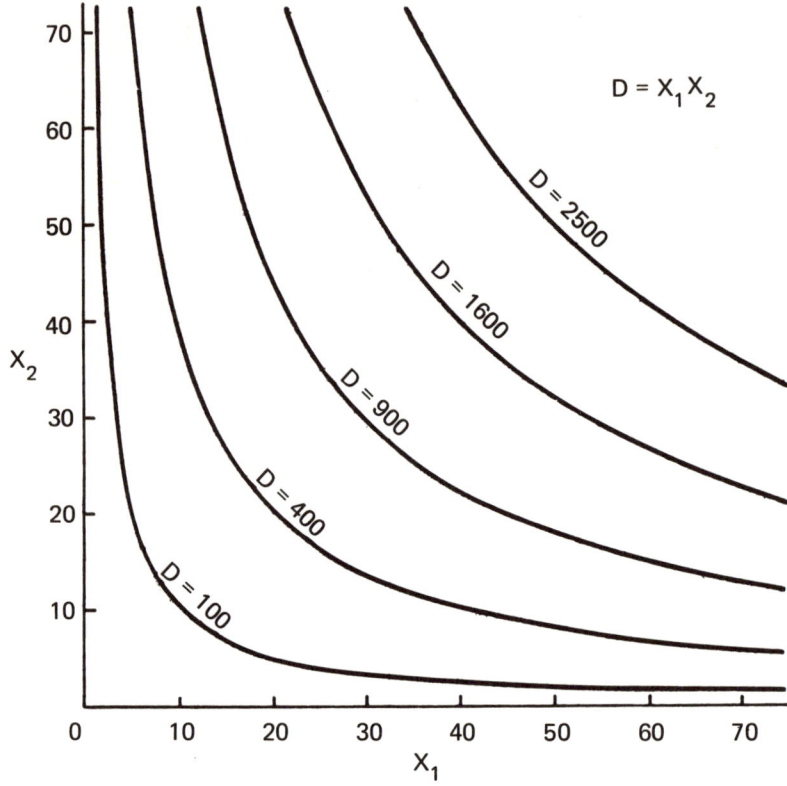

Figure 8. Example of the multiplicative damage function $D = X_1 X_2$ plotted in the damage space.

$$\text{MRS}_{21} = \frac{\partial X_2}{\partial X_1} = -X_1^{\rho-1} [D^\rho - X_1^\rho]^{\frac{1-\rho}{\rho}} = -\left(\frac{X_1}{X_2}\right)^{\rho-1} \quad (42)$$

$$\frac{\partial(\text{MRS}_{21})}{\partial X_1} = -(\rho-1) \frac{X_1^{\rho-2} [X_1^\rho + X_2^\rho]}{X_2^{(2\rho-1)}} = -(\rho-1) \frac{X_1^{\rho-2}}{X_2^{(2\rho-1)}} D^\rho \quad (43)$$

Because $\rho > 1$ and $X_1 \geqslant 0$, $X_2 \geqslant 0$, Equation 42 is always negative, indicating that the equivalence curves have negative slopes, and Equation 43 is always negative, indicating that the marginal rate of substitution decreases (becomes more negative) as X_1 increases, giving curves that are concave to the origin. The marginal damage with respect to X_1 is computed as follows:

$$MD_1 = \frac{\partial D}{\partial X_1} = [X_1^\rho + X_2^\rho]^{\frac{1-\rho}{\rho}} X_1^{\rho-1} = \left(\frac{X_1}{D}\right)^{\rho-1} \quad (44)$$

The criteria for interaction are applied by differentiating Equation 44 with respect to X_2:

$$\frac{\partial^2 D}{\partial X_1 \partial X_2} = -(\rho-1)[X_1^\rho + X_2^\rho]^{\frac{1-2\rho}{\rho}} (X_1 X_2)^{\rho-1}$$
$$= -(\rho-1)(X_1 X_2)^{\rho-1} D^{1-2\rho} \quad (45)$$

Because the result is negative, the function is antagonistic. Notice that the damage function given in Equation 40 is identical in form to the root-sum-power index aggregation function described in Chapter II (page 72).

A special case occurs if $\rho = 2$ in Equation 40, giving the root-sum-square damage function:

$$D = \sqrt{X_1^2 + X_2^2} \quad (46)$$

This function plots in the damage space as a series of concentric circles of radius D centered at the origin (Figure 9). From Equation 42, the slope of each of these equivalence curves is $MRS_{21} = -X_1/X_2$, and the marginal damage is obtained from Equation 41 as $MD_1 = X_1/D$.

As a final example, consider the following damage function:

$$D = X_1^2 + X_2^2 \quad (47)$$

This function plots in the damage space as a circle of radius $\sqrt{D}$ (Figure 10). Initial inspection might suggest that this damage function is antagonistic, because its equivalence curves are concave to the origin. However, the characteristic shapes shown in Figure 6 do not apply in this situation, because the requirement that $D(X_1,0)$ and $D(0,X_2)$ be linear functions of X_1 and X_2, respectively, is not satisfied. Actually, this damage function is not antagonistic, which can be seen by applying the criteria for interaction:

$$MD_1 = \frac{\partial D}{\partial X_1} = 2X_1 \quad (48)$$

$$MD_2 = \frac{\partial D}{\partial X_2} = 2X_2 \quad (49)$$

344 ENVIRONMENTAL INDICES

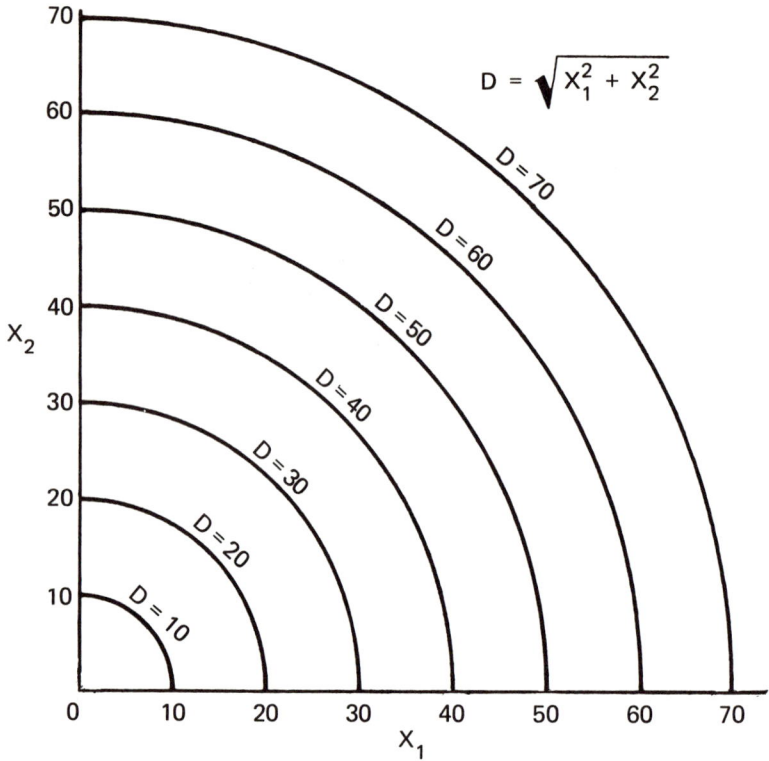

Figure 9. Example of the damage function $D = \sqrt{X_1^2 + X_2^2}$ plotted in the damage space.

$$\frac{\partial(MD_1)}{\partial X_2} = \frac{\partial^2 D}{\partial X_1 \partial X_2} = \frac{\partial(MD_2)}{\partial X_1} = 0 \qquad (50)$$

Although damage D is a nonlinear function of X_1 and X_2, this damage function meets the superposition criterion given by Equation 27. This example illustrates that the shape of an equivalence curve alone is insufficient to determine if synergism, superposition or antagonism applies to a given damage function. Either the general criteria (Equations 18, 19 and 20) or the criteria for interaction (Equations 26, 27 and 28) must be applied for a conclusive test.

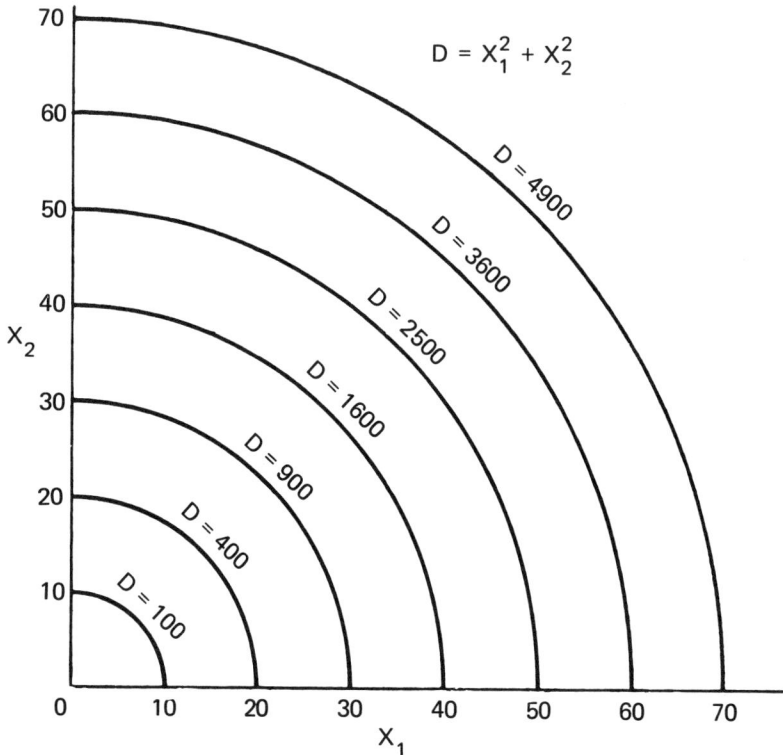

Figure 10. Example of the damage function $D = X_1^2 + X_2^2$ plotted in the damage space.

Discussion

The above analysis describes the properties of theoretical damage functions. The development of real damage functions, as discussed in Chapter I, is a formidable technical challenge that is beset by many experimental problems. If valid damage functions were available, they would provide the necessary rationale for the construction of environmental indices which have a very clearly defined scientific basis. Such an environmental index might be a linear function of the damage equation $I = \alpha D + \beta$ or it might be the damage function itself; that is, $I = D$.

If today's environmental indices were interpreted as damage functions, what properties would they have? The above criteria are useful for analyzing their characteristics. Most existing indices, when subjected to these tests, are found to contain numerous inherent assumptions about the relationships of the pollutants to each other and to overall damage. It is likely that the original index developers were unaware of these assumptions.

Consider, for example, a simple air pollution index, the Ontario API discussed in Chapter III (page 109):

$$\text{API} = 0.2(30.5X_1 + 126X_2)^{1.35} \tag{51}$$

where X_1 = Coefficient of Haze (COH)
X_2 = SO$_2$ concentration (ppm)

If this equation were treated as a damage function, it would be written in the following general form:

$$D = \kappa(\omega_1 X_1 + \omega_2 X_2)^\rho \tag{52}$$

The MRS$_{21}$ is calculated by solving Equation 52 for X_2 and then differentiating:

$$\text{MRS}_{21} = \frac{\partial X_2}{\partial X_1} = -\frac{\omega_1}{\omega_2} = -\frac{30.5}{126} = -0.24 \tag{53}$$

This result implies that, for damage to remain fixed, a decrease of 1 COH unit should be accompanied by an increase of 0.24 ppm SO$_2$. That is, 1 COH unit is "worth" $1/0.24 = 4.1$ ppm SO$_2$ in terms of its impact on damage, and this "exchange rate" is assumed constant. In view of the limited scientific knowledge about the relative effects of these two pollutants on health (see Chapter I), this assumption appears questionable.

The marginal damage, MD$_1$, is computed by differentiating Equation 52 with respect to X_1:

$$\text{MD}_1 = \frac{\partial D}{\partial X_1} = \omega_1 \kappa(\omega_1 X_1 + \omega_2 X_2)^{\rho-1} = \frac{\rho \omega_1 D}{\omega_1 X_1 + \omega_2 X_2} \tag{54}$$

Note that κ does not appear explicitly in the right-hand side of Equation 54 because it is contained in D. The criteria for interaction then are applied by differentiating Equation 54 with respect to X_2:

$$\frac{\partial^2 D}{\partial X_1 \partial X_2} = (\rho-1)\omega_1 \omega_2 \kappa(\omega_1 X_1 + \omega_2 X_2)^{\rho-2} = \frac{\rho(\rho-1)\omega_1 \omega_2 D}{(\omega_1 X_1 + \omega_2 X_2)^2} \tag{55}$$

Because $\rho = 1.35$, $\omega_1 > 0$, and $\omega_2 > 0$, the result is positive, and this equation meets the synergism criterion given by Equation 26. Thus, the Ontario

CONCEPTUAL APPROACHES 347

API is synergistic. The original article in which this index was published says nothing about its synergistic properties, and the authors might be surprised to learn that their index embodies a synergistic relationship.

Earlier, it was noted that specification of a damage function, as in Equation 6, implies that a common measure of damage can be found which applies to all the pollutant variables contained in the damage function. If no common measure of damage is available, the MRS cannot be defined. For example, if one pollutant affects respiratory illness and another pollutant affects the central nervous system, it will be difficult to relate the damage associated with lung function to the damage associated with the central nervous system. The MRS can be defined only if damage can be related to some common measure, such as the mortality rate (for example, the number of deaths per year) or the morbidity rate (for example, the number of days of illness per year).

The MRS deals very explicitly with the relationship of one variable to another. Its existence implies that a specific, defined relationship exists between pairs of variables in the damage function. In the familiar problem of "combining apples and oranges," it means that one apple is "worth" a specific quantity of oranges. If today's environmental indices are interpreted as damage functions, we will discover that the MRS exists for most of them. The principal exceptions are indices which involve the maximum or minimum operator, for which the MRS is not mathematically defined.

Even a simple index consisting of the sum of two pollutant variables causes the MRS to exist. In some indices, the MRS is a constant, but often it is more complex. For indices such as the NSF WQI (see Chapter IV), the MRS exists for all pairs of variables, even though it cannot be written formally as an equation because of the implicit nature of each subindex function. Thus, if the NSF WQI is interpreted as a damage function, BOD can be related, through the MRS, to pH. Similarly, fecal coliforms can be related, through the appropriate MRS, to temperature. These facts illustrate that many intrinsic assumptions are contained within these indices.

Of the candidate index aggregation functions, the maximum and minimum operators appear to embody the fewest intrinsic assumptions about the relationship of one pollutant to another. Only at the point at which the subindices are exactly equal is an explicit assumption made about the relationship between pollutant variables. This assumption is appropriate if, as discussed in Chapter II, the univariate damage function for each pollutant variable is dichotomous (two-state). Because the maximum and minimum operators contain relatively few intrinsic assumptions, they are well suited for use in environmental indices in which detailed specification of the damage functions is lacking and the effects of the pollutants differ. If the damage imparted to an organism by one pollutant is of a distinctly different nature

348 ENVIRONMENTAL INDICES

from the damage imparted by another pollutant, and if no common measure of damage can be identified, the maximum or minimum operator would appear to be the best choice. An even more conservative choice would be to avoid aggregating the subindices at all and present the individual variables in an environmental quality profile (see Chapter I) or in a multivariate display such as PSI bars (see Chapter III). However, the inevitable consequence of this conservative approach is that the presentation becomes richer in information and therefore more complex. Once again, this discussion illustrates the tradeoff between simplicity, with its attendant risk of misinterpretation, and completeness of detail, with its attendant complexity. It is this tradeoff which causes the classic dichotomy of views toward environmental indices (see Chapter I). The data analyst, after considering the audience involved, must always choose the proper balance between these two extremes.

CONCLUSIONS

Few formal environmental indices have been developed in fields outside of the air and water pollution topic areas. The QOL concept has been suggested as a broad framework for reflecting all components of man's total well-being. Attempts to apply the QOL concept to the comparison of U.S. metropolitan areas, although interesting scientific efforts in themselves, have suffered seriously from data limitations and methodological difficulties. In general, making valid comparisons of environmental conditions in different geographical areas is a formidable technical challenge, requiring considerable skill and analytical rigor. The analyst faces serious technical problems, such as the ranking paradox discussed in Chapters I and III, as well as political problems, such as the public controversy that these comparisons inevitably stir.

In contrast with QOL, the concept of environmental damage is an increasing scale system. This chapter has introduced a formal mathematical approach for analyzing the properties of theoretical damage functions. If today's indices are interpreted as damage functions, their properties (for example, synergism, superposition and antagonism) can be explored readily using this technique. Thus, analysis of damage functions can give important insights into the structure, behavior and underlying assumptions of today's environmental indices. This theoretical treatment also helps to reveal the nature of the scientific data that are needed to construct increasingly meaningful environmental indices in the future.

PROBLEMS FOR STUDY

1. Liu's environmental component for determing the QOL in U.S. SMSAs contains 17 variables. (a) What additional variables would you include in the environmental component if data were available? (b) Discuss the data collection problems that might result from inclusion of these additional variables. (c) Discuss the data collection problems that would occur if a QOL system were developed around the 14 environmental categories proposed by Pikul.

2. Consider a dichotomous univariate damage function in which $D_1 = 0$ if X_1 is less than its threshold level of $X_1 = 50$, and $D_1 = 100$ damage units if $X_1 \geqslant 50$. Consider a second dichotomous damage function in which $D_2 = 0$ if X_2 is less than its threshold level of 75, and $D_2 = 100$ damage units if $X_2 \geqslant 75$. (a) Suppose that a bivariate damage function is constructed that is the average of the two individual damage functions:

$$D = \frac{1}{2}(D_1 + D_2)$$

Plot this damage function in the damage space and show that $D = 100$ is represented graphically by an area for which *both* pollutants equal or exceed their threshold levels. [This area is the upper right quadrant of the intersecting lines $X_1 = 50$ and $X_2 = 75$.] (b) Suppose that a bivariate damage function is constructed using the maximum operator:

$$D = \max\{D_1, D_2\}$$

Plot this damage function in the damage space and show that $D = 100$ is represented by a larger area for which *either* pollutant equals or exceeds its threshold level. [This area is the region above the line $X_2 = 75$ and to the right of the line $X_1 = 50$.] Notice that these two cases correspond to the AND and OR functions in computer science.

3. Consider the following damage function:

$$D = \alpha_1 X_1 + \alpha_2 X_2 + \alpha_3 X_1 X_2$$

(a) Plot this function in the damage space for the case in which $\alpha_1 = \alpha_2 = \alpha_3 = 1$. (b) Calculate the MRS_{21} and show that the equivalence curves are negatively sloped and concave to the origin. [Answer: $MRS_{21} = -(D\alpha_3 + \alpha_1\alpha_2)/(\alpha_2 + \alpha_3 X_1)^2$] (c) Calculate the marginal damages. [Answers: $MD_1 = \alpha_1 + \alpha_3 X_2$, $MD_2 = \alpha_2 + \alpha_3 X_1$]. (d) By applying the criteria for interaction, show that this function is synergistic if $\alpha_3 > 0$ and is antagonistic if $\alpha_3 < 0$.

4. Consider the following damage function:

$$D = \sqrt{X_1^{K_1} + X_2^{K_2}}$$

(a) Show that the marginal damage with respect to pollutant variable X_1 is expressed as follows:

$$MD_1 = \frac{K_1}{2} \frac{X_1^{K_1 - 1}}{D}$$

(b) Show that the derivative of the marginal damage with respect to X_2 is expressed as follows:

$$\frac{\partial(MD_1)}{\partial X_2} = \frac{\partial D}{\partial X_1 \partial X_2} = -\frac{1}{4} K_1 K_2 \frac{X_1^{K_1 - 1} X_2^{K_2 - 1}}{D^3}$$

(c) Using the criteria for interaction, show that this damage function is antagonistic if K_1 and K_2 have the same sign and synergistic if K_1 and K_2 have different signs.

5. If $D(X_1, 0)$ and $D(0, X_2)$ are linear functions of X_1 and X_2, respectively, show that equivalence curves for the superposition case must plot in the damage space as straight lines. [Hint: If $D(X_1, 0)$ is a linear function of X_1, then $\partial D/\partial X_1$ is constant for the case $X_2 = 0$. Likewise, $\partial D/\partial X_2$ is constant for the case $X_1 = 0$. For the superposition situation, Equations 23 and 24 are satisfied; therefore, $\partial D/\partial X_1$ is the same for all values of X_2, and $\partial D/\partial X_2$ is the same for all values of X_1. Thus, $\partial D/\partial X_1$ and $\partial D/\partial X_2$ are constant for all X_1 and X_2. From this information, it is possible to define $D(X_1, X_2)$ and to show that $D(X_1, k \cdot X_2) =$ constant, in which k is a constant, giving straight lines in the damage space.]

6. Consider the examples given in Equations 12, 33, 40, 47 and 52. (a) Calculate the MRS_{21} and its derivative for each example. (b) Calculate MD_1 and MD_2 for each example. (c) Apply the criteria for interaction (Equations 26, 27 and 28) to each example in order to establish whether synergism, superposition or antagonism applies. (d) Verify your results by applying the general criteria (Equations 18, 19 and 20) to $D(X_1, 0)$ and $D(0, X_2)$ for each example.

REFERENCES

1. "The Quality of Life Concept: A Potential New Tool for Decision-Makers," U.S. Environmental Protection Agency, Office of Research and Monitoring, Washington, DC (March 1973).
2. Liu, Ben-Chieh. "Quality of Life Indicators in U.S. Metropolitan Areas, 1970: A Comprehensive Assessment," report prepared by Midwest Research Institute for the U.S. Environmental Protection Agency under Grant No. R803049-01-0, Washington, DC (May 7, 1975).
3. Lloyd, Cliff. *Microeconomic Analysis* (Homewood, IL: Richard D. Irwin, Inc., 1967).
4. Henderson, James M., and Richard E. Quandt. *Microeconomic Theory: A Mathematical Approach* (New York: McGraw-Hill Book Co., 1958).
5. Liu, Ben-Chieh. "The Quality of Life in the United States, 1970: Index, Rating, and Statistics," Report No. 816-561-0202, Midwest Research Institute, Kansas City, MO (October 1975).
6. Liu, Ben-Chieh. "Quality of Life Indicators in the U.S. Metropolitan Areas, 1970: Summary," Midwest Research Institute, Kansas City, MO (February 1976).
7. Ott, Wayne R., and David T. Mage. "A General Purpose Univariate Probability Model for Environmental Data Analysis," *Computers and Operations Research*, 3:209-216 (1976).
8. "How U.S. Areas Rated in Quality of Life Study," *Dallas Times Herald*, Dallas, TX (September 21, 1975).
9. "Our Quality of Life Rated Low," *Detroit Free Press*, Detroit, MI (September 21, 1975).
10. "Charlotte 59th in Life Quality Study," *Charlotte Observer*, Charlotte, NC (September 21, 1975).
11. "New Orleans' Quality of Life Ranked 63 of 65 Cities," *Times-Picayune*, New Orleans, LA (September 21, 1975).
12. "Dallas Gets 'Good' Rating," *Morning News*, Dallas, TX (September 21, 1975).
13. "Quality of Life in Topeka Rated 'Outstanding'," *Topeka Capitol-Journal*, Topeka, KS (September 21, 1975).
14. "Study Rates Quality of Life for 243 U.S. Urban Centers," *Chicago Tribune*, Chicago, IL (September 25, 1975).
15. "Livability Good Enough in City for Many To Move Here for It," *Express*, San Antonio, TX (September 25, 1975).
16. "Report Rating San Antonio Livability Blasted," *Express*, San Antonio, TX (September 24, 1975).
17. "Life Quality Listings Doubted," *Omaha World Herald*, Omaha, NB (October 10, 1975).
18. *Congressional Record–House*, September 24, 1975, pp. H 9084-H 9087.
19. "EPA Admits that Survey of U.S. Cities Is Misleading," *Hartford Times*, Hartford, CT (October 3, 1975).
20. "EPA Softens Study's Blow," *Des Moines Register*, Des Moines, IA (October 4, 1975).
21. "EPA Research Chief Regrets If Report Misused," *Birmingham News*, Birmingham, AL (October 10, 1975).

22. "EPA Says M.R.I. Study Wrongly Ranked Cities," *Kansas City Star*, Kansas City, MO (October 6, 1975).
23. Wingo, Lowdon, and Alan Evans. *Public Economics and the Quality of Life* (Baltimore, MD: The Johns Hopkins University Press, 1977).
24. Campbell, Angus, Philip E. Converse, and Willard L. Rodgers. *The Quality of American Life* (New York: Russell Sage Foundation, 1976).
25. Collins, John P. "An Environmental Quality Index for Solid Waste Disposal Systems," presented at the 171st National Meeting of the American Chemical Society, New York, NY, April 7, 1976.
26. Manns, Basil. "Community Noise Modeling," in *Proceedings of the EPA Conference on Environmental Modeling and Simulation*, Wayne Ott, Ed., U.S. Environmental Protection Agency, Washington, DC, EPA-600/9-76-016 (June 1976), pp. 803-807.
27. Meyer, Alvin F. "EPA's Implementation of the Noise Control Act of 1972," *Sound and Vibration* (December 1975), pp. 10-17.
28. "Information on Levels of Environmental Noise: A Requisite to Protect the Public Health and Welfare with an Adequate Margin of Safety," U.S. Environmental Protection Agency, Washington, DC, EPA-550/9-74-004 (March 1974).
29. Pikul, Robert. "Development of Environmental Indices," in *Statistical and Mathematical Aspects of Pollution Problems*, John W. Pratt, Ed. (New York: Marcel Dekker, 1974), pp. 103-121.
30. "'76 EQ Index," *National Wildlife*, National Wildlife Federation, Washington, DC (February-March 1976).
31. Shannon, C.E., and W. Weaver. *The Mathematical Theory of Communication* (Urbana, IL: University of Illinois Press, 1963).
32. Hurlbert, Stuart H. "The Nonconcept of Species Diversity: A Critique and Alternative Parameters," *Ecology* 52(4):577-586 (summer 1971).
33. Hamilton, Martin A. "Indexes of Diversity and Redundancy," *J. Water Poll. Control Fed.* 47(3):630-632 (March 1975).
34. Pielou, E. C. *Mathematical Ecology* (New York: John Wiley & Sons, 1977).
35. Finney, D. J. *Probit Analysis* (London: Cambridge University Press, 1971).
36. Meisel, William S., and Alan Horowitz. "The Selection of Air Quality Indices," report prepared for the U.S. Environmental Protection Agency by Technology Service Corporation, Santa Monica, CA (August 16, 1973).

APPENDIX A

Table A-1
PSI VALUES, IN STEPS OF 5, FOR PSI FROM 5 TO 200

PSI	Gravimetric Units[a]				Volumetric Units[a]			PSI
	CO mg/m^3	O_3 $\mu g/m^3$	SO_2 $\mu g/m^3$	TSP $\mu g/m^3$	CO ppm	O_3 ppm	SO_2 ppm	
5	0.50	8.00	8.00	7.50	0.4500	0.0040	0.0030	5
10	1.00	16.00	16.00	15.00	0.9000	0.0080	0.0060	10
15	1.50	24.00	24.00	22.50	1.3500	0.0120	0.0090	15
20	2.00	32.00	32.00	30.00	1.8000	0.0160	0.0120	20
25	2.50	40.00	40.00	37.50	2.2500	0.0200	0.0150	25
30	3.00	48.00	48.00	45.00	2.7000	0.0240	0.0180	30
35	3.50	56.00	56.00	52.50	3.1500	0.0280	0.0210	35
40	4.00	64.00	64.00	60.00	3.6000	0.0320	0.0240	40
45	4.50	72.00	72.00	67.50	4.0500	0.0360	0.0270	45
50	5.00	80.00	80.00	75.00	4.5000	0.0400	0.0300	50
55	5.50	88.00	108.50	93.50	4.9500	0.0440	0.0410	55
60	6.00	96.00	137.00	112.00	5.4000	0.0480	0.0520	60
65	6.50	104.00	165.50	130.50	5.8500	0.0520	0.0630	65
70	7.00	112.00	194.00	149.00	6.3000	0.0560	0.0740	70
75	7.50	120.00	222.50	167.50	6.7500	0.0600	0.0850	75
80	8.00	128.00	251.00	186.00	7.2000	0.0640	0.0960	80
85	8.50	136.00	279.50	204.50	7.6500	0.0680	0.1070	85
90	9.00	144.00	308.00	223.00	8.1000	0.0720	0.1180	90
95	9.50	152.00	336.50	241.50	8.5500	0.0760	0.1290	95
100	10.00	160.00	365.00	260.00	9.0000	0.0800	0.1400	100
105	10.35	172.00	386.75	265.75	9.3000	0.0860	0.1480	105
110	10.70	184.00	408.50	271.50	9.6000	0.0920	0.1560	110
115	11.05	196.00	430.25	277.25	9.9000	0.0980	0.1640	115
120	11.40	208.00	452.00	283.00	10.2000	0.1040	0.1720	120
125	11.75	220.00	473.75	288.75	10.5000	0.1100	0.1800	125
130	12.10	232.00	495.50	294.50	10.8000	0.1160	0.1880	130
135	12.45	244.00	517.25	300.25	11.1000	0.1220	0.1960	135
140	12.80	256.00	539.00	306.00	11.4000	0.1280	0.2040	140
145	13.15	268.00	560.75	311.75	11.7000	0.1340	0.2120	145
150	13.50	280.00	582.50	317.50	12.0000	0.1400	0.2200	150
155	13.85	292.00	604.25	323.25	12.3000	0.1460	0.2280	155
160	14.20	304.00	626.00	329.00	12.6000	0.1520	0.2360	160
165	14.55	316.00	647.75	334.75	12.9000	0.1580	0.2440	165
170	14.90	328.00	669.50	340.50	13.2000	0.1640	0.2520	170
175	15.25	340.00	691.25	346.25	13.5000	0.1700	0.2600	175
180	15.60	352.00	713.00	352.00	13.8000	0.1760	0.2680	180
185	15.95	364.00	734.75	357.75	14.1000	0.1820	0.2760	185
190	16.30	376.00	756.50	363.50	14.4000	0.1880	0.2840	190
195	16.65	388.00	778.25	369.25	14.7000	0.1940	0.2920	195
200	17.00	400.00	800.00	375.00	15.0000	0.2000	0.3000	200

[a]CO is measured as an 8-hr average; O_3 is a 1-hr average; SO_2 and TSP are 24-hr averages.

354 ENVIRONMENTAL INDICES

Table A-2. PSI VALUES, IN STEPS OF 5, FOR PSI FROM 200 TO 500

PSI	Gravimetric Units[a]				Volumetric Units[a]				Mixed Units[a]			
	CO mg/m³	O₃ µg/m³	SO₂ µg/m³	TSP µg/m³	NO₂ µg/m³	CO ppm	O₃ ppm	SO₂ ppm	NO₂ ppm	TSPXSO₂ 10³ (µg/m³)²	TSPXSO₂ ppm·µg/m³	COH COHs	COHXSO₂ COHs-ppm
200	17.00	400.	800.	375.00	1130.	15.00	0.200	0.300	0.60	65.0	24.820	3.00	0.200
205	17.85	420.	840.	387.50	1187.	15.75	0.210	0.315	0.63	74.8	28.562	3.10	0.230
210	18.70	440.	880.	400.00	1243.	16.50	0.220	0.330	0.66	84.6	32.304	3.20	0.260
215	19.55	460.	920.	412.50	1300.	17.25	0.230	0.345	0.69	94.4	36.046	3.30	0.290
220	20.40	480.	960.	425.00	1356.	18.00	0.240	0.360	0.72	104.2	39.788	3.40	0.320
225	21.25	500.	1000.	437.50	1413.	18.75	0.250	0.375	0.75	114.0	43.530	3.50	0.350
230	22.10	520.	1040.	450.00	1469.	19.50	0.260	0.390	0.78	123.8	47.272	3.60	0.380
235	22.95	540.	1080.	462.50	1526.	20.25	0.270	0.405	0.81	133.6	51.014	3.70	0.410
240	23.80	560.	1120.	475.00	1582.	21.00	0.280	0.420	0.84	143.4	54.756	3.80	0.440
245	24.65	580.	1160.	487.50	1639.	21.75	0.290	0.435	0.87	153.2	58.498	3.90	0.470
250	25.50	600.	1200.	500.00	1695.	22.50	0.300	0.450	0.90	163.0	62.240	4.00	0.500
255	26.35	620.	1240.	512.50	1752.	23.25	0.310	0.465	0.93	172.8	65.982	4.10	0.530
260	27.20	640.	1280.	525.00	1808.	24.00	0.320	0.480	0.96	182.6	69.724	4.20	0.560
265	28.05	660.	1320.	537.50	1865.	24.75	0.330	0.495	0.99	192.4	73.466	4.30	0.590
270	28.90	680.	1360.	550.00	1921.	25.50	0.340	0.510	1.02	202.2	77.208	4.40	0.620
275	29.75	700.	1400.	562.50	1978.	26.25	0.350	0.525	1.05	212.0	80.950	4.50	0.650
280	30.60	720.	1440.	575.00	2034.	27.00	0.360	0.540	1.08	221.8	84.692	4.60	0.680
285	31.45	740.	1480.	587.50	2091.	27.75	0.370	0.555	1.11	231.6	88.434	4.70	0.710
290	32.30	760.	1520.	600.00	2147.	28.50	0.380	0.570	1.14	241.4	92.176	4.80	0.740
295	33.15	780.	1560.	612.50	2204.	29.25	0.390	0.585	1.17	251.2	95.918	4.90	0.770
300	34.00	800.	1600.	625.00	2260.	30.00	0.400	0.600	1.20	261.0	99.660	5.00	0.800
305	34.60	810.	1625.	637.50	2297.	30.50	0.405	0.610	1.22	267.6	102.182	5.10	0.820
310	35.20	820.	1650.	650.00	2334.	31.00	0.410	0.620	1.24	274.2	104.704	5.20	0.840
315	35.80	830.	1675.	662.50	2371.	31.50	0.415	0.630	1.26	280.8	107.226	5.30	0.860
320	36.40	840.	1700.	675.00	2408.	32.00	0.420	0.640	1.28	287.4	109.748	5.40	0.880
325	37.00	850.	1725.	687.50	2445.	32.50	0.425	0.650	1.30	294.0	112.270	5.50	0.900
330	37.60	860.	1750.	700.00	2482.	33.00	0.430	0.660	1.32	300.6	114.792	5.60	0.920
335	38.20	870.	1775.	712.50	2519.	33.50	0.435	0.670	1.34	307.2	117.314	5.70	0.940
340	38.80	880.	1800.	725.00	2556.	34.00	0.440	0.680	1.36	313.8	119.836	5.80	0.960
345	39.40	890.	1825.	737.50	2593.	34.50	0.445	0.690	1.38	320.4	122.358	5.90	0.980
350	40.00	900.	1850.	750.00	2630.	35.00	0.450	0.700	1.40	327.0	124.880	6.00	1.000

[a]CO is measured as an 8-hr average; O₃ and NO₂ are 1-hr averages; SO₂, TSP, TSPXSO₂, COH, and COHXSO₂ are 24-hr averages.

APPENDIX A 355

Table A-2 (Continued)

	Gravimetric Units[a]					Volumetric Units[a]					Mixed Units[a]			
PSI	CO mg/m^3	O$_3$ μg/m^3	SO$_2$ μg/m^3	TSP μg/m^3	NO$_2$ μg/m^3	CO ppm	O$_3$ ppm	SO$_2$ ppm	NO$_2$ ppm	TSPXSO$_2$ 10^3 (μg/m^3)^2	TSPXSO$_2$ ppm·μg/m^3	COH COHs	COHXSO$_2$ COHs-ppm
355	40.60	910.	1875.	762.50	2667.	35.50	0.455	0.710	1.42	333.6	127.402	6.10	1.020
360	41.20	920.	1900.	775.00	2704.	36.00	0.460	0.720	1.44	340.2	129.924	6.20	1.040
365	41.80	930.	1925.	787.50	2741.	36.50	0.465	0.730	1.46	346.8	132.446	6.30	1.060
370	42.40	940.	1950.	800.00	2778.	37.00	0.470	0.740	1.48	353.4	134.968	6.40	1.080
375	43.00	950.	1975.	812.50	2815.	37.50	0.475	0.750	1.50	360.0	137.490	6.50	1.100
380	43.60	960.	2000.	825.00	2852.	38.00	0.480	0.760	1.52	366.6	140.012	6.60	1.120
385	44.20	970.	2025.	837.50	2889.	38.50	0.485	0.770	1.54	373.2	142.534	6.70	1.140
390	44.80	980.	2050.	850.00	2926.	39.00	0.490	0.780	1.56	379.8	145.056	6.80	1.160
395	45.40	990.	2075.	862.50	2963.	39.50	0.495	0.790	1.58	386.4	147.578	6.90	1.180
400	46.00	1000.	2100.	875.00	3000.	40.00	0.500	0.800	1.60	393.0	150.100	7.00	1.200
405	46.58	1010.	2126.	881.25	3038.	40.50	0.505	0.810	1.62	397.9	151.950	7.05	1.215
410	47.15	1020.	2152.	887.50	3075.	41.00	0.510	0.820	1.64	402.7	153.800	7.10	1.230
415	47.73	1030.	2178.	893.75	3113.	41.50	0.515	0.830	1.66	407.6	155.650	7.15	1.245
420	48.30	1040.	2204.	900.00	3150.	42.00	0.520	0.840	1.68	412.4	157.500	7.20	1.260
425	48.88	1050.	2230.	906.25	3188.	42.50	0.525	0.850	1.70	417.3	159.350	7.25	1.275
430	49.45	1060.	2256.	912.50	3225.	43.00	0.530	0.860	1.72	422.1	161.200	7.30	1.290
435	50.03	1070.	2282.	918.75	3263.	43.50	0.535	0.870	1.74	427.0	163.050	7.35	1.305
440	50.60	1080.	2308.	925.00	3300.	44.00	0.540	0.880	1.76	431.8	164.900	7.40	1.320
445	51.18	1090.	2334.	931.25	3338.	44.50	0.545	0.890	1.78	436.7	166.750	7.45	1.335
450	51.75	1100.	2360.	937.50	3375.	45.00	0.550	0.900	1.80	441.5	168.600	7.50	1.350
455	52.33	1110.	2386.	943.75	3413.	45.50	0.555	0.910	1.82	446.4	170.450	7.55	1.365
460	52.90	1120.	2412.	950.00	3450.	46.00	0.560	0.920	1.84	451.2	172.300	7.60	1.380
465	53.48	1130.	2438.	956.25	3488.	46.50	0.565	0.930	1.86	456.1	174.150	7.65	1.395
470	54.05	1140.	2464.	962.50	3525.	47.00	0.570	0.940	1.88	460.9	176.000	7.70	1.410
475	54.63	1150.	2490.	968.75	3563.	47.50	0.575	0.950	1.90	465.8	177.850	7.75	1.425
480	55.20	1160.	2516.	975.00	3600.	48.00	0.580	0.960	1.92	470.6	179.700	7.80	1.440
485	55.78	1170.	2542.	981.25	3638.	48.50	0.585	0.970	1.94	475.5	181.550	7.85	1.455
490	56.35	1180.	2568.	987.50	3675.	49.00	0.590	0.980	1.96	480.3	183.400	7.90	1.470
495	56.93	1190.	2594.	993.75	3713.	49.50	0.595	0.990	1.98	485.2	185.250	7.95	1.485
500	57.50	1200.	2620.	1000.00	3750.	50.00	0.600	1.000	2.00	490.0	187.100	8.00	1.500

[a]CO is measured as an 8-hr average; O$_3$ and NO$_2$ are 1-hr averages; SO$_2$, TSP, TSPXSO$_2$, COH, and COHXSO$_2$ are 24-hr averages.

INDEX

absolute index
   See index
acetycholinesterase   263
acidity   8,203,269
   See Also alkalinity, pH
adaptation   29
aesthetic factors   14-15,201,248-249,269,292,323
aggregation function   50,66-93
   additive forms   66-76,78-82, 91-93
   maximum operator   76-79, 91-93
   multiplicative forms   79-89, 92-94,126
   minimum operator   89-90,92-93
   See Also linear sum, maximum operator, minimum operator, root-mean-square, root-sum-square, root-sum-power, weighted linear sum, weighted product
aggregation of subindices   26,50,74
   See Also aggregation function
Air Pollution Control Association   123
air quality   9-10,13-22,97-198, 323-234
   comparison with water quality   49-50,284,295-297,304
   criteria   28-29,98,170,186
   health effects   16-18,149-150, 170
   monitoring   37-39
   standards   13,19-21,99,110, 120,126,131,136,186

   See Also monitoring, National Ambient Air Quality Standards
Air Quality Control Region (AQCR)   98,103
Air Quality Index (AQI)   107-108,120-121,127
Alabama   124,184,220,222,316, 319,320
Alaska   9-27,294
Alberta, Canada   123,129,134
aldrin   22,240
Alert level
   See episode criteria
alkalinity   199-200,203,221,226, 247,267-269,288,301
alkyl benzene sulfonates   214-216,267
Allegheny County Air Pollution Control Advisory Committee   179-180
aluminum   203,241,269
ambiguity   68,71,78,83,91-93, 299-300
   See Also ambiguous region
ambiguous region   67-68,70-95, 116,120-121
   definition   68
American Lung Association   145
ammonia   203,214-216,226,229, 241,257-258,262,267,286, 290-291,302-303
ammonium   39
Anaheim, California   171-178, 181-182,316
Andrews, Richard N. L.   250-251

antagonism 330-333,336-338, 343,348,350
  definition 330-332,338-339
  diminutive 331,337
  See Also criteria for interaction
antimony 39
arithmetic mean 70,78-79,105-106,119,121,162,174-175, 179,205,210,215,237,247, 270-273,313,315
Anchorage, Alaska 19-20
Arizona 124,316
Arkansas 316
Army Corps of Engineers, U.S. 41
arsenic 39,205,240-241,253
asbestos 17
Association of Local Air Pollution Control Officials (ALAPCO) 145
averaging times, conversion of 110,117-118

Babcock, Lyndon 109-111,120
Baltimore, Maryland 123,128,316
barium 240
Barker, Bruce 285-286,300
Barnett, Stanley M. 121
basic type 131-132,138
Beebe, James A. 292-294
benzene-soluble organics 39
beryllium 17,39,241
beta distribution 261
Beta Function Index 259-263, 266,286,302-303
bicarbonates 203,288
biochemical oxygen demand (BOD) 44,94,203,207,211, 214-219,221,226,230-232, 252,255,267,273,275,278-279,288,290-292,302,347
biological indices 197,263-264
Birmingham, Alabama 316,320
bismuth 39
Bisselle, C. C. 114,120
boron 241,269
Boston, Massachusetts 124,316-317

breakpoint 55-58,65-66,91,93-94,147,159,161,186-187, 189-191,200-201,205,215
  definition 55-56
Brown, Robert M. 59,61,79,88, 94,202-213,225,228-229, 255,265-275,287
  See Also National Sanitation Foundation Water Quality Index
Buchanan, Congressman 319-320
Bureau of Reclamation, U.S. 41,249

cadmium 39,205,240-241,251, 253
calcium 203,269
calculation method 125,130-132,135
calculation mode 125-126,130-132,135,154
calculation of subindices 50,51-66,91
California 38,110,124,128,139, 171-179,181-184,315-316
Camden, New Jersey 171-178, 181-182
Campbell, Angus 322
Canada 108,118-119,123,129-131,134,185-186,251
Canadian Federal-Provincial Committee on Air Pollution 185
Canadian National Index 248, 251,266
carbamates 22,24,240
carbon chloroform extract (CCE) 199-200,203,214-216,267, 301
carbon dioxide 269
carbon monoxide (CO) 13,16, 18-19,20-22,26,39,93,99-191, 326,353-355
  episode criteria values 102
  Federal Reference Method 100-101
  health effects 16
  NAAQS values 99

INDEX 359

PSI subindex function 151
Significant Harm values 104
carbonates 288
Carp River 281-282
case study approach 287-291, 295-296
cautionary statement 169
Central Limit Theorem 260
Chamber of Commerce, Charlotte 319
Chamber of Commerce, San Antonio 318
Charlotte, North Carolina 124, 317,319
chemical oxygen demand (COD) 203,214-216,218-219,267, 286
chemiluminescence 100
Chicago, Illinois 28,114-115, 124,128,174,183,316-318
Chicago Tribune 318
chi-square distribution 260-261
chlordane 238-240
chlorides 7,44,199-200,203,214-216,221,226,229,238,241, 243,247,252,267,269,288,301
chlorophyll-*a* 262
chromium 39,205,240-241
city-county air pollution indices 124-125,128-132
classic dichotomy 7,27,45,247, 348
classification systems 123-126, 197-199,213-215,255,262
  biological 262
  index 123-126,129
  nonparametric 255
  saprobic 263
  stream 197-199,213-215
Clean Air Act 13,37-38,98,185
climate 314,324
Coate, Edwin L. 6
cobalt 39,241
Coefficient of Haze (COH) 33-36,102-190,346,353-355
  episode criteria values 102
  PSI subindex function 155
  relationship to TSP 103,156, 179
  Significant Harm values 104

coefficient of kurtosis 162,174
coefficient of skewness 162,174
coefficient of variation 162,174
Cogger, William J. 291
COH x SO_2 102-104,162,165, 167,190,353-355
  episode criteria values 102
  Significant Harm values 104
coliforms
  See fecal coliforms, total coliforms
Collins, John P. 323
color 7,44,201,203,221,226,230-232,234-235,238,241,247, 267-269
color coding system 169,183, 212,292-293
Colorado 124,128,139,184,278, 280,291-292,316
Colorado River 255,320
Combustion Products Index (CPI) 106-107,120-121,127
Common Air Quality Reporting Format 140-141
comparison of cities 176-180
  See Also Pollutant Standards Index, ranking paradox
comparison of indices 119-122, 123-135,264-273
Compendium of air pollution indices 6,105,122,136,138-140,141,147,170,179,185
Composite Pollution Index 254, 266
computer simulation 260
Congress, U.S. 37,184-185,279, 281,286,301,319-321
Connecticut 124,128-129,184, 316
Continuous Air Monitoring Project (CAMP) 114
control value 256-258,261-262, 303
Converse, Philip E. 322
copper 39,203,241,251
Coughlin, Robert E. 255,266
Council on Environmental Quality (CEQ) 4-5,37,105,114,140, 197,248,288-289,300
Creason, John P. 30

360 ENVIRONMENTAL INDICES

criteria for a uniform air pollution index   135-138,141,147, 186-188
criteria for an ideal water quality index   300
criteria for interaction   338-339, 341,343,346,349-350
   applications   339,341,343,346, 350
   definition   338-339
critical pollutant   154,162,177-178,190,298
   See Also Pollutant Standards Index
Cullen, James J.   121
Cumberland River   237
cumulative frequency   162,175
customizing   295
cyanides   205,240,269
Czechoslovakia   214

dalapon   240
Dallas Times Herald   317
damage function   27-36,120,326-349
   definition   27
   economic   27-28
   examples   339-345
   general criteria   331-332
   general form   326
   graphical representation   332-339
   linear   339-340
   logistic (sigmoid)   326-328
   multiplicative   339-342
   multivariate   329-345
   root-sum-power   341-343
   root-sum-square   343-344
   univariate   326-329
   See Also criteria for interaction, hockey stick function
damage space   334-350
   definition   334-335
Daniels, Congressman Dominick V.   320-321
DDT   22-23,240
decreasing scale   49,53-54,66,79-90,91-93,197,202,215,220, 243,262-266,270-272,297, 317,322,326
   conversion to increasing scale   270
   definition   49
Dee, Norbert   248-251,266-267, 270-273,289
   See Also Environmental Evaluation System
Deininger, Rolf A.   202,224,228-233,259,265-267,270-273
Delaware River   215,277,298-299
Delphi technique   202,225,287
Denver, Colorado   316,320
Department of Environmental Quality, Oregon   287,290-291
Department of Human Resources, Nevada   287
Department of Natural Resources, Georgia   284
Department of Transportation, Illinois   285
descriptor category   126,130-135, 147,149-150,161-162
   definition   126
Detroit, Michigan   107
diminutive   331,334,337
   See Also antagonism, criteria for interaction, damage function
Dinius, S. L.   198,219-222,265-267,270-273,289
   See Also Social Accounting System
dichotomous subindex   59-60,66, 68,71,78,87,91,201,349
   definition   59
   See Also step function
dieldrin   22,240
Dinman, Bertram   29
dissolved oxygen (DO)   7-8,14-15, 44,59-60,94,199-200,203, 205-206,211,214-219,221, 226,230-235,247,250,255, 257-258,267,269,272,279, 286,288,290-292,298,301-303
dissolved solids   44,203,226,230-232,247,267,269,286,288, 302-303

See Also total dissolved solids
District of Columbia 124,128-129,139,316
dose-effect function 27-28,32,36
  See Also damage function
dose-effect information 32-36
drinking water 10,40,304
  See Also public water supply, water uses
Dunnette, David A. 287-291, 300-301,304
Dziewulski, C. 122

eclipsing 70-93,106,111,114,119-121,211,237,245,270-273, 299-300,317,322
  definition 70-71
  region 70-93,120-121
economic indicators 7,309
  See Also Gross National Product
Emergency level
  See episode criteria
endrin 240
energy equivalent noise level 323-324
Energy Resources Co., Inc. 197
England 214
environmental damage function 326-348
  See Also damage function
Environmental Evaluation System (EES) 248,266
environmental impact (EI) 250
environmental indicator 2
  definition 8
Environmental Protection Agency, Illinois 259,286,313,318-319,321
Environmental Protection Agency (EPA), U.S. 5,24,38,40-42, 98,103,105,117,140,143,145, 147,154-156,161-162,168, 170,179,185,248,254,262, 279,284,294,309,313,318-319,321,324
  Chicago Regional Office (Region V) 257

Denver Regional Office (Region VIII) 278,291-293
Seattle Regional Office (Region X) 9-27,117,278, 289,292-294
Environmental Quality (EQ) index 324-325
Environmental Quality Index (EQI) 118-119,120-121, 127,251
Environmental Quality Profile 9-27,117
epidemiology 29,32
episode criteria 20,78,97,101-103,126,131-133,135-137, 141,145,147-155,159,162, 186-189,190
  U.S. values 102
episodes 105,108-109,121
equivalence curves 334-350
  definition 334
equivalent method 99-100,154
  definition 99
Estimated Average Daily Excess Deaths 35
eutrophication 14-15,262,290
Evans, Alan 322
excess deaths 35,326
explicit subindex function 59, 62-64,91,215-216,239-241, 270-273,287-288
  definition 59
  See Also nonlinear subindex function
exponential function 63-64,234-235,270-273
Extreme Value Index (EVI) 114, 116-117,120-121,127

factor analysis 254
Fairbanks, Alaska 19-20
fecal coliforms 94,203,206,211, 221,226,230-232,241,247, 250,263,267,269,272,274, 279,286,288,290-292,298-299,302-303,347
Federal Interagency Task Force on Air Quality Indicators 140-141,143,147

Federal Reference Method (FRM) 99-101,103,137-138,154-155,159
  definition 99
  list of methods 100
Federal Register 101,147,159-160,162,184
Federal Water Pollution Control Act 10,40,254,279
feedback process 3,147,205
Fensterstock, Jack C. 107-108, 120
Ficke, John F. 296
Finney, D.J. 327
fish 198,234,249,251,255,263-264,286,296,324
Fish and Wildlife (FAWL) index 225-228
five-day biochemical oxygen demand
  See biochemical oxygen demand
Flaherty, Theodore V. 121
Florida 124,128,139,184,243-244,316
fluorides 203,226,230-232,239, 241-242,252,267,269
forecasts of air pollution levels
  See Pollutant Standards Index
Forest Service, U.S. 41
FORTRAN 160-161
frequency/severity characteristics 117
  See Also ranking paradox

Gaussian distribution
  See normal distribution
general water quality indices 198-222,249,262,264-265, 267,270-273
Genesee River 219,283
Geological Survey, U.S. 41,296
geometric mean 85-86,93,146, 231,237,271-273
Georgia 124,128,184,277-278, 280,284-285,316
Germany 197,214,252
Gillies, D. K. A. 170
Glasser, Marvin 33-35

Goldstein, Gerald 264
gravimetric units 99,157,159-160, 162,164-165,190-191
grease 203,234-235,267
Great Lakes 277,280,286
Greeley, R. S. 248
Green, Marvin H. 105-106
Greenburg, Leonard 33-35
Green's Index 105-106,120-121, 127
gross national product (GNP) 7, 251,309

Hahn, Gerald J. 261
Hamilton, Martin A. 324
Haney, Johnathan T. 215-219, 220,265,271
  See Also River Pollution Index
hardness 7,44,203,221,226,230-232,247,251-252,267,269
Harkins, Ralph D. 255-263,266, 275,277-278,288-290,300, 302-303,313
Hasselblad, Victor 30
Hawaii 316
heptachlor 240
herbicides 203
Hershaft, A. 27,31
high-volume sampler 38-39,100, 103,107,154-156
  relationship to tape sampler 156
histogram 133,162-163
hockey stick function 30-31,54-55,68,78-79,326
Holzworth, George C. 108
homogeneous 329
Horowitz, Alan 335
Horton, Robert K. 59-60,198-202,215,220,265-275,289, 301
  See Also Quality Index
Houston, Texas 174,181,183,316
Hudson River 215
Hunt, William F., Jr. 29,53,140-141,145-147,156,159,168, 170-171,176,178-181,184-185
Hurlbert, Stuart H. 325

INDEX 363

hydrocarbons (nonmethane) 13, 16,23,98-99,190
  health effects 16-17
  NAAQS values 99
hyperbola 84,341

(I_1,I_2)-plane 67-95,116,119,334
  arithmetic mean 18
  definition 68
  linear sum 69
  maximum operator 77
  minimum operator 90
  root-mean-square 76
  root-sum-power 75
  root-sum-square 74
  weighted linear sum 70,72-73, 81-82,94
  weighted product 84,86
Idaho 9-27,294
ideal gas law 160
Illinois 28,124,183,257,259,262, 277-278,280,284-286,316-318
impairment categories 288-291
Implicit Index of Pollution 198, 213-217,265,270-273
implicit subindex function 59, 91,212,270-273,281,293
  definition 59
  examples 61
  See Also nonlinear subindex function
increasing scale 49,53-54,66,79-80,91-93,119,197,215,219, 244,264,270-273,286,291, 297,326,335,348
  conversion to decreasing scale 270
  definition 49
index 2,5,8,26-27,45,50-51,91-93
  absolute 51,255,259,264
  classification system 97,123-135
  definition 2,8,26-27,45,50-51
  mathematical structure of 50-51
  purposes for using 5,280
  relative 51,259,288
  role 2

  See Also long-term index, short-term index
Index of Industrial Emissions 118-119
Index of Interurban Air Quality 118
Index of Partial Nutrients 255, 266
Index of Specific Pollutants 118
Index of Total Nutrients 255, 266
Index of Well-Being 323
INDEX.PLOT 146,160-163,171, 181
Indiana 124,278,280,316
indicator 2,8-9,18,26-27,45,324
  definition 8,45
  economic 309
  organism 263-264
individual mode
  See calculation mode
information theory 262,325
Inhaber, Herbert 7,118-119,251, 266,289
in-place pollutants 253-254
integrating nephelometer 156
Interstate Commission on the Potomac River Basin 275-277,280
iodine 240
Iowa 184
iron 39,203,214-216,230-232, 241,247,252,267,269
isocapability curves 312
isoquality curves 310-312
Italy 215

Jackson Turbidity Units (JTU)
  See turbidity
Janardan, Konanur G. 259-263, 266,286,300
Japan 254
Jersey City, New Jersey 316,320
Johanson, Edward E. 248,253-254,266
  See Also Pollution Index
Johnson, Jaret C. 248,253-254, 266
  See Also Pollution Index

Johnstown, Pennsylvania 28
Joung, H. M. 254-266

Kansas 184,298-299,301,316
Kansas River Basin 212
Kazen, Congressman 321
Kendall, Maurice 255
Kentucky 124,128,139,184,316
Klamath River 277
Knox County Department of Air Pollution Control 118
Knoxville, Tennessee 118
Kramer, Paul 285-286,300

Lake Michigan 286
land quality 118,251,323
Landwehr, Jurate 79,197,212-213,224,228-233,259,270-273
   See Also National Sanitation Foundation Water Quality Index, Public Water Supply Index
Laredo, Texas 321
Larsen, Ralph I. 110
lead 17,39,205,238,240
Lennox, California 171-178, 181-182
lindane 240
linear subindex function 51-54, 91,109-110,119-125,131-132, 239-241,245,270-273,287-288
linear sum 66-68,71,73,91-92, 110,121,201,286-287
lithium 251
Liu, Ben-Chieh 28,310-322,349
logarithmic function 64
logistic curve 328
   See Also sigmoid curve
lognormal distribution 175,314
long-term index 120
   See Also index
Los Angeles, California 124,128, 171-178,181-182,316
Louisiana 316
Lubore, S. H. 114

Mage, David T. 314
magnesium 203,269
manganese 39,203,214-216,241, 247,267,269
Manns, Basil 323
marginal damage (MD) 327-350
   definition 327
marginal rate of substitution (MRS) 335-350
   definition 327
marginal utility 327
Maryland 124,128,139,316
Mason, Anthony K. 6
Massachusetts 124,184,316-317
mathematical structures 270-273
Mathtech, Inc. 197
Maxim, L. C. 197
maximum mode
   See maximum operator
maximum operator 74,76-79,90-92,121,126,130-135,141, 147,154,297,337,347-349
McAdie, H. G. 170
McClelland, Nina I. 61,202-213, 298,301
   See Also National Sanitation Foundation Water Quality Index
McDuffie, Bruce 198,215-219, 220,265-267,270-273,289-290
   See Also River Pollution Index
McGuire, Terry 110
Measure of Undesirable Respirable Contaminants (MURC) 107, 120-121,127
median 117-118,146,254
Meisel, William S. 335
mercury 22,24,240,251,253-254
mesotrophic 262
methoxychlor 240
methylene blue active substances (MBAS) 239,241
Metro Nashville Wastewater Treatment Plant 237
Mexico 187
Mexico City 187
Michigan 124,128,278-282,316-317
microeconomic theory 310,326

Midwest Research Institute
 (M.R.I.) 313-322
Miller, M. E. 108
Miller, Terry L. 117-118,120
minimum operator 89-90,92-93,
 95,297,347-348
Minnesota 124,128-129,134,316
Mississippi 184
Missouri 124,316-317
MITRE Air Quality Index (MAQI)
 114-117,120-121,127
MITRE Corp. 114-117,247-251,
 313-314
 See Also MITRE Air Quality
  Index, Prevalence Duration
  Intensity index
Mohawk River 216
molybdenum 39,240
monitoring 36-45,79,176,179,183
 air monitoring station 37-38,
  79,176,179,183,190-191
 air pollution 37-39
 continuous 38-40
 discrete 39
 water monitoring station 41-
  44
 water pollution 39-45
Montana 278,280,291-292
Montreal, Canada 123
morbidity rate 28,347
mortality rate 28,32-36,326,347
Morton, J. 27,31
Murawski, Thadeus J. 36

Nagda, Niven L. 111
Nakamura, Takakazu 254
National Academy of Sciences
 (NAS) 4,10,238
National Air Quality Index
 (NAQI) 114
National Ambient Air Quality
 Objectives 186
National Ambient Air Quality
 Standards (NAAQS) 13,18-
 22,29,36-37,53-55,66-69,71,
 73,75,78-79,91,93,97-191,
 296
 primary and secondary 13,98

National Bureau of Standards
 140
National Environmental Policy
 Act 4
National Eutrophication Survey
 262
 See Also eutrophication
National Oceanic and Atmospheric
 Administration 140,145
National Planning Priorities Index
 (NPPI) 249-250,266
National Sanitation Foundation
 Water Quality Index (NSF
 WQI) 198,202-213,220,224-
 233,250,255,259,265-267,
 270-273,277-284,287-292,
 294-295,297-299,300-302,
 347
National Weather Service, U.S.
 138,170
National Wildlife Federation 324-
 325
nationally uniform index 5-6,76,
 97,104,122,134-136,140-189,
 294-301
 See Also Pollutant Standards
  Index
Nebraska 316
Nelson, William C. 30
Nemerow, Nelson L. 224,244-
 247,265-267,270-273,289,
 302
 See Also Pollution Index
Nevada 254-255,277,284,286-287
New England Interstate Water
 Pollution Control Commis-
 sion 277-278,280-281
New Jersey 121,129,139,171-
 179,181-183,316,320
New Mexico 124
New York 27-28,33-36,124,129,
 139,219,247,278-279,283,
 316
New York City, New York 27-
 28,33-36,124,128,316
New York State Water Quality
 Surveillance Network 219
New Zealand 214
Newark, New Jersey 163,171-
 178,181-182,316

nickel 39,241
nitrates 17,39,94,203,208,211, 214-216,226,230-232,235, 252,267,269,286,290-292, 302
nitrite 239,241-242,267
nitrogen 44,218,241,247,262,267, 286,292,302
nitrogen dioxide (NO_2) 13,16,21, 39,99-191,353-355
  episode criteria values 102
  Federal Reference Method 100
  health effects 16
  NAAQS values 99
  PSI subindex graph 151
  Significant Harm values 104
nitrogen oxides
  See nitrogen dioxide
noise 9-10,24-26,263,313-317, 322-324
  indicator 323-324
Noise Control Act 24
Noll, Kenneth E. 110
nomogram 112,162-165
nondispersive infrared (NDIR) 101
nonpoint source 11-13,45,253, 285
nonlinear subindex function 59-64,91-94,119-121,125,130-131,212,215,234-235,270-273,284,287-288
nonlinearity 87-89,93-94,129-131
  See Also nonlinear subindex function
nonparametric 255,261
nonvolatile suspended solids 218
normal distribution 174,260,314, 327
North Carolina 124,316,319
North Dakota 291-292
nutrients 44,269

Oak Ridge Air Quality Index (ORAQI) 110-114,120-121, 126-127,131,134
Oak Ridge National Laboratory 110

obvious pollution 200-201
  See Also aesthetic factors
O'Connor, Michael Fredrick 224-229,233,265-267,271,290, 304
  See Also Fish and Wildlife index, Public Water Supply index
odor 44,201,234-235,238,323-324
Office of Management and Budget 41
Ohio 124,128-129,139,178-180, 284,316-317
Ohio River 201,277
Ohio River Water Sanitation Commission (ORSANCO) 199,201
oil 201,203,267
Oklahoma 278-279,316
oligotrophic 262
Ontario Air Pollution Index (API) 108-109,120-121,127,346-347
Ontario, Canada 108-109,123, 134,170
Oregon 9-27,125,128,184,277-278,280,284,287-291,294, 304,316
Oregon Water Quality Index (OWQI) 288-291
organophosphates 22-24,240,252, 288
Orlando, Joseph A. 197,263-264
Ott, Wayne 76,105,122-147,156-167,170-181,185-187,213, 268-269,273-297,314
oxidants
  See photochemical oxidants
ozone
  See photochemical oxidants

parabola 60-63,66,234,239-241, 270-273
pararosaniline colorimetry 100, 154
Parker, Frank L. 61-62,93,222, 224,234-237,265,270-273

particulates 17-19,21,98,156,263, 330
  See Also total suspended particulates
Pavanello, R. 65,213
Payne, Marshall L. 291
peakedness
  See coefficient of kurtosis
Pennsylvania 28,124,128,179-180, 184,315-317
permanganate 214-216
Pesarin, F. 65,213
pesticides 9-10,13-15,22-24,44, 203,205,210,226-227,229, 240,250,263,292,324
  persistent and nonpersistent 22
pH 5,14-15,44,59-61,65-66,93-94,199-200,203,207,211, 214-216,226,230-235,241, 245-247,250,267,269,272-273,279,288,290-291,301-302,347
phenols 203,226,230-232,241, 257-258,267
phosphates 203,208,211,218-219, 226,232,235,252,255,267, 279,286,288,302
phosphorus 44,94,262,292
photochemical oxidants 13,17-19,21,39,98-191,353-355
  episode criteria values 102
  Federal Reference Method 100
  health effects 17
  NAAQS values
  PSI subindex function 152
  Significant Harm values 104
  UNIPEX subindex function 188
Pielou, E. C. 263,325,327
Pikul, Robert 114,324-325,349
PINDEX 109-110,120-121,127
Pittsburgh, Pennsylvania 125, 315-317
planning indices 198,247-254, 264,266
point source 11-12,45
Poland 214
Pollutant Standards Index (PSI) 76,79,97,104-105,135,141, 143-189,297-298,341
  adoption 184-189
  breakpoints 149-150
  calculation 154,156-162
  cautionary statements 149-150,169
  comparison of cities 176-183
  computer calculation 156-162
  coefficients 157-158
  critical pollutant 154,171,178, 190
  cumulative frequencies 162-163,175
  data used in 154-156
  descriptive language 149-150
  descriptor categories 149-150, 161-162
  development process 143-148
  forecasting 170-171
  histogram 163
  international adoption 185-189
  logarithmic probability plots 175
  nomograms 162-165
  numerical ranges 149-150,161
  performance 171-184
  PSI Bars 180-184,348
  reporting 168-171
  statistical properties 171-176
  structure 147-154
  subindex functions 151-153
  tables for calculating 157-158,353-355
  time series plot 171-173
  TV reports 169
  U.S. Adoption 184-185
pollutant variable, definition 8
Pollution Index (PI) 254,266, 285-286
Pollution Potential Index 248, 266
polychlorinated biphenyls (PCB) 11
Portland, Oregon 21-22,123, 125,128,316
potassium 203,269
power subindex function 59,63, 105,270-273
  See Also nonlinear subindex function

368 ENVIRONMENTAL INDICES

Prati, L.   65,198,213,220,265-267,270-273,289-290
See Also Implicit Index of Pollution
preferred index characteristics   134,141,147
Prevalence Duration Intensity (PDI) index   247-249,266
principal component analysis   255,266
Priority Action Index (PAI)   249,266
probit analysis   327
prospects for a nationally uniform water quality index   294-300
PSI Bars   180-184
See Also Pollutant Standards Index
public water supply   224-234, 237-244,265,296,304
Public Water Supply (PWS) index   225-234,265

Quality Index (QI)   198-202, 215,220,265,270-273,301
Quality of Life (QOL)   309-323,326,348-349
   in metropolitan areas   312-322
   other QOL approaches   322-323

radiation   22-23
See Also radioactivity
radioactivity   14-15,44,203,229, 269,324
radium-226   238-240
Raisin River   282
Rand Corporation   202
random variable   171,260
range   174
ranking paradox   20,45,175-176, 181,191,348
ranking procedure   255-258, 260-263
reference measurement prinicple   99-100

Reidy, M.   122
Reisa, James J.   140
relative index
See index
Rhine River   252
Rhode Island   316
Rich, T. A.   106,120
River Pollution Index   215-219, 265
Rodgers, Willard L.   322
root-mean-square   75-76,119-121,245,251,270-273
root-sum-power   71-75,91-92,341
root-sum-square   72-75,115-117, 120-121,343-344
Rosen, Richard H.   197
Rubin, Edward S.   179-180
running average   154

Sacramento, California   315-316
salinity   269
San Antonio, Texas   316,318
saturation level   31,35
Sayers, William T.   41-43,268-269
Schaeffer, David J.   259-263,266, 286,300
Schimmel, Herbert   34,36
Seattle, Washington   174,183,316-317
Secchi Disk Transparency   235,262
Section 305(b) report   279-286,301
segmented linear function   54-59, 91,93,125,130-132,134-135, 151-153,155,159,186-188, 245,249,270-273
   definition   56-57
segmented nonlinear function   64-66,93-94,158-159,234-235, 249,270-273
   definition   65
selenium   205,240
severe events list   292
Shannon, C. E.   325
Shapiro, Samuel S.   261
Shea, G.   27,31
Sheen, Jackson R.   287
Shenfeld, L.   108-109,120
Shoji, Hikaru   254,266
short-term index   120-122
See Also index

Short Time Averaging Relationships to Air Quality Standards (STARAQS) 117-118,120-121,127
sigmoid curve 31
See Also logistic curve
Significant Harm level 55,97, 103-104,136,141,145,147, 155,187,190-191
  published values 104
silica 203,269
silvex 240
slope 52,56,60,71,80,85,159, 327,336-337,341-343
  definition 52
Smith, Raymond 145
Smoke Shade 103
See Coefficient of Haze
Snake River 255
Social Accounting System 198, 219-222,265
social indicators 309
sodium 203,269
Sodium Absorption Ratio 241
Soiling Index 103
See Coefficient of Haze
solid waste 9-10,24-25,313-317,323-324
Solid Waste Environmental Quality Index (SWEQI) 323
South Dakota 291-292
Soviet Union 214
species diversity 263,325
specific conductance 44,199-200,203,218,221,241,255, 267,269,301
specific-use water quality indices 198,222-247, 262,264-265,267,270-273
Sprenger, Lester D. 291
standard deviation 162,174, 179,256-258,303,313
Standard Metropolitan Statistical Area (SMSA) 28,312-322,349
Standardized Urban Air Quality Index (SUAQI) 141-143,146
  development process 142

State and Territorial Air Pollution Program Administrators (STAPPA) 145
State Implementation Plan (SIP) 98,101,103,161
  definition 98
statistical approaches 198,254-264,266
step function 59-60,66,78,91, 201,238,270-273,286,347, 349
  definition 59
  dichotomous 59-60,66,68,71, 78,91,238,347,349
  staircase 59-60,91,201,270-273,286
Steubenville-Pittsburgh-Wheeling 135-136,179-180
Stoner, Jerry D. 224,237-249, 265-267,270-273,302
Stones River 237
STORET 42,45
strip chart 38
strontium 240
subindex 8,50-66
  definition 8,50
  See Also linear subindex function, nonlinear subindex function, segmented linear function, segmented nonlinear function
sulfates 17,39,203,226,241, 247,267,269
sulfur dioxide 8,13,18-19,21, 28,33-36,39,55-59,98-191, 313-314,322,330,346,353-355
  episode criteria values 102
  Federal Reference Method 100
  health effects 18
  NAAQS values 99
  PSI subindex function 152
  Significant Harm values 104
sulfur oxides 18
See Also sulfur dioxide
Sumitomo, Hisashi 224,244-247,265,270-273,302
See Also Pollution Index
summary of index structures 91-93

370 ENVIRONMENTAL INDICES

superposition 329-333,336-339, 344,348,350
  definition 329-332,338-339
  See Also criteria for interaction, damage function
suspended solids 7,44,214-216, 235,247,269
synergism 18-32,330-332,336-339,341,344,346-348,350
  definition 330-332,338-339
  See Also criteria for interaction

Talley, Wilson K. 321
tape sampler 33,103,155-156
  relationship to high-volume sampler 103,156
  See Also Coefficient of Haze
temperature 7,14-15,44,93-94, 200-201,203,209,211,218-219,226,230-232,234-236, 247,255,267,269,278-279, 288,301-302,347
Tennessee 124,128,184,237,316
Tennessee River 213
Tennessee Valley Authority 41,213
Texas 124,128,183-184,243-244,316-318
theoretical damage functions 29-32,326-348
  See Also damage function
threshold limiting value (TLV) 29-31,34,54,79,349
Thom, Gary 76,105,122-147, 170,185-187
Thomas, William A. 113,264
tin 39
titanium 39
total coliforms 7,44,199-200, 203,218,221,235,252,267-269,274,288,292,301
total dissolved solids (TDS) 14-15,199-200
  See Also total solids
total environmental indices 324-325
total organic carbon 205

total solids 94,203,210-211, 226,232,267,279,290-291, 298,302
total suspended particulates (TSP) 17,19,21,38-39,99-191,313-314,322,353-355
  episode criteria values 102
  Federal Reference Method 100
  health effects 17-18
  NAAQS values 99
  PSI subindex function 153
  relationship to COH 103
  Significant Harm values 104
toxic substances 11,14-15,30, 201,205,210,215,226,229, 238,240,250,253-254,263, 269
Tozer, Ronald G. 202
Train, Russell E. 3
Treatment Index (TI) 286
trend analysis 5,120,279-281, 301
Trend Monitoring Index (TMI) 284-285
trichloroacetic acid (TCA) 240
tri-state metropolitan area 179-180
Trophic Index 262
  See Also eutrophication
Truett, J. B. 248,266,289
  See Also Prevalence Duration Intensity index
TSP x SO_2 102-104,136,141, 147-158,178,341,353-355
  episode criteria values 102
  PSI subindex function 153
  Significant Harm values 104
turbidity 7,40,44,94,203,206, 209,211,226,230-232,234-236,247,250-251,267,269, 279,288,292,298,302

uniform administrative limits 97-104
uniform index
  See nationally uniform index, Pollutant Standards Index

unimodal 61,91,234-235,245
    definition 61
UNIPEX system 97,186-189, 191
United Nations 187
Utah 291,316
utility theory 311,323

vanadium 39,241
variable weighting 284-285
variance 256-258,303
    See Also standard deviation
Virginia 124,128,184,316
visibility 8,118,121,128-129, 156,313-317
volumetric units 158-160,162, 166-167

Wakarusa River 298-299,301-302
Wallace, Lance A. 162,165-168
Walski, Thomas M. 61-62,93, 222,224,234-237,265-267, 270-273,289
Warning level
    See episode criteria
Washington, DC 114-115,124, 128-129,316
Washington state 9-27,124,128, 183,294,316-317
water monitoring station 41-43
Water Pollution Index (WPI) 297-299
water quality 9-13,49-50,313-317,323-324
    comparison with air quality 284,295-297,304
    criteria 29,239,297
    geographical variation 295-296
    monitoring 39-45
    standards 39-40,291,295, 297
water uses 40,198,222-247, 268-269,296
    fish and wildlife 223,225-228,269

human contact 244-247
indirect contact 244-247
industrial water supply 223, 268-269
irrigation and agriculture 237-244,268-269
public water supply 223-234, 237-244,269,296
recreation 234-237
remote contact 244-247
variables for 269
Weaver, W. 325
weighted linear sum 68-72, 79-82,85,91-94,211,226-227,247,249,270-273,290, 297,315,317,322
    definition 68
weighted product 82-89,90, 92-94,211,259,270-273, 285,297
West Virginia 184
Wilcox, William H. 264
Willamette River Basin 287-291
Wingo, Lowdon 322
Wisconsin 124,184,316
Working Group to Develop an Air Quality Index 143, 145-146,184
World Health Organization 186
Wrightington, Robert B. 197, 263-264
Wyoming 278,280,291-292

Yamamoto, Takeo 254
Yodo River System 254
York, Pennsylvania 28
Young, J. W. S. 186
Yu, Eden Siu-hung 28

zinc 39,205,241,251-252
Zoeteman, B. C. J. 248,251-253,266
    See Also Pollution Potential Index